The R.A.M.S. Library of Alchemy

Volume 40

The Holy Guide

by

John Heydon

R.A.M.S. Publishing Company

The Holy Guide

John Heydon

The Holy Guide

by

John Heydon

Produced by

Restorers of Alchemical Manuscripts Society

R.A.M.S. Publishing Company

R.A.M.S. Publishing Company
7309 East 102nd Street
Kansas City Missouri 64134

The R.A.M.S. Library of Alchemy, Volume 40:
The Holy Guide
Copyright © 2018 Althea Productions LLC

All rights reserved. No part of this publication may be reproduced or transmitted in any form or by any means, electronic or mechanical, including but not limited to any information storage and retrieval system, without written permission from Althea Productions LLC.
Reviewers may quote brief passages.

First Edition 2015
Second Edition 2018

ISBN-13: 978-1726241274
ISBN-10: 1726241270

Image Processing by Philip N. Wheeler

Printed in the United States of America

Table of Contents

Disclaimer 10

Introduction 11

The Holy Guide 14

The Preface...19

To the Reader..73

Chemical Dictionary 85

Book I. Chapter I. 100

Chapter II. 108

Book II. Chapter I. 126

Chapter II. 130

Chapter III. 134

Chapter IV...140

Chapter V. 144

Chapter VI. 149

Chapter VII. 159

Chapter VIII. 163

Chapter IX. 169

Chapter X. 181

Chapter XI 185

Chapter XII. 188

Chapter XIII. 194

Chapter XIV. 204

Chapter XV. 214

Chapter XVI. 217

Chapter XVII. 229

The Holy Guide, Part 2 242

Book III. Chapter I. 245

Chapter II. 276

Chapter II. 280

Chapter III. 284

Chapter IV. 287

Chapter VI. 290

Chapter VII. 300

Chapter VIII. 305

Chapter IX. 307

Chapter X 312

Chapter XII. 315

Chapter XIII. 319

Chapter XIV. 322

Chapter XV. 329

Chapter XVI. 336

Chapter XVII. 342

Chapter XVII. 348

Chapter XVIII. 366

Chapter XIX. 384

Book IV. Chapter I. 405

Chapter II. 417

Chapter III. 428

Chapter IV. 445

BOOK V. 457

Book V. Chapter I. 460

Chapter II. 476

Chapter III. 496

Chapter IV. 508

Chapter V. 516

Chapter VI. 524

Chapter VII. 534

Chapter VIII. 539

Chapter IX. 551

Book VI. 565

An Apologue for an Epilogue. 568

The Rosie Cross Uncovered: The Sixth Book. 571

The Rosie Crucian Prayer 595

A Word from the Publisher 597

The R.A.M.S. Library of Alchemy 598

Colophon 600

Dedicated to Hans W. Nintzel,

Founder of the

Restorers of Alchemical Manuscripts Society

(R.A.M.S.)

Disclaimer

Liability: The publisher does not warrant or assume any legal liability or responsibility for the accuracy, completeness, or usefulness of any information, apparatus, product, or process disclosed. The publisher makes no representation as to the accuracy or completeness of the contents of this book and specifically disclaims any implied warranty of merchantability or fitness for a particular purpose. No warranty may be created or extended by written sales materials or sales representatives. You should obtain professional consultation where appropriate. The publisher shall not be liable for any loss of profit or other commercial or personal damages, including but not limited to special, incidental, consequential, or other damages.

This book is sold for informational purposes only. Neither the publisher nor the editor shall be held accountable for the use or misuse of the information in this book.

Introduction

Philip N. Wheeler

John Heydon (1629 – 1667 or 1670) was an English Rosicrucian, alchemist, and attorney. He published more than ten books during the last twelve years of his life.

In this work is a chapter called "Chemical Dictionary," which contains a useful set of definitions of various terms used in Alchemy. That chapter includes much excellent advice for students of Alchemy, such as, "Try not at first experiments of great cost, or great difficulty, for it will be a great discouragement to thee, and thou will be very apt to mistake."

There is an interesting section on sounds and harmonies: "We have also Sound-houses, where we practise and demonstrate all sounds, and their Generation." (near pg. 55).

The figure near the end of Part I incorporates the image of a serpent (dragon) eating its own tail: Ouroboros. This image is included in various forms in other Alchemical works, including "The Golden Chain of Homer" and "Lamspring's Process," both available in The R.A.M.S. Library of Alchemy (Volumes 33 and 28).

The illustrations were not included in the original R.A.M.S. edition. They were added from the 1662 printed edition. This 2nd edition includes Part I and Part II of the original text.

John Heydon

Portrait engraved by Thomas Cross.

The Holy Guide

LEADING THE WAY TO THE WONDER OF THE WORLD

A complete Physician teaching the knowledge of all things, Past, Present and yet to Come, viz., of Pleasure, Long Life, Health, Youth, Blessedness, Wisdom and Virtue, and to Cure, Change and Remedy all Diseases in both Young and Old.

WITH Rosie Crucian MEDICINES, WHICH
ARE VERIFIED BY A PRACTICAL EXAMINATION
OF PRINCIPLES IN THE GREAT WORLD,
AND FITTED FOR THE EASY UNDERSTANDING,
PLAIN PRACTICAL USE, AND BENEFIT OF
MEAN CAPACITIES.

By John Heydon, Gent., Φιλοτομος, a servant of God, and a Secretary of Nature.

"And he took the golden Calf which they had made, and burned it £n the Fire, and ground it to powder, and strewed it upon the Water, and made the Children of Israel drink of it". EXO. 32, V. 20.

LONDON. Printed by T.M. and are to be sold by Thomas Whittlesey at the Globe in Cannon Street near London-Stone, and at all Booksellers and Shops. 1682.

To the truly Noble
(by all Titles)
Sr. Richard Temple,
Baronet, etc.

External, Internal and Eternal happiness be wished, Honored Sir,

I do observe that every man naturally desires a Superiority, to have Treasures of Gold and Silver, and to seem great in the eyes of the world; God indeed Created all things for the use of man, that he might rule over them, and acknowledge therein the singular and Omnipotence of God, and give him thanks for his benefits; honor him and praise him: But there are few men who look for these things otherwise then by spending their days idly, they would enjoy them without any previous labor and danger; neither do they look them out of that place, where God has treasured them up, who expects also that man should seek for them there, and to those that seek, will he give them: But there is not any that labors for a possession in that place, and therefore these Riches are not found: For the way to this place, and the place itself, has been unknown for a long time, and it is hidden from the greatest part of the world. But notwithstanding it be difficult, and laborious to find out this way and place; yet the place should be sought after; But it is not the will of God to conceal anything from those that are his; and therefore in this last age, before the final judgement comes, all these things shall be manifested to those that are worthy: As he himself (though obscurely, lest it should be manifested to the unworthy) has spoken in a certain place; there is nothing covered that shall not be revealed, and hidden that shall not be known; and therefore being a servant of God, and secretary of nature, we do declare the will of God to the World, which we have also already performed and published in Italy and England; but most men, either revile or contemn our *Harmony of the World*, and *Temple of Wisdom* etc., or else waving the spirit of God, they expect the Proposals thereof from us, supposing we will straight away teach them how to make Gold by Art, or furnish them with ample treasures, whereby they may live pompously in the face of the world, swagger, and make wars, turn Usurera, Gluttons and Drunkards, live unchastely, and defile their whole life with several other sins; all which things are contrary to the blessed will of God; these men should have learnt from those Ten Virgins (whereof five that were foolish demanded Oil for their Lamps, from those five that were wise) how that the case is much otherwise; It is expedient that every man should labor for the treasure by the assistance of God, and his own particular search and industry. But the perverse intentions of these Fellows we understand out of their writings, by the singular Grace and Revelation of God, we do stop our ears, and wrap ourselves, as it were in clouds, to avoid the bellowing and howling of those men, who in vain cry out for Gold. And hence indeed it comes to pass, that they brand us with infinite Calumnies and Slanders, which notwithstanding we do not refute, but God in his good time will judge them for it. But after that we had well known (though unknown to you) and perceived by all your writing, how diligently you are to peruse the holy Scripture and seek the true knowledge of God: We Honor you Sir Richard above thousands, and magnify thus much to you, not, but that you know as much as our self: But as a token of our good will, that may make you mindful of us. There is a Mountain situated in the midst of the Earth, or Center of the World, which is both small and great. It is soft, and also above measure hard and stony. It is far off, and, near at hand, but by the providence of God invisible. In it are hidden most ample treasures, which the world is not able to value. This mountain by envy of the Devil, who always opposes the glory of God, and the happiness of man, is compassed about with every cruel Beast and other ravenous

Birds, which make the way thither both difficult and dangerous. And therefore hitherto, because the time is not yet come, the way thither could not be sought after, nor found out, but now the way is to be found by those that are worthy, but notwithstanding by every man's self-labor and endeavors. To this mountain, you shall go in a certain Night (when it comes) most long and most dark, and see that you prepare yourself by Prayer. Insist upon the way that leads to the Mountain, but ask not of any man where the way lies: Only follow your *Holy Guide*, who will offer himself to you, and will meet you in the way, but you shall not know him. This Guide will bring you to the Mountain at Midnight, when all things are silent and dark. It is necessary that you arm yourself with heroic courage, least you fear those things that will happen, and so fall back: You need no Sword, Horse and Pistols, etc., nor any other bodily weapons, only call upon God sincerely and heartily; When you have discovered the Mountain, the first Miracle that will appear, is this, a most vehement and very great wind that will shake the Mountain, and shatter the Rocks in pieces; you shall be encountered also by Lions and Dragons, and other terrible Beasts, but fear not any of these things, be resolute and take heed that you return not, for your *Holy Guide* that brought you thither, will not suffer any evil to befall you. As for the treasure, it is not yet discovered, but it is very near, after this wind will come an Earthquake that will overthrow those things, which the wind has left, and make all flat; But be sure that you fall not off; the Earthquake being past, there shall follow a fire, that will consume the Earthly Rubbish, and discover the treasure; but as yet you cannot see it: After all these things, and near the day break, there shall be a great Calm, and you shall see the Day Star arise, and the dawning will appear, and you shall perceive a great treasure; the chiefest things and most perfect that are there are written of at large in this Book. These medicines being used, as your *Holy Guide* shall teach you, will make you young when you are old, healthful, long lived, wise and virtuous; and you shall perceive no disease in any part of your body, by means of the things taught in this Book, you shall find Pearls of that Excellency, which cannot be imagined: But do not you arrogate anything to yourself, because of our present power, but be contented with that which the *Holy Guide* shall communicate to you, praise God perpetually for this his gift, and have a specially care that you use not for worldly Pride; but employ it in such works, which are contrary to the world; use it rightly, and enjoy it so, as if you had it not; live a temperate life, and beware of all sin, otherwise the *Holy Guide* will forsake you, and you shall be deprived of this happiness: For, know this of a truth, whosoever abuses what he learns from his *Guide*, and lives not exemplary purely, and devoutly before men, he shall lose this benefit, and scarce any hope will there be left, ever to renew it afterwards. Thus, craving pardon for my boldness, but you may partly thank yourself; You taught me this familiarity: And now I humbly present myself, Sir;

	Your affectionate
March 15.	Servant,
2h. 45.	
P.M.	John Heydon.
1661.	

The Preface.

We travelled from Sydmouth (where we had continued by the space of a whole year) for London and Spain by the South Sea, taking with us Victuals for twelve months; and had good Winds from the East, though soft and weak, for five months space, and more. But then the wind came about, and settled in the West for many days, so as we could make little or no way, and were sometimes in purpose to turn back. But then again there arose strong and great Winds from the South, with a point East, which carried us up, (for all that we could do) towards the North: By which time our Victuals failed us, though we had made good spare of them. So that finding ourselves in the midst of the greatest wilderness of Waters in the World, without Victuals; we gave ourselves for lost men, and prepared for Death. Yet we did lift up our hearts and voices to God above, who shows his wonders in the Deep; Beseeching him of his mercy, that as in the beginning he discovered the Face of the Deep, and brought forth Dry land; So, he would now discover land to us, that we might not perish. And it came to pass; that the next day about evening, we saw within a kenning[1] before us, towards the North, as it were thick Clouds, which did put us in some hope of Land; Knowing how that part of the South Sea was utterly unknown, and might have Islands and Continents, that hitherto had not come to light; Therefore we bent our course thither, where we saw the appearance of Land, all that evening.

And in the Dawning of the next day, we might plainly discern that it was a Land; flat to our sight and full of Boscage[2], which made it show

---

[1] Verified spelling match with 1662 printed edition; apparently indicates an estimate of distance. -pnw

[2] Massed trees or shrubs. -pnw

the more Dark: And after an hour and a half sailing, we entered into a good Haven, being the Port of a fair City; not great indeed, but well built, and that gave a pleasant view from the Sea: And we thinking every minute long, till we were on Land, came close to the shore, and offered to land: But straightaways we saw divers of the people, with Bastons in their hands, (as it were) forbidding us to land; Yet without any cries or fierceness, but only as warning us off, by signs that they made. Whereupon being not a little discomforted, we were advising with ourselves, what we should do. During which time, there made forth to us a small boat, with about eight persons in it: Whereof one of them had in his hand a Tipstaff of a yellow Cane, tipped at both ends with green, who came aboard our ship, without any show of distrust at all. And when he saw one of our number present himself somewhat afore the rest, he drew forth a little Scroll of Parchment (somewhat yellower than our Parchment, and shining like the Leaves of Writing Tablets, but otherwise soft and flexible) and delivered it to our foremost man. In which Scroll were written in ancient Hebrew, and in ancient Greek, and in good Latin of the School, and in Spanish, these words, "Land ye not, none of you; And provide to be gone from this Coast, within sixteen days, except you have further time given you. Meanwhile, if you want fresh Water, or Victual, or help for your Sick, or that your ship needs repair, write down your wants, and you shall have that which belongs to Mercy." This Scroll was signed with a Stamp of Cherubim Wings, not spread, but hanging downwards: And by them a Cross. This being delivered, the Officer returned, and left only a Servant with us to receive our Answer. Consulting hereupon amongst ourselves, we were much perplexed. The denial of Landing, and hasty warning us away, troubled us much: On the other side, to find that the People had Language's, and were so full of humanity, did comfort us not a little. And above all, the sign of the Cross to that Instrument, was to us a great rejoicing, and as it were a certain presage of Good. Our Answer was in the Spanish tongue, that for our ship, it was well; For we had rather met with calms, and contrary winds than any tempests. For our sick, they were many, and in some cases very ill: So that if they were not permitted to Land, they ran in danger of their lives. Our other wants we set down in particular, adding; that we had some little store of Merchandise, which if it pleased them to deal for, it might supply our wants, without being chargeable unto them. We offered some reward in Pistols to the Servant, and a piece of Crimson Velvet to be presented to the Officer: But the Servant took them not, nor would scarce look upon them; And so, left us, and went back in another boat, which was sent for him.

About three hours after we had dispatched our Answer, there came towards us, a Person (as it seemed) of place. He had on him a Gown with wide sleeves, of a kind of Water Chamolot, of an excellent green Color, far glossier than ours: His under apparel was green Azure; And so was his Hat, being in the form of a Turban, daintily made, and not so huge as the Turkish Turbans; And the Locks of his Hair came down below the Brims of it. A reverend Man was he to behold. He came in a Boat, gilt in some part of it, with four Persons more only in that Boat; And was followed by another Boat, wherein were some twenty. When he was come within a Flight-shot of our Ship, Signs were made to us, that we should send forth some to meet him upon the Water; which we presently did in our Ship-boat, or Skiff, sending the principal Men amongst us save one, and four of our Number with him. When we were come within six yards of their Boat, they called to us to stay, and not to approach further; which we did. And thereupon the Man, who I before described, stood up, and with a loud voice, in Spanish, asked; Are ye Christians? We answered; we were; fearing the less, because of the Cross we had seen in the Subscription. At which answer the said person lift up his Right hand towards Heaven, and drew it softly to his mouth, (which is the gesture they use, when they thank God;) And then said; If ye will swear, (all of you) by the Merit of the Savior, that ye are no Pirates; Nor have shed blood, lawfully, nor unlawfully, within forty days past, you may have license to come to Land. We said, we were all ready to take that Oath, whereupon one of those that were with him, being (as it seemed) a Notary, made an Entry of this all. Which done, another of the same Boat, after his Lord had spoken a little to him, said aloud; My Lord would have you know that it is not of Pride, or greatness, that he comes not aboard your Ship; but for that, in your Answer, you declare, that you have many sick amongst you, he was warned by the Conservator of Health, of the City, that he should keep a distance. We were his humble Servants: and accounted for great Honor, and singular Humanity towards us, that which was already done; But hoped well, that the Nature of the sickness, of our Men, was not infectious. So, he returned; And a while after came the Notary to us aboard our Ship; holding in his hand a Fruit of that Country, like an Orange, but of color between Orange-Tawney and Scarlet, which cast a most excellent Odor. He used it (as it seemed) for a preservative against Infections. He gave us our Oath; By the Name of Jesus and his Merits: And after told us, that the next day, by six of the clock in the Morning, we should be sent to, and brought to the stranger's house, (so he called it) where we should be accommodated of things both for our whole and for our sick. So, he left us; And when we offered him some Pistolets, he smiling said; He must not be twice paid for one Labor: Meaning (as I take it) that he had salary sufficient of the State for his Service. For (as I after learned) they call an Officer that takes Rewards, twice paid.

The next Morning early, there came to us the same Officer, that came to us at first with his Cane, and told us; He came to conduct us to the Strangers house; And that he had prevented the hour, because we might have the whole day before us, for our business. For (said he) if you Will follow mine advice, there shall first go with me some few of you, and see the place, and how it may be made convenient for you; And then you may send for your Sick, and the rest of your Number, which ye will bring on Land. We thanked him, and said, that this care which he took of desolate strangers, God would reward. And so, six of us went ashore with him: And when we were landed, he went before us, and turned to us, and said; he was our Servant, and our guide. He led us through three fair streets; And all the way we went, there were gathered some people on both sides, standing in a row; but in so civil a fashion, as if it had been, not to wonder at us, but to welcome us; And divers of them, as we passed by them, put their arms a little abroad; which is their Gesture, when they bid any welcome. The stranger's house is a fair and spacious house, built of brick, of somewhat a bluer Color than our brick; And with handsome windows, some of glass, some of a kind of Cambrick oiled. He brought us first into a fair Parlor above stairs, and then asked us; What number of Persons we were? And how many sick? We answered, we were in all (sick and whole) 250 Persons, whereof our sick, there were seventeen. He desired us to have patience a little, and to stay till he came back to us; which was about an hour after; And then he led us to see the Chambers, which were provided for us, being in number 250. They having cast it (as it seemed) that four of those Chambers, which were better than the rest, might receive four of the principle Men of our company; and lodge them alone by themselves; and the rest were to lodge us. The Chambers were handsome and cheerful Chambers, and furnished civilly. Then he led us to a long Gallery like a Porture, where he showed us all along the one side, (for the other side was but Wall and Window) Seventeen Cells, very neat ones, having partitions of Cedar wood. Which Gallery and Cells, being in all 900, (many more than we needed) were instituted as an Infirmary; for sick persons. And he told us withal, that as any our sick waxed well, he might be removed from his Cell to a Chamber: For which purpose, there were set forth ten spare Chambers, besides the number we spoke before. This done, he brought us back to the Parlor, and lifting up his cane a little, (as they do when they give any Charge or Command) said to us: Ye are to know, that the custom of the Land requires, that after this day, and tomorrow, (which we give you for removing of your people from your ships) you are to keep within doors for three days. But let it not trouble you, nor do not think yourselves restrained, but rather left to your rest and ease. You shall want nothing, and there are six of our people appointed to attend you, for any business you have abroad. We gave him thanks, with all affection and respects, and said; God surely is manifested in this Land. We offered him also twenty Pistolets; But he smiled, and only said; What? Twice paid! And so, he left us. Soon after our Dinner was served in; which was right good Viands, both for bread, Meat, Wine, etc. Better than any Collegiate Diet that I have known in Europe. We had also drink of three sorts, Ale, Beer, Cider, all wholesome and good; Wine of the Grape, another drink of Grain, such as is with us our Mum, but more clear: And a kind of Berry like the Pear juice, made of a fruit of that Country; A wonderful pleasing and refreshing Drink Besides, there were brought in to us, great store of those Scarlet Oranges, for our Sick; which (they said) were an assured Remedy for sickness taken at sea. There was given us also a Box of small gray, or whitish Pills, which they wished our sick should take, one of the Pills, every night before sleep; which (they said) would hasten their recovery. The next day, after that our trouble of Carriage and Removing of our men and goods, out of our ship, was somewhat settled and quiet, I thought good to call our Company together, and when they were assembled, said unto them; My dear friends, let us know ourselves, and how it stands with us. We are Men cast on Land as Jonas was, out of the Whales Belly, when we were as buried in the

Deep: And now we are on Land, we are but between Death and Life; For we are beyond, both the old World, and the New; And whether ever we shall see Europe, God only knows. It is a kind of Miracle has brought us hither: And it must be little less, that shall bring us hence. Therefore, in regard of our Deliverance past, and our danger present, and to come, let us look up to God, and every man reform his own ways. Besides, we are come here amongst a Christian People, full of Piety and Humanity: Let us not bring that Confusion of face upon ourselves, as to show our vices, or unworthiness before them. Yet there is more. For they have by Commandment, (though in form of Courtesy) Cloistered us within these walls, for three days; who knows, whether it be not, to take some taste of our manners and conditions? And if they find them bad, to banish us straight ways; if good, to give us further time. For these men, that they have given us for attendance, may withal have an eye upon us. Therefore, for God's love, and as we love the weal of our Souls and Bodies, let as so behave ourselves, as we may be at peace with God, and may find grace in the eyes of this People. Our Company with one voice thanked me for my good Admonition, and promised me to live Soberly and Civilly, and without giving any the least occasion of Offence. So, we spent our three days joyfully and without care, in expectation what would be done with us, when they were expired. During which time, we had every hour joy of the amendment of our sick; who thought themselves cast into some Divine Pool of Healing; They mended so kindly and so fast, as you may read in our *Temple of Wisdom*.

The morrow after our three days were past, there came to us a new Man, that we had not seen before, clothed in Azure, as the former was, save that his Turban was white, with a small red Cross on the Top. He had also a Tippet of fine Linen. At his coming in, he did bend to us a little, and put his arms broad. We of our parts saluted him in a very lowly and submissive manner; As looking that from him, we should receive Sentence of Life, or Death. He desired to speak with some few of us: Whereupon six of us only stayed, and the rest avoided the room. He said; I am by Office Governor of this House of Strangers, and by Vocation I am a Christian Priest, and of the order of the Rosie Cross; and therefore, am come to you to offer you my service, both as strangers, and chiefly as Christians. Some things I may tell you, which I think you will not be unwilling to hear. The State has given you License to stay on Land, for the space of six weeks: And let it not trouble you, if your occasions ask further time, for the Law in this point is not precise; And I do not doubt, but myself shall be able to obtain for you, such further time, as may be convenient. Ye shall also understand, that the Strangers House, is at this time Rich, and much aforehand; For it has laid up Revenue these 36,000 years: For so long it is since any Stranger arrived in this part. And therefore, take ye no care; the State will defray you all the time you stay: Neither shall you stay one day the less for that. As for any Merchandize ye have brought, ye shall be well used, and have your return, either in Merchandize, or in Gold and Silver: For to us it is all one. And if you have any other Requests to make, hide it not. For ye shall find, we will not make your Countenance to fall, by the Answer ye shall receive. Only this I must tell you that none of you must go above a Julo, or Karan (that is with them, a Mile and a half) from the walls of the City, without special leave. We answered, after we had looked a while one upon another, admiring this gracious and parent-like usage; That we could not tell what to say:

For we wanted words to express our thanks; And his Noble free Offers left us nothing to ask. It seemed to us, that we had before us a picture of our Salvation in Heaven: For we that were a while since in the jaws of Death, were now brought into a place where we found nothing but

Consolations. For the Commandment laid upon us, we would not fail to obey it, though it was impossible but our Hearts should be enflamed to tread further upon this happy Holy Ground. We added, that our Tongues should first cleave to the Roofs of our Mouths, ere we should forget, either his Reverend Person, or this whole Nation, in our Prayers. We also most humbly besought him, to accept of us as his true Servants, by a just Right as ever men on earth were bounden; laying and presenting, both our Persons, and all we had, at his feet. He said, He was a Priest, and looked for a Priest's reward; which was our Brotherly love, and the good of our Souls and bodies. So, he went from us, not without tears of tenderness in his eyes; And left us also confused with joy and kindness, saying amongst ourselves, that we were come into a Land of Angels, which did appear to us daily, and present us with Comforts, which we thought not of, much less expected.

The next day about 10 of the Clock, the Governor came to us again, and after Salutation, said familiarly; That he had come to visit us; And called for a Chair, and sat him down; And we being some ten of us, (the rest were of the meaner sort, or else gone abroad) sat down with him. And when we were set, he began thus. We of this Island of Apanua or Christie in Arabia (for so they call it in their language) have this, that by means of our Solitary Situation, and of the Laws of Secrecy, which we have for our Travelers, and our rare admission of Strangers, we know well most parts of the Habitable World, and are ourselves unknown. Therefore, because he that knows least, is fittest to ask questions, it is more reason, for the Entertainment of the time, that ye ask me questions, than that I ask you. We answered: That we humbly thanked him, that he would give us leave so to do: And that we conceived by the taste we had already, that there was no worldly thing on Earth, worthier to be known than the State of that happy

Land. But above all (we said) since that we were met from the several Ends of the World, and hoped assuredly, that we should meet one day in the Kingdom of Heaven (for that we were both parts Christians) we desired to know (in respect that Land was so remote, and so divided by vast and unknown Seas, from the Land, where our Savior walked on Earth) who was the Apostle of that Nation, and how it was converted to the Faith? It appeared in his face, that he took great contentment in this question in the first place; For it shows that you first seek the Kingdom of Heaven; And I shall gladly, and briefly, satisfy your demand.

About twenty years after the Ascension of our Savior, it came to pass, that there was seen by the people of Dancar (a City upon the Eastern Coast of our Island) within Night, (the Night was cloudy and calm) as it might be some mile into the Sea, a great Pillar of Light; Not sharp but in form of a Column or Cylinder, rising from the Sea, a great way up towards Heaven; and on the top of it was seen a large Cross of Light, more bright and resplendent than the body of the Pillar. Upon which so strange a spectacle, the people of the City gathered a space together upon the Sands to wonder; And so, after put themselves into a number of small Boats, to go nearer to this marvelous sight. But when the Boats were come within (about) 60 yards of the Pillar, they found themselves all bound, and could go no farther; yet so as they might move to go about, but might not approach nearer. So, as the Boats stood all as in a Theater, beholding this Light as a Heavenly Sign. It so fell out, that there was in one of the Boats, one of the wise Men of the Society of the Rosie Crucians, whose House or College (my good Brethren) is the very Eye of this Kingdom, who having a while attentively and devoutly viewed, and contemplated this Pillar and Cross, fell down upon his face; and then he raised himself upon his knees, and lifting up his hands to Heaven, made his prayers in this manner.

Lord God of Heaven and Earth; you have vouchsafed of thy Grace, to those of our Order, to know thy works of Creation and the Secrets of them; And to discern (as far as appertains to the Generation of Men) Between divine Miracles, Works of Nature, Works of Art, and Impostures, and Illusions of all sorts. I do here acknowledge and testify before this people, that the Thing which we now see before our eyes, is thy Finger, and a true Miracle. And for as much as we learn in our Books that thou never works Miracles but to a Divine and excellent End (for the Laws of Nature are thine own Laws, and thou exceeds them not but upon great cause), we most humbly beseech thee, to prosper this great Sign; And to give us the Interpretation and use of it in Mercy; which you do in some part secretly promise, by sending it unto us.

When he had made his Prayer, be presently found the Boat he was in, moveable, and unbound; whereas all the rest remained still fast; And taking that for an assurance of Leave to approach, he caused the Boat to be softly, and with silence, rowed towards the Pillar. But ere he came near it, the Pillar and Cross of Light broke up, and cast itself abroad, as it were, into a firmament of many Stars; which also vanished soon after, and there was nothing left to be seen but a small Ark or Chest of Cedar, dry, and not wet at all with water, though it swam. And in the Fore-end of it which was towards him, grew a small green Branch of Palm; And when the Rosie Crucian had taken it with all reverence into his Boat, it opened of itself, and there were found in it a Book and a Letter; Both written in fine Parchment, and wrapped in Sindons of Linen. The Book contained all the Canonical Books of the Old and New Testament, according as you have them; (For we know well what the Churches with you receive;) And the Apocalypse itself; And some other Books of the New Testament, which were not at that time written, were nevertheless in the Book. And for the Letter, it was in these words.

I, John, a Servant of the Highest, and Apostle of Jesus Christ, was warned by an Angel, that appeared to me in a vision of Glory, that I should commit this Ark to the floods of the Sea. Therefore, I do testify and declare unto that people where God shall ordain this Ark to come to Land, that in the same day, is come unto them Salvation and Peace, and good Will, from the Father, and from the Lord Jesus.

There was also in both these writings, as well the Book, as the Letter, wrought a great Miracle, Conforms to that of the Apostles in the Original gift of Tongues. For there being at that time, in this Land, Hebrews, Persians, and Indians, besides the Natives, every one read upon the book and the Letter, as if they had been written in his own Language. And thus, was this Land saved from Infidelity, (as the Remains of the Old World was from Water) by an Ark, through the Apostolical and Miraculous Evangelism of Saint John. And here he paused, and a Messenger came, and called him from us. So, this was all that passed in that Conference.

The next day the same Governor came again to us, immediately after dinner, and excused himself, saying: That the day before, he was called from us somewhat abruptly, but now he will make us amends, and spend time with us, if we held his Company and Conference agreeable. We answered, that we held it so agreeable and pleasing to us, as we forgot both dangers past, and fears to come, for the time we heard him speak; And that we thought an hour spent with him, was worth years of our former life. He bowed himself a little to us, and after we were set again he said; Well, the Questions are on your part. One of our numbers said after a little pause; That there was a Matter we were no less desirous to know, then fearful to ask, least we might presume too far. But encouraged by his rare Humanity towards us, (that could scarcely think ourselves strangers, being his vowed and professed Servants) we would take the Hardiness to propound it: Humbly beseeching him, if he thought it not fit to be answered, that he would pardon it, though he rejected it. We said, We well observed those his words, which he formerly spoke, that this happy Island, where we now stood, was known to few, and yet knew most of the Nations of the World; which we found to be true, considering they had the Languages of Europe, and knew much of our state and business; And yet we in Europe, (notwithstanding all the remote Discoveries, and Navigations of this last Age) never heard any of the least Inkling or Glimpse of this Island. This we found wonderfully strange; For that all Nations have Enter—knowledge one of another, either by Voyage into Foreign Parts, or by Strangers that come to them:

And though the Traveler into a Foreign Country, does commonly know more by the eye, than he that stays at home can by relation of the Traveler; Yet both ways suffice to make a mutual knowledge, in some degree, on both parts. But for this Island, we never heard tell of any Ship of theirs, that had been seen to arrive upon any shore of Europe; No nor of either the East or West Indies, nor yet of any ship of any other part of the World that had made return from them.

And yet the Marvel rested not in this: For the Situation of it (as his Lordship said) in the Secret Conclave of such a vast Sea might cause it. But then that they should have knowledge of the Languages, Books, Affairs of those that lye such a distance from them, it was a thing we could not tell what to make of; For that it seemed to us a condition and propriety of Divine Powers and Beings, to be hidden and unseen to others, and yet to have others open, and as in a light to them. At this speech the Governor gave a gracious smile and said; That we did well to ask pardon for this Question we now asked: For that it imported, as if we thought this Land, a Land of Magicians, that sent forth Spirits of the Air into all ports, to bring them news and intelligence of other Countries, it was answered by us all, in all possible humbleness, but yet with a Countenance taking knowledge, that we knew he spoke it but merrily, That we were apt enough to think, there was somewhat Supernatural in this Island, but yet rather as Angelical, than Magical. But to let his Lordship know truly, what it was that made us tender and doubtful to ask this Question, it was not any such conceit, but because we remembered, he had given a Touch in his former speech, that this Land had Laws of Secrecy touching strangers. To this he said you remember it aright:

And therefore, in that I shall say to you, I must reserve some particulars that it is not Lawful for me to reveal; but there will be enough left to give you satisfaction.

You shall understand that which perhaps you will scarcely think credible, that about three thousand years ago, or somewhat more, the Navigation of the world (specially for remote voyages) was greater than at this day. Do not think with yourselves, that I know not how much it is increased with you within these six score years: I know it well; and yet I say, greater then, than now:

Whether it was, that the example of the Ark, that saved the remnant of men from the Universal Deluge, gave men confidence to adventure upon the Waters; Or what it was: But such is the Truth. The Phoenicians, and specially the Tyrians, had great Fleets. So, had the

Carthaginians their Colony, which is further West. Toward the East the shipping of Egypt, and of Palestine was likewise great. China also, and the great America, which have now but Junks, and Canoes, abounded then in tall ships. This Island, (as appears by faithful Registers of those times) had then fifteen hundred strong Ships, of great content. Of all this, there is with you sparing Memory, or none; But we have large knowledge thereof.

At that time, this land was known and frequented by the ships and Vessels of all the nations before named; (And as it cometh to pass) they had many times men of other Countries, that were no Sailors, that came with them: As Persians, Chaldeans, Egyptians and Grecians. So as almost all Nations of Might and Fame resorted hither: Of whom we have some Stirps, and little Tribes with us, at this day. And for our own ships, they went on sundry Voyages, as well to your straights, which you call the Pillars of Hercules, as to other parts in the Persian and Mediterranean Seas; As to Paguin, (which is the same with Cambaline) and Quinzy, upon the Oriental Seas, as far as to the Borders of the East Tartary.

At the same time, and an age after, or more, the Inhabitants of the Holy Land did flourish. For though the Narration and description, which is made by a great Man with you, that the Descendants of Neptune planted there; and of the Magnificent Temple, Palace, City and Hill; See my Rosie Crucian *Infallible Axiomata*, and the manifold streams of goodly Navigable rivers, (which as so many Chains environed the same Site, and Temple;) And the several Degrees of Ascent, whereby men did climb up to the same, as if it had been a *Scala Caeli*, be all Poetical and Fabulous: Yet so much is true, that the said Country of Judea, as well that of Peru then called Goya, as that of Mexico then named Tyrambel, were mighty and proud Kingdoms, in Arms,

Shipping, and Riches: So mighty, as at one time (or at least within the space of ten years), they both made two great expeditions: They of Tyrambel through Judea to the Mediterranean Sea; and they of Goya through the South Sea upon this our Island: And for the former of these, which was into Europe, the same Author amongst you (as it seems), had some relation from his Beata, whom he recites: See the *Harmony of the World*, lib. 1, The Preface which indeed is an introduction to the work. For assuredly such a thing there was. But whether it was the ancient Athenians that had the glory of the Repulse, and Resistance of those Forces, I can say nothing: But certain it is, there never came back, either Ship, or Man, from that Voyage. Neither had the other Voyage of those of Goya upon us, had better fortune, if they had not met with Enemies of great clemency. For the King of this Island, (by name Phroates who was raised three times from death to life;) a wise Man, and a great Warrior; Knowing well both his own strength, and that of his Enemies; handled the matter so, as he cut off their Land-forces from their ships; and entoyled both their Navy, and their Camp with a greater power than theirs, both by Sea and land: And compelled them to render themselves without striking stroke: And after they were at his mercy, contenting himself only with their Oath, that they should no more bear Arms against him, dismissed them in safety. But the Divine Revenge overtook not long after those proud enterprises. For within less than the space of one Hundred years, the Island was utterly lost and destroyed: Not by a great Earthquake, as your man says, (For that whole Tract is little subject to Earthquakes) But by a particular Deluge or Inundation; those Countries having, at this day, far greater Rivers, and far greater Mountains, to pour down Waters, than any part of the Old World. But it is true, that the same Inundation was not deep, not past forty feet, in most places, from the ground; So that, although it destroyed man and beast generally, yet some few wild Inhabitants of the Wood escaped.

Birds also were saved by flying to the high Trees and Woods. As for Men, although they had Buildings in many places, higher than the Depth of the Waters, yet that Inundation, though it were shallow, had a long continuance; whereby they of the Vaile, that were not drowned, perished for want of food, and other things necessary. So as marvel you not at the thin Population of America, nor at the rudeness and ignorance of the people; for you must account your inhabitants of America as a young People; Younger a thousand years, at the least than the rest of the world; For that there was so much time, between the Universal Flood, and this Particular Inundation. For the poor Remnant of Human Seed, which remained in their Mountains, Peopled the Country again slowly, by little and little; And being simple and savage People, (Not like Noah and his Sons which was the chief family of the Earth) they were not able to leave Letters, Arts, and Civility, to their Posterity; And having likewise in their Mountainous Habitations been used, (in respect of the extreme cold of those Regions,) to clothe themselves with the Skins of Tigers, Bears, and great Hairy Goats, that they have in those parts; When after they came down into the Valley, and found the intolerable heats which are there, and knew no means of lighter apparel; they were forced to begin the Custom of going naked, which continues at this day. Only they take great pride and delight in the Feathers of Birds, that came up to the high Grounds, while the Waters stood below. So, you see, by this main Accident of time, we lost our traffic with the Americans, with whom, of all others, in regard they lay nearest to us, we had most commerce. As for the other parts of the World, it is most manifest, that in the ages following, (whether it was in respect of Wars, or by a natural revolution of time,) Navigation did everywhere greatly decay; and especially, far voyages, (the rather by the use of Galleys, and such vessels as could hardly brook the Ocean,) were altogether left and omitted. So then, that part of intercourse, which could be from other Nations, to sail to us, you see how it has long since ceased; Except it were by some rare Accident, as this of yours.

But now of the Cessation of that other part of intercourse, which might be by our saying to other Nations, I must yield you some other cause. For I cannot say, (if I shall say truly) but our Shipping, for Number, Strength, Mariners, Pilots, and all things that appertain to Navigation, is as great as ever: And therefore, why we should sit at home, I shall now give you an account by itself; And to will draw nearer, to give you satisfaction, to your principal Question.

There reigned in this Island, about nineteen hundred years ago, a King, whose memory of all others we most adore; Not Superstitiously, but as a divine instrument, though a mortal man: His name was Eugenius Theodidactus, you may read this at large in our Idea of the Law: and we esteem him as the Law-giver of our Nation. This King had a large heart, inscrutable for good; and was wholly bent to make his Kingdom and people happy. He therefore taking into consideration, how sufficient and substantive this Land was to maintain itself without any aid (at all) of the Foreigner; being 56,000 Miles in circuit and of rare Fertility of Soil, in the greatest part thereof; And finding also the shipping of this country might plentifully set on work, both by fishing and by transportations from Port to Port, and likewise by sailing into some small Islands that are not far from us, and are under the Crown and Laws of this State: And recalling into memory, the happy and flourishing estate, wherein this Land then was, so as it might be a thousand ways altered to the worse, but scarce any one way to the better; though nothing wanted to his Noble and Heroical intentions, but only as far as Human foresight might reach to give perpetuity to that, which was in his time so happily established. Therefore, amongst his other fundamental Laws of this Kingdom, he did ordain the

Interdicts and Prohibitions, which we have touching of Strangers; which at that time (though it was after the calamity of America) was frequent; Doubting novelties, and commixture of Manners. It is true, the like Law, against the admission of strangers without License, is an ancient law, in the Kingdom of China, and yet continued in use. But there it is a poor thing: Read our Book called *The Fundamental Element of Moral Philosophy, Policy Government and Laws*. And has made them a curious ignorant, fearful, foolish nation. But our Law-giver made his Law of another temper. Read our Book called *The Idea of the Law*, etc. For first, he has preserved all points of humanity, in taking Order, and making provision for the relief of strangers distressed, whereof you have tasted. At which Speech (as reason was) we all rose up, and bowed ourselves. He went on. That King also still desiring to join humanity and policy together; and thinking it against humanity, to detain strangers here against their wills; And against policy, that they should return, and discover their knowledge of this state, he took this course: He did ordain, that of the strangers, that should be permitted to Land, as many at all times might depart as would; but as many as would stay, should have very good conditions, and means to live, from the state. Wherein he saw so far, that now in so many ages since the Prohibition, we have memory not of one ship that ever returned, and but of thirteen persons only, at several times, that chose to return in our Bottoms. What those few may have reported abroad, I know not. But you must think, whatsoever they have said, could be taken where they came, but for a Dream. Now for our Travelling from hence into parts abroad, our Law giver thought fit altogether to restrain it; read our *Idea of Government*, etc. So is it not in China. For the Chinese sail where they will, or can; which shows, that their Law of keeping out strangers, is a Law of Pusisanimitie, and fear. But this restraint of ours, has only one Exception, which is admirable; preserving the good which comes by communicating with strangers, and avoiding the hurts: And I will now open it to you. And here I shall seem a little to digress but you will by and by find it pertinent. Ye shall understand, (my dear friends) that among the excellent acts of that King, one above all has the preeminence, it was the Erection and Institution of an Order, or Society, which we can The Temple of the Rosie Cross; The noblest Foundation, (as we think) that ever was upon the earth; And the Lanthorne of this Kingdom. It is dedicated to the study of the works, and Creatures of God. Some think it bears the Founders name a little corrupted, as if it should be F.H.R.C. his house. But the Records write it, as it is spoken. So, as I take it to be denominate of the King of the Hebrews which is famous with you, and no stranger to us: For we have some parts of his works, which with you have lost; namely that Rosie Crucian M which he wrote of all things past, present or to come; And of all things that have life and motion. This makes me think, that our king finding himself to Symbolize, in many things, with that King of the Hebrews (which lived many years before him) honored him with the Title of this Foundation. And I am the rather induced to be of this opinion, for that I find in ancient Records, this Order or Society of the Rosie Cross is sometimes called the Holy House, And sometimes the College of the Six days Works? Whereby I am satisfied, that our Excellent King had learned from the Hebrews, that God had created the World, and all that therein is, within six days; And therefore he instituting that house, for the finding out of the true Nature of things, (wherewith God might have the more Glory in the workmanship of them, and men the more fruit in the use of them), did give it also that second name. But now to come to our present purpose; When the King had forbidden, to all his people, Navigation into any part, that was not under his Crown, he had nevertheless this Ordinance: That every twelve years there should be set forth, out of this Kingdom, two ships appointed to several voyages; That in either of these Ships, there should be a mission of three of the Fellows, or Brethren of the holy house; whose errand was only to give us Knowledge of the Affaires and State of those Countries, to which they were designed; And especially of the Sciences, Arts, Manufactures, and Inventions of all the world; And withal to bring unto us, Books, Instruments and Patterns, in every kind: That the ships after they had landed the Brethren of the Rosie Cross should return; And that the Brethren R.C. should stay abroad till the new Mission. These ships are not otherwise fraught, then with store of Victuals, and good quantity of Treasure to remain with the Brethren, for the buying of such things, and rewarding of such persons as they should think fit. Now for me to tell you how the Vulgar sort of Mariners are contained from being discovered at land; And how they that must be put to shore for any time, Color themselves under the name of other Nations, and to what places these voyages have been designed: and what places of Rendezvous are appointed for the new missions, and the like circumstances of the practique, I may not do it; Neither is it much to your desire. But thus, you see, we maintain a Trade, not for Gold, Silver, or Jewels; nor for Silk, nor for Spices, nor any other commodity of matter; But only for Gods first Creature, which was Light: To have Light (I say) of the Growth of all Parts of the World. And when he had said thus, he was silent; and so were we all. For indeed we were all astonished, to hear so strange things so probably told. And he perceiving, that we were willing to say somewhat, but had it not ready, in great courtesy took us off, and descended to ask us questions of our voyage and fortunes, and in the end concluded, that we ought to do well, to think with ourselves, what time of stay we would demand of the state; And bade us not to scant ourselves; for he would procure such time as we desired. Whereupon we all rose up, and presented ourselves to kiss the skirt of his Tippet, but he would not suffer us, and so took his leave. But when it came once amongst our people, that the State used to offer conditions to strangers that would stay, we had work enough to get any of our Men to look to our Ship; and to keep them from going presently to the Governor, to crave conditions. But with much ado we restrained them, till we might agree what course to take.

We took ourselves now for freemen, seeing there was no danger of our utter Perdition; And lived most joyfully, going abroad, and seeing what was to be Seen, in the City and places adjacent, within our *Tedder*; And obtaining acquaintance with many of the City, not of the meanest Quality; at whose hands we found such humanity, and such a freedom and desire, to take strangers, as it were, into their bosom, as was enough to make us forget all that was dear to us, in our own Countries; and continually we met with many things, right worthy of Observation and relation: As indeed, if there be a Mirror in the World worthy to hold men's eyes, it is that Country. One day there were two of our Company bidden to feast of the fraternity, and as they call it; a most Natural, Pious, and Reverend custom it is, showing that nation to be compounded of all Goodness. This is the manner of it. It is granted to any Man, that shall live to see thirty persons, descended of his body, alive together, and all above three years old, to make this Feast, which is done at the cost of the State. The Father of the fraternity, whom they call the R.C. two days before the Feast, taketh to him three of such friends as he likes to choose; And is assisted also by the Governor of the City, or place where the feast is celebrated; and all the Persons of the family of both Sexes, are summoned to attend him. These two days the Rosie Crucian sits in consultation, concerning the good estate of the Fraternity. There, if there be any discord or suits between any of the Fraternity, they are compounded and appeased. There, if any of the family be distressed or decayed, order is taken for their Relief, and competent means to live. There, if any be subject to vice, or take ill Courses, they are reproved and Censured. So likewise, direction is given touching Marriages, and the Courses of life, which any of them should take, with divers other the like Orders and Advices. The Governor assists, to the end to put in Execution, by his Public Authority, the Decrees and Orders of the Tirsan, if they should be disobeyed; though that seldom needs; such reverence and obedience they give, to the order of Nature. The Tirsan doth also then ever chose one man from amongst his Sons, to live in house with him; Who is called, ever after the Son of the Vine. The reason will hereafter appear. On the Feast day, the Father or Tirsan comes forth after Divine Service, into a large Room, where the Feast is celebrated; Which room has a half pace at the upper end. Against the wall, in the middle of the Half-pace, is a Chair placed for him, with a Table and Carpet before it. Over the Chair is a State, made Round or Oval, and it is of Ive; a lye somewhat whiter than ours, like the leaf of a Silver Asp, but more shining; For it is green all Winter. And the State is curiously wrought with silver and silk of divers Colors, braiding or binding in the Ivy; And is ever of the work of some of the Daughters of the family; and tailed over at the top, with a fine net of silk and silver. But the substance of it, is true lye; whereof, after it is taken down, the Friends of the Family, are desirous to have some Leaf or Sprig to keep. The Tirsan comes forth with all his Generation or Linage, the Males before him, and the Females following him; and if there be a Mother, from whose body the whole Lineage is descended, there is a Traverse placed in a loft above, on the right hand of the Chair, with a privie Door, and a carved window of Glass; leaded with Gold and blue, where she sits, but is not seen. When the Tirsan is come forth, he sits down in the Chair; and all the Linage place themselves against the wall, both at his back, and upon the return of the Half-pace, in order of their years, without difference of Sex, and stand upon their feet. When he is set, the Room being a ways full of company, but well-kept and without disorder, after some pause, there comes in from the lower end of the Room, a Taratan (which is as much as a Herald); And on either side of him two young Lads; Whereof one carries a scroll of their shining yellow Parchment: And the other a Cluster of Grapes of Gold, with a long Foot or Stalk. The Herald, and Children, are clothed with mantles of Seawater green Satin; But the Heralds Mantle is streamed with Gold, and has a Traine. Then the Herald with three curtsies, or rather inclinations, comes up as far as the Half-pace; And there first taketh into his Hand the Scroll. This Scroll is the King's Charter, containing Gifts of Revenue and many Privileges, Exemptions, and points of honor, granted to the Father of the Fraternity; And it is ever styled and directed, to Such a one, our well-beloved friend and Creditor: Which is a Title proper only to this Case. For they say, the King is Debtor to no man, but for Propagation of his subjects. The Seal set to the Kings Charter, is R.C. and the Kings image embossed or molded in Gold; And though such Charters be expedited of Course, and as of Right; yet they are varied by discretion according to the Number and Dignity of the Fraternity. This Charter the Herald read aloud, and while it is read, the father or Rosie Crucian stands up, supported by two of his sons, such as he chooses. Then the Herald mount the half-pace, and deliver the Charter into his Hands; and with that there is an Acclamation, by all that are present, in their Language, which is thus much, Happy are the people of Apamia. Then the Herald taketh into his hand from the other child, the cluster of Grapes, which is of Gold; both the stalk and the grape. But the Grapes are daintily Enameled; And if the males of the Holy Island be the greater number, the Grapes are enameled Purple, with a little Sunset on the top; If the females, then they are enameled into a greenish yellow, with a Crescent on the top. The Grapes are in number as many as there are Descendants of the Fraternity. This Golden cluster, the Herald delivers also to the Rosie Crucian, who presently delivers it over to that Son that he had formerly chosen to be in House with him; who bears it before his Father as an ensign of Honor, when he goes in public ever after; and is thereupon called The Son of the Vine. After this Ceremony ended, the Father or Rosie Crucian retires; and after some time, comes forth again to Dinner, where he sits alone under the State, as before; and none of his descendants sit with him, of what degree or dignity soever, except he happens to be of the Holy House. He is served only by his own children, such as are Male; who perform unto him all service of the table upon the knee; and the Women only stand about him, leaning against the wall. The room below the half pace, has Tables on the sides for the Ghosts that are bidden; Who are served with great and comely orders; and towards the end of Dinner (which in the greatest feasts with them, lasts never above an hour and an half) there is an Hymn sung, varied according to the Invention of him that composed it; (for they have excellent Poesy) but the Subject of it is, (always) the praises of Adam, and Noah, and Abraham, whereof the former two peopled the world, and the last was the father of the faithful. Concluding ever with a thanksgiving for the Nativity of our Savior Jesus Christ, in whose Birth the Births of all are only blessed. Dinner being done, the R. Crucian returns again; And having withdrawn himself alone into a place where he makes some private Prayers, he comes forth the third time, to give the Blessing with all his descendants, who stand about him as at the first. Then he called them forth by one and by one, by name, as he pleases, though seldom the Order of Age be inverted. The person that is called (The Table being before removed) kneels down before the chair, and the Father lays his hand upon his head or her head, and gives the blessings in these words: Son of the Holy Island (or Daughter of the Holy Island), thy Father says it; The man by whom you have breath and life, speaks the word; The Blessing of the Everlasting Father, the Prince of Peace and the Holy Dove, be upon thee, and make the days of thy Pilgrimage good and many. This he says to every One of them; And that done, if there be any of his Sons of eminent Merit and Virtue, (so they be not above two) he calleth for them again, and says, laying his arm over their shoulders, they standing; Sons, it is veil ye are borne, give God the praise, and persevere to the end. And withal delivers to either of them a Jewel, made in the figure of an ear of wheat, which they ever after do wear in the front of their Turban, or Hat. This done, they fall to Music and dances, and other Recreations, after their manner, for the rest of the day. This is the full order of that Feast of the Rosie Cross.

By that time, six or seven days were spent, I was fallen into straight Acquaintance with a Merchant of that City, whose Name was Nicholes Walford, and his man; Sede John Booker; He was a Jew and Circumcised: For they have some few Strips of Jews yet remaining amongst them, whom they leave to their own religion: Which they may the better do, because they are of a far differing Disposition from the Jews in other parts. For whereas they hate the Name of Christ; And have a secret inbred Rancor against the people amongst whom they live; These (contrariwise) give unto our Savior many high Attributes, and love the Nation of Chassalonia extremely. Surely this man of whom I speak, would ever acknowledge that Christ was born a Virgin; And that he was more than a man: And he would tell how God made him Ruler of the Seraphems which guard his Throne; read the *Harmony of the World*. And they call him also the milken way Emepht[3] and the Eliah of the Messiah, and many other high Names; which though they be inferior to his Divine Majesty, yet they are far from the language of other Jews. And for the Country of Apamia, the Holy Island or Chassalonia, for it is one place this man would make no end of commending it; Being desirous by tradition amongst the Jews there to have it believed, that the People thereof were of the generations of Abraham, by another son, whom they call Machoran; And that Moses

---

[3] Spelled exactly as in the 1662 edition. -pnw

by a secret Cabala, read *The Temple of Wisdom* lib. 4 ordained the Laws of Bensalem which they now use; and that when the Messiah Should come, and sit in his Throne at Jerusalem, the King of Chassanlonia, should sit at his feet, whereas other Kings should keep a great distance. But yet setting aside the Jewish Dreams, the man was a wise man, and learned, and of great Policy, and excellently seen in the Laws and customs of that Nation. Amongst other discourses, one day, I told him, I was much affected with the Relation I had, from some of the Company, of their customs in holding the Feast of the Fraternity: For that (me thought) I had never heard of a Solemnity, wherein Nature did so much preside. And because Propagation of families proceeds from the Nuptial copulation, I desired to know of him, what Laws and customs they had concerning Marriage; And whether they kept Marriage well; And whether they were tied to one wife; For that where Population is so much affected, and such as with them it seemed to be, there is commonly Permission of Plurality of Wives. To this he said; You have Reason for to commend that excellent Institution of the Feast of the Family. And indeed, we have Experience, that those Families that are partakers of the blessing of that Feast, do flourish and prosper ever after, in an extraordinary manner. But hear me now, and I will tell you what I know. You shall understand, that there is not under the Heavens, so chaste a Nation, as this of Apamia; Nor so free from all Pollution, or foulness. It is the Virgin of the World. I remember I have read in one of your European Books, of a holy Hermit amongst you, that desired to see the Spirit of Fornication, and there appeared to him a little foul ugly Æthiope. But if he had desired to see the Spirit of Chastity of the Holy Island, it would have appeared to him in the likeness of a fair beautiful Cherubim. For there is nothing, amongst Mortal men fairer and more admirable than the chaste Minds of this people. Know therefore, that with them there are no Stewes, no dissolute Houses, no courtesans, nor anything of that kind. Nay they wonder (with detestation) at you in Europe, which permit such things. They say ye have put marriage out of office: For marriage is ordained a remedy for unlawful concupiscence; And natural concupiscence seems as a Spirit to marriage. But when men have at hand a remedy more agreeable to their corrupt will, marriage is almost expulsed. And therefore, there are with you seen infinite men that marry not, but chase rather a libertine and impure single life, than be yoked in marriage; And many that do marry, marry late, when the Prime and strength of their years is past. And when they do marry, what is marriage to them, but a very bargain, wherein is sought alliance, or Portion, or Reputation, with some desire (almost indifferent) of Issue; and not the faithful Nuptial union of man and wife, that was first instituted? Neither is it possible, that those that have cast away so basely, so much of their Strength, should greatly esteem children, (being of the same matter) as chaste Men do. So likewise, during Marriage is the case much amended, as it ought to be if those things were tolerated only for necessity? No, but they remain still a very affront to marriage. The haunting of those dissolute places, or resort to Curtizans, are no more punished as married Men, than in Bachelors; And the depraved custom of change and the delight in Meretricious Embracement, (where sin is turned into Art,) makes Marriage a dull thing, and a kind of imposition, or Tax. They hear you defend these things, as done to avoid greater Evils; As adulteries, Deflowering of Virgins, unnatural lust, and the like: But they say, this is a preposterous Wisdom; and they call it Lot's offer, who to save his guests from abusing, offered his daughters: Nay they say further, that there is little gained in this; for that the same vices and appetites do still remain and abound; unlawful lusts being like a furnace, that if you stop the flames altogether, it will quench; But if you give it any vent, it will rage. As for masculine Love, they have no touch of it, and yet there are not so faithful and inviolate friendships in the world again, as are there; and to speak generally (as I said before), I have not read of any such Chastity in any people, as theirs: and their usual saying is, That whosoever is unchaste, cannot reverence himself: And they say, That the Reverence of a man's self is, next to religion, the chiefest Bridle of all vice. And when he had said this, the good Jew paused a little; Thereupon, I was far more willing to hear him speak on, than to speak myself; yet thinking it descent, that upon his pause of speech, I should not be altogether silent, said only thus; That I would say to him, as the Widow of Serepta said to Elias, that he was come to bring to memory our sins; and that I confess the Righteousness of Apamia was greater than the Righteousness of Europe. At which speech he bowed his head, and went on in this manner. They have also many wise and excellent laws touching Marriage. They allow no Polygamy. They have ordained that none do intermarry or contract, until a month be past from their first Interview. Marriage without consent of Parents they do not make void, but they *mulct* it in the inheritors: For the children of such Marriages are not admitted to inherit, above a third part of their Parents Inheritance. I have read in a Book of one of your Men, of a Famed commonwealth, where the Married couple are permitted, before they contract, to see one another Naked. This they dislike: For they think it a Scorn to give a refusal after so Familiar knowledge: But because of many hidden defects in men and Woman's bodies, they have a more civil way: For they have near every Town, a couple of Pools, (which they call Adam and Eves Pools,) where it is permitted to one of the friends of the Man, and another of the Friends of the Woman, to see them severally bathe Naked.

And as we were thus in conference, there came one that seemed to be a messenger, in a rich Huke, that spoke with the Jew; Thereupon he turned to me, and said You will pardon me, for I am commanded away in haste. The next morning, he came to me again, joyful as it seemed, and said; There is word come to the Governor of the City that one of the Fathers of the Temple of the Rosie Cross or Holy House, will be here this day Seven-night: We have seen none of them these dozen years. His coming is in State; But the cause of his coming is secret. I will provide you, and your fellows, of a good standing, to see his Entry. I thanked him and told him I was most glad of the news. The day being come he made his entry. He was a man of middle stature and age, comely of person, and had an Aspect as if he pitied Men. He was clothed in a Robe of fine black Cloth, with wide sleeves, and a Cape. His under garment was of excellent white linen, down to the foot, girt with a girdle of the same; and a Lindon or Tippet of the same about his neck. He had gloves that were curious, and set with Stones; and Shoes of Peach-colored Velvet. His neck was bare to the shoulders. His Hat was like a Helmet, or Spanish Montera; and his Locks curled below it cresently: They were of Color brown. His Beard was cut round, and of the same Color with his hair, somewhat lighter. He was carried in a rich Chariot, without wheels, Litter-wise; With two horses at either end, richly trapped in blue Velvet Embroidered, and two Footmen on each side in the like attire. The Chariot was all of Cedar gilt and adorned with Chrystal; save that the Fore-end had Panels of Sapphire, set in borders of Gold; And the hinder-end the like of Emeralds of the Peru Color. There was also a Son of Gold, Radiant upon the Top, in the Midst; And on the Top before, a small Cherub of Gold with wings Displayed. The Chariot was covered with cloth of Gold tissued upon Blue. He had before him fifty attendants young men, all in white Satin loose coats to the Mid Leg; and Stockings of white Silk; and shoes of blue Velvet; and Hats of blue Velvet; with fine Plumes of divers colors, set round like Hat-bands. Next before the Chariot, went two Men bare-headed, in Linen Garments down to the foot, girt, and shoes of blue Velvet; Who carried the one a Crosier, the other a Pastoral staff like a Sheep-hook:

Neither of them of Metal, but the crosier of Palmwood, the Pastoral Staff of Cedar. Horsemen he had none, neither before, nor behind his Chariot: as it seemed to avoid all tumult and trouble behind his Chariot, went all the Officers and Principals of the Companies of the City. He sat alone upon cushions, of a kind of excellent Plush, blue; And under his foot curious Carpets of Silk of divers Colors, like the Persian, but far finer. He held up his bare hand, as he went, as blessing the people, but in silence. The street was wonderfully well kept; So that there was never any Army had their men stand in better battel-Array, than the people stood.

The windows likewise were not crowded, but everyone stood in then, as if they had been placed. Then the show was past, the Jew said to me; I shall not be able to attend you as I would, in regard of some charge the city has laid upon me, for the entertaining of this Rosie Crucian, Three days after the Jew came to me again, and said; Ye are happy men; for the Father of the Temple of the Rosie Cross taketh notice of your being here, and commanded me to tell you, that he will admit all your company to his presence, and have private conference with one of you, that ye shall choose: And for this has appointed the next day after tomorrow. And because he means to give you his blessing, he has appointed it in the Forenoon. We came at our day, and hour, and I was chosen by my fellows for the private Access. We found him in a fair Chamber, richly hanged, and carpeted under foot, without any degrees to the State. He was set upon a low Throne richly adorned, and a rich cloth of State over his head, of blue Satin Embroidered. He was alone, save that he had two Pages of honor, on either Hand one, finely attired in White. His under Garments were the like that we saw him wear in the Chariot; but instead of his Gown, he had on him a Mantle with a Cape, of the same fine black, fashioned about him. Then we came in, as we were taught, we bowed low at our first entrance; And when we were come near his Chair, he stood up, holding forth his hand ungloved, and in posture of blessing; and we every one of us stooped down and kissed the Hem of his Tippet. That done, the rest departed, and I remained. Then he warned the Pages forth of the Room, and caused me to sit down beside him, and spoke to me thus in the Spanish Tongue.

God bless thee, my Son, I will give thee the greatest Jewel I have: For I will impart unto thee, for the love of God and Men, a Relation of the true State of the Rosie Cross. Son, to make you know the true state of the Holy House, I will keep this order. First, I will set forth unto you the end of our Foundation. Secondly, the preparations and instruments we have for our Works. Thirdly, the several employments and functions whereto our fellows are assigned. And Fourthly, the Ordinances and Rights which we observe.

The end of our Foundation is the Knowledge of Causes, and Secret Motions of Things; And the enlarging of the bounds of Kingdoms to the effecting of all Things possible.

The Preparations and Instruments are these. We have large and deep Caves of several depths; The deepest are sunk, 36,000 Feet: And some of them are dug and made under great Hills and Mountains: So that if you reckon together the depths of the Hill, and the depth of the Cave, they are (some of them) above seven miles deep. For we find, that the depth of a Hill, and the depth of a Cave from the flat, is the same Thing; Both remote alike, from the Sun and Heavens Beams, and from the open Air. These Caves we call the Lower Regions, and we use them for all Coagulation, Induration, Refrigeration, and Conservations of Bodies. We use them likewise for the imitation of natural Mines; and the producing also of new artificial Metals, by compositions and materials which we use, and lay therefore many years. We use them also sometimes (which may seem strange), for curing of some Diseases, and for prolongation of life, in some Hermits that choose to live there well accommodated of all things necessary, and indeed live very long; By whom also we learn many things; Read our *Temple of Wisdom*.

We have Burials in several earths, where we put diverse Cements, as the Chinese do their Bercellane, but we have them in greater Variety, and some of them finer. We have also great varieties of composts, and Soils, for the making of the earth fruitful.

We have High Towers; the highest about half a mile in height; and some of them likewise set upon high Mountains; So that the vantage of the Hill with the Tower, is in the highest of them three miles at least. And these places we call the upper Region, accounting the Air between the highest places, and the lower, as a middle Region. We use these Towers, according to their several Heights, and Situations, for Insolation, Refrigeration, Conservation; And for the View of divers Meteors, as Winds, Rain, Snow, Hail, and some of the fiery Meteors also. And upon them, in some places, are Dwellings of Hermits, whom we visit sometimes, and instruct what to observe. Read our *Harmony of the World*.

We have great Lakes, both Salt and fresh; whereof we have use for the fish and fowl. We use them also for burials of some natural bodies: For we find a difference in things buried in earth, or in Air, below the Earth, and things buried in the Water. We have also Pools, of which some do strain fresh water out of Salts; And other by Art do turn fresh Water into salt. We have also some Rocks in the midst of the Sea; and some Bays upon the shore for some works, wherein is required the Air and Vapor of the Sea, we have likewise violent streams and cataracts which serve us for many Motions:

And likewise, Engines for multiplying and Enforcing of winds to set also on-going divers Motions.

We have also a number of Artificial Wells, and fountains made in imitation of the natural Sources and Baths, as tincted upon Vitriol, Sulphur, Steel, Brass, Lead, Nitre, and other Minerals. And again, we have little Wells for infusion of many things, where the waters take the virtue quicker and better, than in Vessels, or the Basins, and amongst them we have a Water, which we call water of Paradise, being, by that we do to it, made very Sovereign for health, and prolongation of Life; As you shall read in this Book.

We have also great and spacious houses, where we imitate and demonstrate Meteors, As Snow, Hail, Rain, some artificial rains of bodies, and not of water, thunders, lightnings; Also, generations of bodies in air, as frogs, flies and divers others.

We have also certain chambers, which we call chambers of Health, where we qualify the air as we think good and proper for the cure of divers diseases, and preservation, of Health.

We have also fair and large Baths, of several mixtures, for the restoring of man's body from arefaction[4]: and others for the confirming of it in Strength of Sinews, vital parts, and the very juice and substance of the body.

---

[4] Spelled exactly as in the 1662 printed edition. -pnw

We have also large and various Orchards; see the Epistle to the Harmony of the World, and Gardens, wherein we do not so much respect beauty, as variety of ground and Soil, proper for diverse Trees, and Herbs: And some very spacious, where Trees and Berries are set, whereof we make divers kinds of drinks, besides the Vineyards. In these we practice likewise all conclusions of Grafting, and inoculating, as well of wild trees, as fruit trees, which produces many effects, and we make (by Art) in the same Orchards, and Gardens, Trees and Flowers to come earlier, or later than their Seasons; and to come up and bear more speedily than by their natural course they do. We make them also by Art greater much than nature; and their fruit greater, and sweeter, and of differing taste, smell, Color, and figure, from their nature. And many of then we so order as they become of medicinal use.

We have also means to make divers plants rise by mixtures of Earth without Seeds: And likewise, to make divers new Plants, differing from the Vulgar; and to make one Tree or Plant turn into another.

We have also parks, and enclosures of all sorts of beasts, and birds; which we use not only for view or rareness, but likewise for dissections, and trials, that thereby we may take light what may be wrought upon the body of man. Wherein we find many strange effects; as continuing life in them, though divers parts, which you account vital, be perished, and taken forth; Resuscitating of some that seem dead in appearance; and the like. We try also all Poisons, and other medicines upon them, as well of Chyrurgery as Physick. By art likewise we make them greater or taller than their kind is; and contrary wise dwarf them and stay their growth. We make them more fruitful and bearing than their kind is; and contrary wise barren and not Generative. Also, we make them differ in Color, shape, activity, many ways. We find means to make commixtures and copulations of divers kinds, which have produced many new kinds, and them not barren, as the general opinion is. We make a number of kinds of Serpents, worms, flies, fishes, of putrefaction, whereof some are advanced (in effects) to be perfect creatures, like beasts, or birds, and have Sexes, and do propagate. Neither do we this by chance, but we know beforehand, of what matter and commixture, what kind of those creatures will rise.

We have also particular Pools, where we make trials upon fishes, as we have said before of beasts, and birds.

We have also places for breeding and generation of those kinds of Worms and Flies which are of special use; such as are with you your Silk worms, and Bees.

I will not hold you long with recounting of our brew-houses, bake-houses, and kitchens, where are made divers drinks, breads and meats, rare and of special effects. Wines we have of Grapes, and drinks of other Juices, of fruits, of Grains, and of roots, and of mixtures with honey, sugar, manna, and fruits dried, and decocted:

Also, of the tears or wounding of trees; And of the Pulp of Canes. And these drinks are of several ages, some to the age or last of forty years. We have drinks also brewed with several Herbs, and roots, and Spices; Yea with several fleshes, and white-meats; whereof some of the Drinks are such as they are in effect meat and drink both: So that divers, especially in age, do desire to live with them, with little or no meat, or Bread. And above all we strive to have drinks of extreme thin parts, to insinuate into the body, and yet without all biting, sharpness, or fretting; insomuch as some of them, put upon the back of your Hand, will, with a little stay, pass through to the palm, and yet taste mild to the mouth. We have also waters, which we ripen in that fashion, as they become nourishing: So that they are indeed excellent Drink: And many will use no other. Breads we have of several Grains, Roots and Kernels: Yea and some of flesh, and fish, dried with divers kinds of Leavenings, and seasonings: So that some do extremely move appetite; some do nourish so as divers do live of them, without any other meat, who live very long. So far meats, we have some of them so beaten, and made tender, and mortified, yet without all corrupting, as a weak heat of the Stomach will turn them into good Chylus[5]; as well as a strong heat would meat otherwise prepared. We have some meats also, and breads, and drinks, which taken by men, enable them to sailing after and some other, that used make the very flesh of man's bodies sensibly more hard and tough, and their strength far greater then otherwise it would be.

We have Dispensatories, or shops of Medicines: wherein you may easily think, if we have such variety of Plants and living Creatures, more than you have in Europe, (for we know what you have) the Simples, Drugs, and Ingredients of Medicines, must likewise be in so much the greater Variety. We have them likewise of diverse ages, and long Fermentations. And for their preparations, we have not only all manner of Exquisite distillations, and separations, and especially by gentle heats and Percolations through diverse Strainers, yea and substances, but also exact forms of compositions, whereby they incorporate almost, as they were natural simples.

---

[5] Spelled exactly as in the 1662 printed edition. -pnw

We have also divers Mechanical Arts, which you have not, and stuffs made by them; as papers, linen, silks, Tissues; dainty works of feathers of wonderful luster; excellent Dyes, and many others:
And shops likewise, as well for such as are not brought into vulgar use amongst us, as for those that are. For you must know, that of the things fore cited many of them are grown into use throughout the Kingdom; But yet, if they did flow from our invention, we have of them also for Patterns and Principles.

We have also furnaces of great diversities, and that keep great Diversity of heats: Fierce and Quick; strong and constant; soft and mild, blown quite dry, moist, and the like. But above all we have heats in imitation of the Suns and heavenly bodies heats, that pass divers inequalities, and (as it were) Orbs, Progresses and returns, whereby we produce admirable effects. Besides we have heats of dungs; and of bellies and maws of living Creatures, and of their bloods, and bodies and of Hayes and herbs, laid up moist; of lime unquenched, and such like. Instruments also which generate heat only by Motion. And further, places for strong insulations; and again, places under the earth, which by nature, or Art, yield heat. The divers heat we use, as the nature of the operation, which we intend, requires.

We have also perspective houses, where we make demonstrations of all lights, and radiations: And of all colors and out of things uncolored and transparent, we can represent unto you all several colors: Not in Rainbows (as it is in Gems, and Prisms), but of themselves single. We respect also all multiplications of light, which we carry to great distances, and make so sharp, as to discern small points and lines. Also, all colorations of light, all delusions and deceits of the sight in figures, Magnitudes, Motions, colors.

All demonstrations of Shadow; we find also diverse means yet unknown to you, of producing light, originally from divers bodies. We procure means of seeing objects afar off as in the heaven, and remote places: and represent things near as a far off; and things a far off as near, making feigned distances. We have also help for the sight, far above Spectacles and Glasses in use. We have also glasses and means to see small and minute bodies, perfectly and distinctly; as the shapes and Color of small flies and worms, Grains and flaws in Gems which cannot otherwise be seen, observation in Urine and Bloods, not otherwise to be seen. We make artificial Rainbows, Halo's, and circles about light. We represent also all manner of Reflections, Refractions, and Multiplications of visual beams of objects.

We have also precious stones of all kinds, many of them of Great beauty, and to you unknown: Crystals likewise, and glasses of divers kinds; and amongst them some of Metals Vitrificated, and other materials, besides those of which you make Glass. Also, a number of Tassels, and imperfect minerals, which you have not. Likewise, Loadstones of prodigious virtue: And other rare stones, both natural and artificial. We have also Sound-houses, where we practice and demonstrate all sounds, and their Generation. We have Harmonies (read the *Harmony of the World*), which you have not, of quarter sounds, and lesser kinds of sounds. Divers instruments of Music likewise to you unknown, some sweeter than any you have; together with bells, and rings that are dainty and sweet; See my book of Geomancy and Telesmes; Lib. 4, chapter the 9th. We represent small sounds as great and deep; likewise, great sounds, Extenuate and sharp; we make diverse Tremblings and Warblings of Sounds, which in their original are entire. We represent all articulate sounds and Letters; read my Cabbala or Art by which Moses showed so many signs in Egypt, and the voices and motes of beasts and birds. We have certain helps, which set to the ear do further the hearing greatly. We have also divers Strange and artificial Echoes, Reflecting the voice many times, and as it were tossing it: And some that give back the voice lower than it came, some shriller, and some deeper; yea some rendering the voice differing in the letters or articular sound, from that they receive. We have also means to convey Sounds in Trunks and pipes in strange lines, and distances.

We have also perfume houses; wherewith we join also practices of taste. We multiply smells which may seem strange. We Imitate smells, making all smells to breathe out of other Mixtures than those that give them. We make divers imitations of taste likewise, so that they will deceive any man's taste. And in this Temple of the Rosie Cross we contain also a Confecture-House where we make all sweetmeats, dry and moist; and diverse pleasant Wines, Milks, Broths, and Salads, in far greater variety than you have.

We have also Engine-houses, where are prepared Engines and instruments for all sorts of motions. There we imitate and practice to make Swifter Motions than any you have, either out of your Muskets, or any Engine that you have, and to make them, and multiply them more easily, and with small force, by wheels, and other means: And to make them Stronger and more violent, than yours are, exceeding your greatest Canons, and Basilisks. We represent also Ordinance and instruments of War, and Engines of all Kinds: And likewise, new Mixtures and Compositions of Gunpowder, Wildfire burning in water, and unquenchable. Also, Fireworks, read my book of *Telesmes*, How Moses did so many Miracles, Joshua made the Sun stand still, and Elijah called down fire from Heaven; of all variety, both for pleasure and use. We imitate also flights of birds: we have some degrees of flying in the Air: Read *The Familiar Spirit*. We have ships and boats for going under water, and brooking of Seas; Also, swimming girdles and supporters. We have divers curious clocks, and other like notions of return: And some perpetual Motions. We imitate also motions of living Creatures, by Images of men, beasts, birds, fishes, and Serpents. We have also a great number of other various motions, strange for Equality, fineness and subtilty.

We have also a Mathematical palace, where are represented all instruments, as well of Geometry, as Astronomy, Geomancy, and Telesmes, viz; astromancy and geomancy exquisitely made.

We have also houses of deceits of the senses, where we represent all manner of feats of Juggling, False Apparitions, Impostures, and illusions and their fallacies. And surely you will easily believe, that we that have so many things truly natural, which induce admiration, could in a world of particulars deceive the senses, if we would disguise those things, and labor to make them seem more miraculous.

But we do hate all impostures and lies: Insomuch as we have severally forbidden it to all our brethren, under pain of ignominy and fines, that they do not show any natural work or Thing abhorred or swelling; but only pure as it is; and without all affectation of strangeness.

These are (my Son) the Riches of the Rosie Crucians; read our *Temple of Wisdom*.

For the several employments and Offices of our fellows, we have twelve that sail into foreign Countries, under the Names of other Nations, but our seal is R.C. and we meet upon the day altogether (for our own we conceal); Who bring us the books, and Abstracts, and

Patterns of experiments of all other parts. These we call merchants of light.

We have three that Collect the experiments which are in all Books. These we call depredators.

We have three that Collect the experiments of all Mechanical Arts; And also of Liberal Sciences; and also of Practices which are not brought into Arts. These we call Mystery Men.

We have three that try new experiments, such as themselves think good. These we call Pioneers or Miners.

We have three that draw the Experiments of the former four into Titles, and Tables, to give the better light for the drawing of observations and axioms out of them. These we call compliers.

We have three that bend themselves, looking into the experiments of their fellows, and cast about how to draw out of them things of use and practice for man's life, and Knowledge, as well for works, as for strange demonstrations of causes, means of natural divinations, and the easy and clear discovery, of the parts of Bodies. These we call Dowry men or benefactors.

Then after diverse meetings and consults of our whole number, to consider of the former labors and collections, we have three that take care, out of them, to direct new Experiments of a higher light, more penetrating into nature than the former. These we call Lamps.

We have three others that do execute the Experiments so directed, and report them. These we call Inoculators.

Lastly, we have three that raise the former Discoveries by experiments, into greater observations, axioms and aphorisms. These we call Interpreters of Nature.

We have also, as you must think, Novices and Apprentices, that the succession of the former employed men, of our fraternity of the Rosie Cross do not fail; Besides, great number of servants and attendants, men and women. And this we do also: We have Consultations, which of the Inventions and Experiences, which we have discovered, shall be published, and which not: And take all an Oath of Secrecy, for the concealing of those which we think fit to keep secret: Though some of those we do reveal sometimes to the State, and some not. Read our *Temple of Wisdom*.

For our Ordinances and Rites: We have two very long and fair Galleries in the Temples of the Rosie Cross; Is one of these we place patterns and samples of all manner of the rarer and excellent inventions: In the other we place the Statues of all principal Inventors. There we have the Statues of the West-Indies:

Also the Invention of Ships; and the monk that was the Inventor of Ordinance, and of Gunpowder: The inventor of Music: The inventor of letters, the inventor of Printing: The inventor of Observations of Astronomy, Astromancy, and Geomancy: The inventor of Works in metal: The inventor of Glass: The inventor of Silk of the Worms: The inventor of Wine: The inventor of Corn and bread: The inventor of Sugars: And all these, by more certain tradition, than you have. Then have we diverse inventors of our own, of excellent Works; which since you have not seen, it was too long to make Descriptions of them; And besides, in the right Understanding of those Descriptions, you might easily err. For upon every invention of value, we erect a Statue to the Inventor, and give him a liberal and honorable reward. These Statues are, some of brass, some of Marble, and Touchstone; some of Cedar and other special woods guilt and adorned, some of Iron, some of Silver, some of Gold, telesmatically made.

We have certain Hymns and Services, which we say daily, of Loud and Thanks to God, for his marvelous works: And Forms of Prayers, imploring his aid and blessing, for the illuminations of our labors, and the turning of them into good and holy Uses.

Lastly, we have Circuits or Visits of divers principal Cities of the Kingdom; where, as it comes to pass, we do publish such News, profitable inventions, as we think good. And we do also declare natural Divinations of Diseases, Plagues, Swarms of hurtful creatures, Scarcity, Tempests, Earthquakes, great Inundations, Comets, Temperature of the Year, and divers other things; And we give Counsel thereupon, what the people shall do, for the Prevention and Remedy of them.

And when he had said this: He desired me to give him an account of my life, and observations of my youth, that he might report it to the brethren of the Rosie Cross. I was descended from a noble family of London in England being born of a complete tall stature, small limbs, but in every part proportionable, of a dark, flaxen hair, it curling as you see in the Effigies: And these Figures of Astrology, at the time I was born:

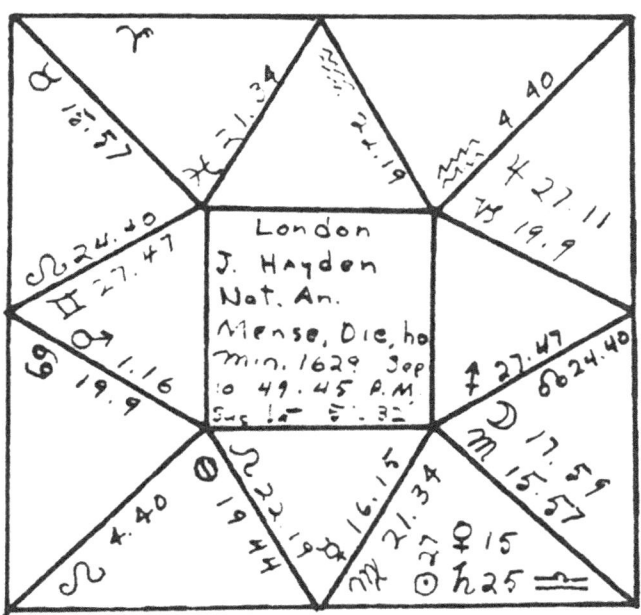

This is also the Character of my Genius Malhitriel, and Spirit Taphza Bnezelthar Thaseraphimarah:

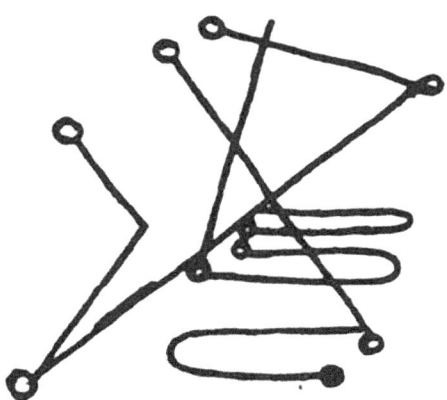

I had the small Pox and Rickets very young; Ascendant to Conjunction, Mars, and Sol to the quartile of Saturn: I was at Tardebick in Warwickshire near Hewel where my Mother was born, and there I learned, and so careful were they to keep me to the book and from danger, that I had one purposely to attend me at school and at home. For indeed my Parents were both of them honorably descended; they put me to learn the Latin-tongue to one Mr. George Linacre the Minister of the Gospel at Golton; of him I learned the Latin and Greek, perfectly, and then was fitted for Oxford. But the Wars began, and the Sun came to the body of Saturn and frustrated that design; And whereat you are pleased to stile me a noble-natured sweet Gentleman, You see my Nativity: Mercury, Venus and Saturn are strong, and by them, the Dragons head and Mars, I judge my behavior full of vigor, and acknowledge my Conversation austere: In my devotion I love to use the civility of my knee, my hat, and hand, with all those outward and sensible motions; which may express or promote invisible devotion: I followed the Army of the King to Edgehill: and commanded a troop of Horse; but never violated any man, etc. Nor defaced the memory of Saint or Martyr: I never killed any man willfully, but took him prisoner and disarmed him; I did never divide myself from any man upon the difference of opinion; or was angry with his judgement for not agreeing with me in that from which perhaps; within a few days I should dissent myself: I never regarded what Religion any man was of, that did not question mine. And yet there is no Church in the world, whose every part so squares into my Conscience, whose Articles, Constitution; and Customs seem so consonant unto reason, and as it were framed to my particular devotion, as this whereof I hold my belief, The Church of England, to whose Faith I am a sworn subject; and therefore in a double Obligation, subscribe unto her Articles, and endeavor to observe her Constitutions: whatsoever is beyond, as points indifferent, I observe according to the rules of my private reason, or the humor and fashion of my devotion; neither believing this, because Luther affirmed it, or disproving that, because Calvin has disvouched it: Now as all that die in the War, are not termed soldiers, so neither can I properly term all those that suffer in matters of Religion Martyrs. And I say, there are not many extant, that in a noble way fear the face of Death less than myself: yet from the moral duty I owe to the Commandment of God, and the natural respects that I tender unto the conservation of my Essence and being, I would not perish upon a Ceremony, politique points or indifferency: nor is my belief of that intractable temper, as not to bow at their obstacles or connive at matters wherein there are not manifest impieties: the leaves therefore and ferment of all, not only civil, but Religious actions, is wisdom, without which, to commit ourselves to the flames is homicide, and I fear, but to pass through one fire into another: I behold as a Champion with pride the spoils and Trophies of my victory over my enemies, and can with patience embrace this life, yet in my best Meditations do often desire death: I honor any man that contemns it, nor can I love any that is afraid of it; this makes me naturally love a soldier that will follow his Captain. In my figure you may see I am naturally bashful: yet you may read my qualities in my countenance. About the time I travelled into Spain, Italy, Turkey, and Arabia, the Ascendant was then directed to the Trine of the Moan, Sextile of Mercury, and Quartile of Venus. I studied Philosophy and writ this Treatise, and my *Temple of Wisdom*, etc. Conversation, Age, or Travel has not been able to affront or enrage me; yet I have one part of the modesty which I have seldom discovered in another, that is (to speak truly) I am not so much afraid of Death as ashamed thereof: It is the very disgrace and ignominy of our natures, that in a moment can so disfigure us that our beloved friends stand afraid and start at us; the birds and beasts of the field that before in a natural fear obeyed us, forgetting all allegiance begin to prey upon us. This very thought in a storm at sea has disposed and left me willing to be swallowed up in the abyss of waters; wherein I had perished, unseen, unpitied, without wondering eyes, tears of pity, lectures of morality, and none had said, *Quantum autatug abalio*! Not that I am ashamed of the anatomy of my parts, or can accuse nature for playing the pupil in any part of me, or my own vitious life for contracting any shameful disease upon me, whereby I might not call myself a complete bodied man free from all diseases, sound, and I thank God in perfect health: I write my *Harmony of the World*, when they were all at discord, and saw many revolutions of Kingdoms, Emperors, Grand Signiours, and Popes: I was twenty when this book was finished, but me thinks I have outlived myself, and begin to weary of the Sun, although the Sun now applies to a Trine of Mars; I have shaken hands with delight and know all is vanity, and I think no man can live well once, but he that could live twice, yet for my own part I would not live over my hours past, or begin again the minutes of my days, not because I have lived them well, but for fear I should live them worse; at my death I mean to take a total adieu of the world, not caring for the burden of a Tombstone and Epitaph, nor so much as the bare memory of my name to be found anywhere, but in the universal Register of God, I think God that with joy I mention it, I was never afraid of Hell, nor never grew pale at the description of Sheol or Tophet, etc. because I understand the policy of a Pulpit, and fix my contemplations on heaven I write The Rosie Crucians *Infallible Axiomata* in four books, and study not for my own sake only, but for theirs that study not for themselves; and in the Law I began to be a perfect Clerk: I write the *Idea of the Law* etc. for the benefit of my friends and practice in the Kings Bench; I envy no man that knows more than myself, but pity them that know less. For Ignorance is rude, uncivil, and will abuse any man as we see in Bayliffs, who are often killed for their imprudent attempts; they'll forge a Warrant and fright a fellow to fling away his money, that they may take it up; the Devil that did but buffet St. Paul, plays methinks at sharp with me. To do no injury nor take none, was a principle, which to my former years and impatient affection, seemed to contain enough of morality, but my more settled years and Christian constitution have fallen upon severer resolutions. I hold there is no such thing as injury, and if there be, there is no such injury as revenge, and no such revenge as the contempt of an injury. There will be those that will venture to write against my doctrine, when I am dead, that never durst answer me when alive: I see Cicero is abused by Cardan, who is angry at Tully for praising his own daughter: and Origanus is so impudent, that he adventures to forge a position of the heavens, and calls it Cornelius Agrippa's Nativity: and they say, Agrippa was born to believe lies, and broach them: is not this unworthiness to write such lies, and show such reasons for them! His Nativity I could never find: I believe no man knows it: but by a false figure thus they scandalize him. Mercury they make combust and in Quartile to Jupiter, and the Moon in Opposition to him and Sol: The Dragons tail they place upon the ascendant; they will have Saturn and Mars disposi18of the Moon, to signify his manners, being in Sextile of the Moon, and Trine of Mercury, and in Opposition from Angles, and the Ascendant evilly beheld by both of them; specially by Saturn; indeed they have made him a Noble person, Agrippa a base fellow by this figure.

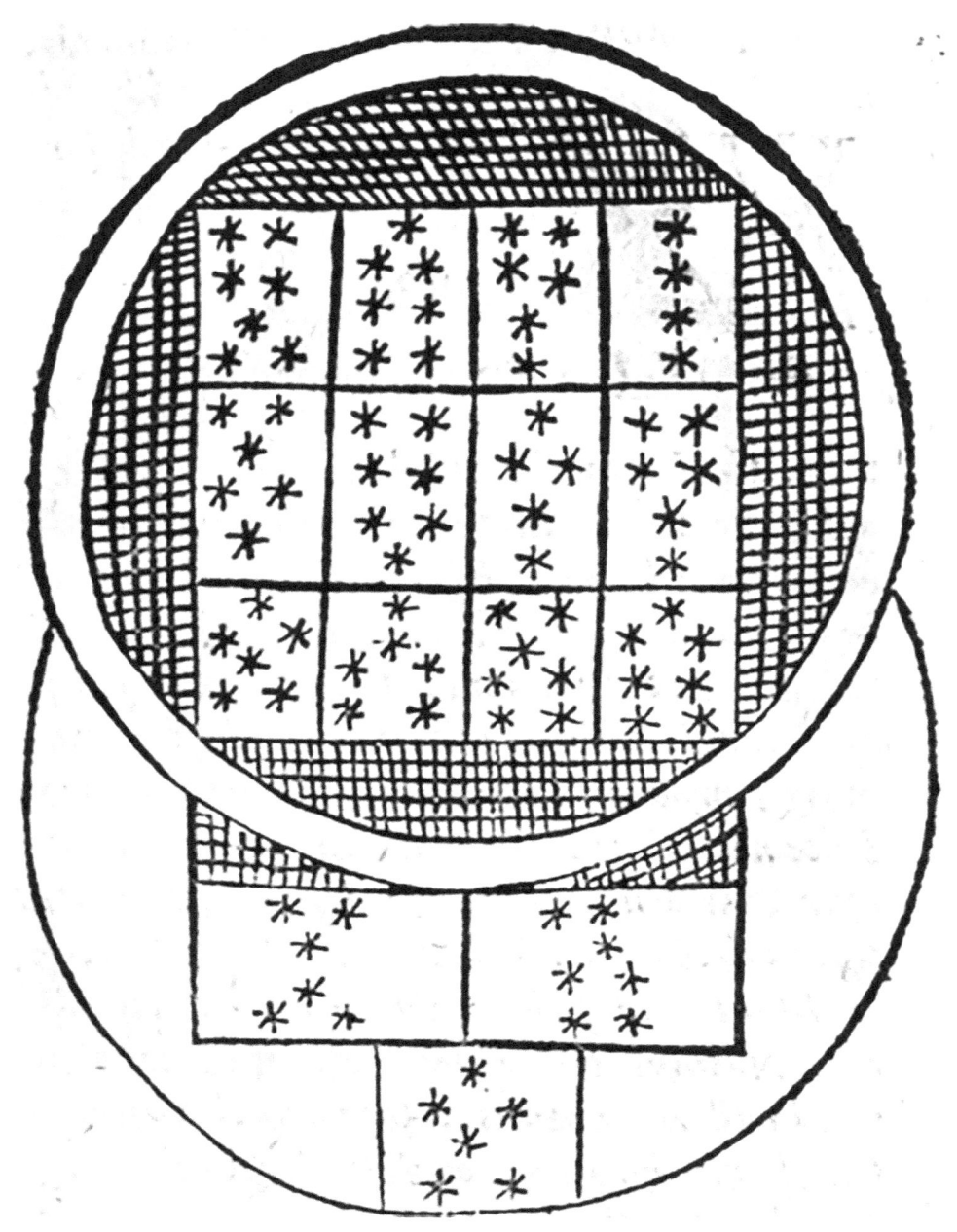

And so, they may use me; But behold the Scheme of my Nativity in Geomancy, and the Character of my Spirit,

Taphzabnezeltharthaseraphimarah, projected by a learned Lord for the honor of birth: now let any Astrologer, Geomancer, Philosopher, etc. judge my Geniture, the Figures are right according to the Exact time of my birth, rectified by Accidents, and verified by the effects of Directions. Now in the midst of all my endeavors, there is but one thought that directs me, that my acquired parts must perish with myself, nor can be legacied amongst my dearly beloved and honored Friends; I do not fall out, or contemn a man for an Error, or conceive why a difference in Opinion should divide an affection: For a modest reproof or dispute, if it meet with discreet and peaceable Natures, doth not infringe the laws of Charity in all Arguments: So much there is of Passion, so much there is of Nothing to the purpose; For then reason like my Hound Lilly spends or calls out aloud, and makes the woods echo upon a false scent: Expecting Poolah to join with him, but Froster, Joyce, Jolliboy, and a white Bitch hunt in their Couples another way, and follow their game first started When the Mid-heaven was directed to the trine of the Moon, I wrote another book and entitled it, *The Fundamental Elements of Philosophy, Policy, Government and the Laws*, etc. After this time, I had many misfortunes, and yet I think there is no man that apprehends his own miseries less than himself, and no man that so nearly apprehends another's. I could lose an Arm without a tear, and with few groans, me thinks, be quartered into pieces: Yet can

I weep seriously with a true passion, to see the merciless Rebels in England forge a debts against the King's most loyal Subjects, purposely to put them in the Marshalsey, or other Houses of Hell, to be destroyed in prison, and starved, or killed by the Keepers, and then two or three poor old women as many shillings shall persuade the Crowner and the people to believe, the men died of Consumptions. It is a barbarous part of humanity to add unto any afflicted party's misery, or endeavor to multiply in any man a passion, whose single nature is already above his patience: Thus, was the greatest affliction of Job, and those oblique expostulations of his friends a deeper injury than the down-right blows of the Devil, etc. The Ascendant to the Quartile of Saturn, and part of Fortune to the Sextile of the Moon came next; and it is true, I had loved a Lady in Devonshire, but when I seriously perused my Nativity, I found the seventh House afflicted, and therefore never resolve to marry; for behold I am a man, and I know not how: I was so proportioned and have something in me, and will be after me; and here is the misery of a sane life; He eats, drinks and then sleeps today that he may do so again tomorrow, and this breeds Diseases, which brings Death. For all flesh is grass. And all those creatures we behold, are but the Herbs of the field digested into flesh in them, or more remotely carnified in our selves: we are devourers not only of men, but of ourselves, and that not in an Allegory, but a Positive truth; for all this mass of flesh, which we behold, came in at our mouths; this frame we look upon, has been upon our trenchers: and we have devoured ourselves, and what are we? I could be content that we might raise each other from death to life as Rosie Crucians do, etc. without Conjunctions, or that there were any way to perpetuate the world without this trivial and vain way of Coition, as Dr. Brown calls it: It is the most foolish act a wiseman commits all his life; nor is there anything that will more deject his cold imagination, than to consider what an odd error he has committed:

had my Stars favored me, I might have been happy in that sweet Sex: Then I consider the love of Parents, the affections of Wives and Children, and they are all dumb dreams, without reality, truth, or constancy; for first, there is a strong bond of affection between us and our Parents; yet how easily dissolved: the Son betakes himself to a woman, forgetting his Mother in a Wife, and the womb that bare him, in that that shall bear his Image: This woman blessing him with Children, his affection leaves the level it held before, and sinks from his bed to his Issue and Picture of posterity, where affections hold no steady mansion: they growing up in years desire his end, or applying themselves to a woman, take a lawful way to love another better than themselves. Thus, I perceive a man may be buried alive, and behold his grave in his Issue. And many take pleasure to be such fools. I remember also that this Quartile of Saturn imprisoned me at a Messengers house, for contending with Cromwell, who maliciously commanded I should be kept close in Lambeth-house, as indeed I was two years; my person he feared, and my tongue and pen offended him, because amongst many things, I said particularly, such a day he would die, and he died: It is very true Oliver opposed me all his life, and made my Father pay Seventeen hundred pounds for his Liberty: Besides, they stole under pretense of sequestering him, two thousand pounds in Jewels, Plate, etc. and yet the Kings noblest servants suffer upon suspicion of Debt: A plot that carries a fairer pretense to persuade the ruder wits all is well, when the King and his best friends are abused: but why should I trouble myself! I do not, believe me, it is not hoping of a place, or a sum of money, or a Commission that I look for; I shall peaceably enjoy my friend, serve God, honor my King and love the Bishops, and few men know who I am.

I look upon France as I do upon the Bear-garden, the Dogs are always quarrelsome; and what is the difference betwixt a man and a beast?

The one is virtuous, learned and wise; the other is rich, proud and foolish; yet indeed the first is most rich, for he studies long life, happiness, health, youth and riches, etc. and enjoys it: Yet I know some will be spectators of this rude Rabble, etc. suddenly dyes an enemy to Reason, Virtue and Religion; and there are a multitude of these, a numerous piece of wonder; and this I observe when they are taken asunder, from men, and the respectable Creatures of God; but confused together, make a Monster more prodigious than any Beast is in the Tower (as Doctor Browne says). It is no breach of charity to call these Fools, as objects of contempt and laughter; and it is the style the Rosie Crucians have afforded them, set down by Solomon in holy Scripture, and a point of our faith to believe so. Neither in the name of multitude do I only include the base and minor sort of peoples; there is a rabble even amongst the Gentry, a sort of Plebian heads, whose fancy moves with the same wheel as these; men in the same level with Mechanics, though their fortunes do somewhat guild their infirmities, and their purses compound for their follies. But as in casting account three or four men together come short in account of one man placed by himself below them: So neither are a troop of these ignorant Doradoes of that true esteem and value as many a forlorn Person, whose condition doth place them below their feet; and there is a Nobility without Heraldry, a natural dignity, whereby one man is ranked with another, another filed before him, according to the quality of his desert, and preeminence of his good parts: though the corruption of these times, and the Byass[6] of present practice wheel another way; thus it was in the first and primitive Commonwealth, and is yet in the integrity and cradle of well-ordered policies, till corruption gutted ground under desires, laboring after that which wiser considerations contemn, every Fool having liberty to amass and heap up riches, and

---

[6] Exact spelling used in the 1662 printed edition. -pnw

they a license or Faculty to do or purchase anything: When the Moon was directed to the Quartile of Sol, and the M.C. to the Opposition of Sol, I was by the phanatick Committee of Safety committed to prison, and my books burnt: yet I would not entertain a base design, or an action that should call me Villain, for all the Riches in England; and for this only do I love and honor my own Soul, and have methinks two arms, too few to embrace myself, my conversation is like the Suns with all men, and with a friendly Aspect to do good and bad.

Methinks there is no man bad, and the worst best, that is, while they are kept within the circle of those qualities, wherein there is good: The method I should use in distributive Justice, I often observe in Commutation, and keep a Geometrical proportion in both, whereby becoming equal to others, I become unjust to myself, and subrogate in that common Principle, Do unto others as you would be done unto the self; yet I give no Alms to satisfy the hunger of my Brother, but to fulfill and accomplish the will and command of God, this general and indifferent temper of mine, doth nearly dispose me to this noble virtue amongst these millions of vices I do inherit and hold from Adam. I have escaped one, and that a mortal enemy to Charity, the first and father sin, not only of man, but of the Devil, Pride: a vice whose name is comprehended in a Monosyllable, but in its nature not circumscribed with a world; I have escaped it in a condition that can hardly avoid it: These petty acquisitions and reputed perfections that advance and elevate the conceits of other men, add no feather unto mine: And this is the observation of my life, I can love and forgive, even my enemies. And when I had said this, he stood up and I kneeled down, and he laid his right hand upon my head, and said, God bless thee my Son, and God bless these Relations, which we have made: I give thee leave to publish them for the good of other Nations: for we are here in God's bosom, a land unknown. And so, he left me. Having assigned a value of about 2000 L. in Gold for a bounty to me and my fellows: For they give great largesse where they come upon all occasions.

John Heydon.

London, from my House in Spittlefields near Bishopgate, next door to the Red Lion, April the 3rd. 1662.

To the Reader.

Gentlemen,

It is thought good to let you know Mr. John Heydon has written many Books, viz. *The Harmony of the World*, *The Temple of Wisdom*, *The Holy Guide*, and The *Wiseman's Crown*; being of affinity, they are to be read together; but in his Preface and other places, speaking of the Person of Nature, and her occult mysterious Truths, he is not understood, as appears by the Knight of the Lobster, being one of a Rabble, who oppose and oppress this Noble Philosopher with a most clamorous insipid Ribaldry; but behold with what an admirable patience our Author heard this report, and answered thus, *Heautontimoreumenon*, and when one told him (of William Lilly being a Laborer or Ditchers Son, born at Diseworth in Liecestershire, and afterwards brought up by one Palyna Taylor in the Strand) how he had abused him with scandalous words, replied, I would not tread upon a Worm, the King of Sweden's sycophantic Ape, let the Ass pass.

It seems in Rome and other parts of Italy his books are highly esteemed by very many persons of honor and worth, and eminent for their skill in these studies; these with Cardinal Ursinus, the Marquess Deffuentes, the Duke of Lorrain, the Prince of Condie, and a Colonel of Spanish Cavalieroes, Thomas Revell, and one Cardinal Antonio by name, have been forward of their own accord to put more honor upon our Author then he in modesty will own; the former, with some other Cardinals and Jesuits, have endeavored to convert him to the Roman Catholic Faith, but all in vain.

The Extract of Colonel Thomas Revell's Letter to Mr. John Heydon.

SIR, etc.,

I beseech you receive, etc. now from a Person who much honors your eminent Learning and Humanity, and would eagerly embrace an occasion to give you most ample testimony of the esteem I have for you, etc. I had your Idea of the Law and Government, etc. The Harmony of the World, but a friend in Rome has borrowed them from me, and since dyed; so I fear my Books are lost: I have once your Fundamental Elements of Moral Philosophy, Policy, Government and Laws, which alone, although your other Labors were not taken in to make up the value, may equal you with the best deservers in Philosophy: I was here advertised of many other Pieces as you writ, etc. Sir, I wish all prosperity to your deserving's, and humbly thank you for the fair admittance you have given me to the acquaintance and friendship of Mr. John Gadbury; be pleased to send his Books with yours, etc. These as memorials of your Loves and Friendships I shall preserve, as a tenderness due to things so estimable; and believe, Sir, you have power at your pleasure to command yours, etc.

Tho. Revell.

The last Letter that came to him, was sent from Colonel Revell, etc. humbly intreating him to have returned him, 1. *The Harmony of the World*. 2. *The Temple of Wisdom*. 3. *The Wiseman's Crown*. 4. *The Fundamental Elements of Moral Philosophy, Policy, Government and Laws*. 5. *The Idea of the Law, Government and Tyranny*. And those excellent Pieces of the Learned Mathematician Mr. John Gadbury, viz.

1. *His Astronomical Tables*. 2. *Caelestis Legatus*. 3. *The Doctrine of Nativities*. 4. *Natura Prodigiorum, Nuncius Astrologicus*. 5. *The King of Sweden's Nativity*. 6. *The Nativity of King Charles*. 7. And the *Examples of Nativities*.

The first Letter was dated from Madrid the ninth of April, 1662, the second was dated the fourth of March from Pozzolo, the third from Fiorenza, the fourth from Venetia, the fifth from Ancona, the sixth from Bisignano, with great respects and honorable salutations from the Learned of those parts of Italy and Spain.

The Learned beyond Sea like these Books never the worse, but much the better (because though every English Reader of Lilly and Mother Shipton understand them not, they do.)

And now let us speak a word or two concerning our Author and his Books, which in many places you may perceive to differ in stile, etc. Our Author writ some of those admirable experienced Truths when he was very young, even before the Wars began in England; and afterwards followed the Army of the King, in which he obtained great honor. Lastly, he revised his first work, and added many things for the interpreting of Nature, and the producing of great and marvelous works for the benefit of Men: And as the Sun tips the Clouds by day, and the Moon the tops of the high Woods by night, with light; so our Author enlightens the Minds of Clowns with knowledge, and they cast dirt at him for his labor: to this he uses no spleen, but shines upon them! And is not this a goodly Age of People the while? A true Servant and Secretary of God and the Mysteries of Nature, is not apprehended by our dull Sermon-sayers, or Jews in the Juggling-box: These with some other fond fools, and some pitiful fine things called Courtiers, curse our Author because they cannot understand him. He smiles at the impudent assaults of vain-glorious humours, and beholds their Antics and Rants, as if they had been trained amongst Apes. But to close all, he loves his Countrymen, even the rudest, and prescribes safe and effectual Medicines for the cure of their Bodies, and gives Laws fit to be observed; he corrects the errors of our Laws, and teaches the best Forms of Government in Church and State, as you may read in his Book, and so I refer you to them.

S.F.

The Holy Guide

To his Esteemed Friend Mr. John Heydon, on his Holy Guide,

and other his Learned Industrious Labors, already Published.

Renowned Eugenius! Famous above all! A Prince, in Physiques! Most Seraphical! The Art's Great Archer! Never shooting wide; Yet Hitt'st the White best, in thy Holy Guide. (took,[7]

Good God! What Pains have learned Physicians for Cleansing Physiques [strange perturbed] Brook? But as their Crooked Labors did Destroy Our hopes, Thy Guide directs the Ready way.

Hippocrates, Great Galen, and Senertus, Rhenodens, Paracelsus, and Albertus, Grave Gerrard, and Ingenious Parkinson, Dead Culpeper, and Living Thomlinson, have all Done well, but ah! they mist Road, thou'st Chalked out, Thou Dear Servant of God; And therefore 'tis no wonder, if they Vary From thee: Great Natures (Highborn) Secretary!

'Tis you alone, who has taught the way to bliss: Tis you alone, that knows what it is: 'Tis you, who has Raked fruitful Egypt ore For Medicines; And Italy for more: And in Arabia thy Collecting Brains, to do us good, has taken wonderous Pains.

This having done, if Critiques will not bow to thy Great Learning,

πεγεϛ · σπανδάλυ

---

[7] Open parenthesis, exactly as this appears in the 1662 printed edition. -pnw

It shall to unto them surely prove: And this Essay of thy Sublime Mysteries, shall make them sure unto the Wise Minerva, yet shall be ignorant of thy Pantarva. But hold! Where am I? Sure, you have set a spell on me, because I can't praise, thy doings well: Release me, Good Eugenius! and the Crowns, shall stand on no brows, but thy Learned Own, Poets, no more, lay claim unto the Bayes! `Tis Heydon shines alone with Splendid Rays: Follow his Guide, he teaches you most sure; Let any make the Wound; `Tis he, must Cure.

For he directs the Well-grown, Old, and Young, to live Rich, Happy, Healthy, Noble, Strong.

John Gadbury,
Φιλμαθηματιχ ☉

To the Reader on the behalf of my such honored Friend the Author Mr. John Heydon.

A Labyrinth, doth need a clue to find
The passage out, and a Daedalian mind
May do strange works, beyond the Vulgar's reach,
And in their understandings make a breach.
It's often seen, when men of pregnant parts,
Study, Invent, and promulgate rare Arts,
Or unknown secrets, how they puzzle those
That Understand them not; Their Yea's their No's,
Are put to Non-plus; Tutors then they lack

To drive them forward, or to bring them back.
How many Learned men (in former ages)
In all the Sciences were counted Sages?
And yet are scarcely understood by men,
Who daily read them o're, and o're again!
Some can recount things past, and present some,
And some would know of things that are to come.
Some Study pleasure, some would faine live long,
Some that are old, would faine again be young;
This Man doth toyle, and work, to purchase wealth,
That man gets sickness studying for his health;
This man would happy be, That Wisdom have:
All are at loss, and every man doth crave;
None is content, but each man wants a Guide,
Them to direct when they do step aside.
Since, this is thus, Our Author took pain
To lead us in, and bring us out again;
Now who is pleas'd, in him for to confide,
In these Discoveries, Here's his Holy Guide.
Pray what can more improve the Commonwealth,
Then the discovery of the way to Health?

The Paradox is made a certain truth,
An Ancient man may dye in his prime of youth.
What wonder is it if he goes aside
The Path, which will not take The Holy Guide!

> Raptim Script.
> 9 Junij 1662.
> John Booker,

Illustrissimis, & were Renatis Fratribue,
I.H.R.C. πρωτοτόκων,
Ecclesiae in tumultuoso hoc
Seculo Apostolis Pacificis,
Salutem à Centro
Salutis.

Quae magna Coeli maenia, & tractae Maris,
Terraeque fines, fiquid aut ultra set, capit;
Mens ipsa tandem Capitur: Omnia hactenus
Quae nosse potuit, nota am primum est fibi;
Accede, Lector, disce quis demum sies;
Tranquilinam jecoris agnoscas tui
Qui propius haeret nil tibi, & nil tam procul.
Non hic Scholarum frivole, aut cassi Loga,
Quales per annos fortè plus septem Legit,
Ut folle pleno prodeat, Rixae Artifex;
Vanasq; merces futili lingua crepet,
Sed sancta Rerum pondera, & fensus graves;
Quale, parari decuit, ipsa cut fuit
Fingenda Ratio, & vindici suo adstitit;
Panduntur omnes Machinae gyri tuae;
Animaeq; vertes, Trocheae, cunei Rotae;
Qua concitetur Arte, quo sufflamine
Sistatur illa rursus & constet sibi.
Nec, si Fenestram Pectori humano suam
Aptasset ipse Momus, inspiceret Magis.
Hic cerno Levia Affectuum vestigia,
Gracilesq; sensus Lineas; video quibus
Nebantur alis blanduli Cupidines,

Quibusq; stimulis urgeant Irae graves.
Hic Dolores, & voluptates fuos
Produnt recessus; ipse nec timor latet;
Has Norit artes quisquis in foro veiit
Animorum habenas flectere, & populos Cupit
Aptis Ligatos nexibus jungi fibi.

Hic Archimedes publicus figat pedem,
Siquando regna machinis Politicis
Surgere satagit, & feras gentes ciet,
Imiq; motum sedibus Mundum quatit;
Facile domabit cuncta, qui menti imperat;
Consultor audax, & Promethaei potens
Facinoris Anime! Quis tibi dedit Deus
Haec intueri saeculis longe abdita;
Oculosq; Luce tinxit ambrosia Tuos!
Tu mentis omnis, at Tuae Nuila est capax;
Hac Iaude solus fruere: Divinum est opus
Animam creare: Proximum huic, oftendere.

      T.H.A.M. Coil. Ex. Oxon.

To his Ingenious Friend Mr. John Heydon, on his Book Intitled The Holy Guide.

The Ancient Magi, Druids, Cabbalists,
The Brachmians, Sybil's, and Gymnosophists
Withal that Occult Arts haberdash,
And make so many mancies, do but trash
By retail vend, and may for Peddlers go:
Your Richer Merchandise doth make them foe.
The Stagarite must with his Manual
Of Elements, Galen of humour call
In all their suit, or your New Art,
Without them, makes their good old cause to smart.
Vulgar Physicians cannot look for more
Patients, then such which do need heliborne:
When Rosie Crucian Power can revive
The dead, and keep old men in youth alive,
Had you not called your work *The Holy Guide*,
It would have puzzled all the world beside
To have Babtiz'd it with a Name so fit
And Adequate to what's contained in it;
Should it be styled the Encyclopedia
Of Curious Arts, or termed a Mystery
In Folio, or be named the Vatican
Reduced unto an Euchiridion,
Or all the Hermae in a Senary,
The Urin and Thummin of Philosophy,
The Art of Hieroglyphics so revealed,
And like the Apocalypse they are concealed.
Or the Orthodox Paradox, or all

Discovered, which men still a wonder call;
Or the Magna Charta of all Sciences,
And be that names it cannot call it less,
The Book and Title might have well agreed;
Yet men have questioned if unto their Creed

They should have put your Article, but Now
The name of Holy none dare disallow.
When so much learning doth in one exist
Heydon, not Hermes, shall be Trismegistus.
And if the Right Reverend of Levies Tribe
Do Hallow it, I cannot but subscribe.

                        Myself your Friend and Servant.

                        Tho. Fyge.

To the most Excellent Philosopher and Lawyer Mr. John Heydon, upon the Holy Guide.

Hail you (admired Heydon) whose great parts
Shine above envy, and the common Arts,
You kin to Angels, and Superior Lights,
(A spark of the first fire) whose Eagle flights
Trade not with Earth, and grossness, but do pass
To the pure Heavens, & make your God your glass,
In whom you see all forms, and so do give
These rare discoveries, how things move and live;
Proceed to make your great design complete,
And let not this rude world our hopes defeat.
Oh, let me but by this the dawning light
Which streams upon me through your three pil'd
Pass to the East of truth,
'till I may see night Mans first fair state,
when sage Simplicity,
The Dove and Serpent, Innocent and wife
Dwell in his breast, and he in Paradise;
There from the Tree of knowledge his best bought,
We pluck a Garland for this Authors brows,
Which to succeeding times Fame shall bequeath,
With this most just Applause, Great Heydon's wreath.

J.W.

A
Chemical Dictionary,[8]
Or,
An Explanation of the hard words and terms of Art which are used in *The Holy Guide*.

A.

Acquisitio, Amissio, Album, are figures of Geomancy, ♈ or Aries, a figure of Astromancy, and they are names that signify the Nature of the Medicines, and you shall find them in Order; there being used 19 Figures of Astromancy, viz. ♄ Saturn, ♃ Jupiter, ♂ Mars, ☉ Sol, ♀ Venus, ☿ Mercury, ☽ The Moon or (Luna), ♉ Taurus, ♊ Gemini, ♋ Cancer, ♌ Leo, ♍ Virgo, ♎ Libra, ♏ Scorpio, ♐ Sagittary, ♑ Capricorn, ♒ Aquarius, ♓ Pisces, And 16, (Carcer, Tristitia, Fortuna Major, Fortuna Minor, Puer, Puella, Conjunctio, Rubeus, Lettitia, Caput Draconis, Cauda Draconis, Populus, via) of Geomancy.

Amalgamation is a Calcining or Corroding Metals with Quicksilver, and it is done thus; take any Metal except Iron, beaten into leaves, or very small Powder, mixt with about eight parts of Quicksilver (which may the better be done, if both be heated first) that they may become one uniform Mass, evaporate the Quicksilver over the fire, and Metal will be left in the bottom as a thin calx.

---

[8] Also refer to Volume 21 of The R.A.M.S. Library of Alchemy, *Alchemical Symbols*. - pnw

Aqua is the water.

Aries and Aquarius signs in Astromancy.

Acquisitio, Amissio, and Albus figures of Geomancy.

Astromancy is Heavenly knowledge, and Geomancy is Earthly knowledge.

C.

Calcination is a reducing anything into Calx, and making it friable; and it may be done in two ways, by firing, by reducing into ashes, by reverberating, by Corrosion, by Amalgamation, Precipitation, Fumigation, or vaporation, Cementation or stratification. Caput Draconis, Cauda Draconis, Conjunctio, and Carcer figures of Geomancy.

Circulation, is when any liquor is so placed in digestion, that it shall rise up and fall down, and rise up and fall down, and so do continually, and thereby become more digested and mature, for which use for the most part we use a Pelican.

Clarification, is the separating of the gross feces from any decoction or juice; and it is done three ways, by the white of an Egg, by digestion, by filtration.

Cure all, is Aurum Potable.

Coagulation, is the reducing of any liquid thing to a thicker substance by evaporating the humidity.

Cohobation, is the frequent abstraction of any liquor poured oft—times on the feces from whence it was distilled, by distillation. Cancer and Capricorn, figures in Astromancy.

Congealation, is when any liquor being decocted to the height, is afterwards by setting into any cold place turned into a transparent substance like unto Ice.

Corrosion, is the Calcining of bodies by corrosive things.

Cure the great is the Pantarva.

D.

Decantation, is the pouring off any liquor which has a settling, by inclination.

Deliquim, the dissolving of a hard body into a liquor, as salt, or the powder of any calcined matter, etc. in a moist, cold place.

Descension, is when the essential juice dissolved from the matter to be distilled doth descend, on fall downward.

Despumation, is the taking off the froth that floats on the top with a spoon or feather, or by percolation.

Distillation, is the extracting of the humid part of things by virtue of heat, being first resolved into a vapor, and then condensed again by cold. Thus, it is generally taken; but how more particularly, I shall afterward show.

Digestion, is a concocting, or maturation of crude things by an easy and gentle heat.

Dissolution, is the turning of bodies into a liquor by the addition of some humidity.

Dulcoration, or Dulcification, is either the washing off the salt from any matter that was calcined therewith, with warm water, in which the salt is dissolved, and the matter dulcified: or it is sweetening of things with sugar, or honey, or syrup.

E.

Elevation, is the rising of any matter in manner of fume, or vapor, by virtue of heat.

Evaporation, or Exhalation, is the vaporing away of any moisture.

Exaltation, is when any matter doth by digestion attain to a greater purity.

Expression, is the extracting of any liquor by the hand, or by press.

Extraction, is the drawing forth of an essence from a corporeal matter by some fit liquor, as spirit of wine; the feces remain in the bottom.

F.

Fermentation, is when anything is resolved into itself, and is rarified, and ripened: whether it be done by any ferment added to it, or by digestion only.

Fortune Major a figure of Geomancy.

Filtration, is the separation of any liquid matter from its feces by making it run through a brown paper made like a tunnel, or a little bag of woolen cloth, or through shreds.

Fixation, is the king of any volatile spiritual body endure the fire, and not fly away, whether it be done by often reiterated distillations, or sublimations, or by the adding of some fixing thing to it.

Fortuna Minor a Figure of Geomancy.

Fumigation, is the calcining of bodies by the same sharp spirits, whether vegetable or mineral, the bodies being laid over the mouth of the vessel wherein the sharp spirits are.

G.

Gemini, A Figure of Astromancy.

H.

Humectation, or Irrigation, is a sprinkling of moisture upon anything.

I.

Imbibition, is when any dry body drinks in any moisture that is put upon it.

Impregnation, is when any dry body has drank in so much moisture that it will admit of no more.

Incorporation, is a mixtion of a dry and moist body together, so as to make a uniform masse of them.

Infusion, is the putting of a hard matter into liquor, for the virtue thereof to be Extracted.

Insolation, is the digesting of things in the Sun.

J.

Jupiter, a planet of Astromancy.

L.

Levigation, is the reducing of any hard matter into a most fine powder.

Letitia, of Geomancy.

Leo and Libra figures of Astromancy.

Liquation, is a melting or making any thing fluid.

Lutation, is either the stopping of the orifices of vessels, that no vapor passes out, or the coating of any vessel to preserve it from breaking in the fire.

M.

Mars, Mercury, and the Moon, Planets of Astromancy.

Maceration, is the same as Digestion.

Maturation, is the exalting of a substance that is immature and crude, to be ripened and concocted.

Menstrum, is any liquor that serves for the extracting the essence of anything.

P.

Precipitation, is when bodies corroded by corrosive spirits, either by the evaporating of the spirits remain in the bottom, or by pouring something upon the spirits, as oil of Tartar, or a good quantity of water, do fall to the bottom.

Puer, Puella & Populus, figures of Geomancy.

Pisces a figure in Astromancy.

Purification, is a separation of any Liquor from its feces, whether it be done by clarification, filtration, or digestion.

Putrefaction, is the resolution of a mixt body into itself, by a natural gentle heat.

Q.

Quintessence, is an absolute, pure end well digested medicine, drawn from any substance, either animal, vegetable or mineral.

R.

Rubeus, a figure of Geomancy.

Rectification, is either the drawing of the flegme from the spirits, or of the spirits from the flegme, or the exaltation of any Liquor by a reiterated distillation.

Reverberation, is the reducing of bodies into a Calx; by a reflecting flame.

S.

Saturn, Sol, Scorpio, and Sagittary, Planets and signs in Astromancy.

Solution, is a dissolving or attenuating of bodies.

Stratification, is a strewing of corroding powder on plates of metal, by course.

Sublimation, is an elevating, or raising of the matter to the upper part of the vessel by way of a subtle powder.

Subtiliation, is the turning of a body into a Liquor, or into a fine powder.

T.

Taurus, a figure of Astromancy.

Transmutation, is the changing of a thing in substance, Color and quality.

V.

Venus, Virgo, figures of Astromancy,

Volatile, is that which flyeth the fire.

Rules to be considered in Rosie Crucian Medicines.

1.

Make choice of a fit place in your house for the furnace, so that it may neither hinder anything, nor be in danger of the falling of anything into it that shall lye over it: for a forcing Furnace, it will be best to set it in a chimney, because a strong heat is used to it, and many times there are used brands which will smoke, and the fire being great the danger thereof may be prevented, and of things of a malign and venerate quality being distilled to such a Furnace, the fume or vapor, if the glass should break may be carried up into the chimney which otherwise will fly about the room to thy prejudice.

2.

In all kinds of Distillations, the vessels are not to be filled too full; for if you distill Liquors, they will run over; if other solider things, the one part will be burnt before the other part be at all worked upon; but fill the fourth part of copper vessels; and in rectifying of spirits fill the vessel half full.

3.

Let those things which are flatulent, as wax, rosin, and such like, as also those things which do easily boil up, as honey, be put in a lesser quantity, and be distilled in greater vessels, with the addition of salt, sand, or such like.

4.

There be some things which require a strong fire, yet you must have a care that the fire be not too vehement for fear their nature be destroyed.

5.

You must have a care that the lute with which vessels are closed, do not give vent and alter the nature of the Liquor, especially when a strong fire is to be used.

6.

Acid Liquors have this peculiar property, that the weaker part goes forth first, and the stronger last, but in fermented and Liquors the spirit goes first, then the flegme.

7.

If the Liquor retain a certain Empyreuma, or snatch of the fire, thou shall help it by putting it into a glass close stopped, and so exposing it to the heat of the Sun, and now and then opening the glass that the fiery impression may exhale, or else let the glass Stand in a cold moist place.

8.

When you put water into a seething Balneum, wherein there are glasses, let it be hot, or else you will endanger the breaking of the glass.

9.

When you take an earthen or glass vessel from the fire, expose it not to the cold air too suddenly for fear it should break.

10.

If thou wouldst have a Balneum as hot as ashes, put sand or sawdust into it, that the heat of the water may be therewith kept in, and made more intense.

11.

If you would make a heat with horse-dung, the manner is this; viz., make a hole in the ground, then lay one course of horse-dung a foot thick, then a course of unslaked lime half a foot thick, then another of dung, as before; then set your vessel, and lay round it lime and horse-dung mixt together; press it down very hard; you must sprinkle it every other day with water, and when it ceases to be hot, then take it out and put in more.

12.

Note that always sand or ashes must be well sifted; for otherwise a coal or stone therein may break your glass.

13.

The time for putrefaction of things is various; for if the thing to be putrefied be vegetables and green, less time is required, if dry, a longer; if Minerals, the longest of all. Thus, much note, that things are sooner putrefied in cloudy weather then in fair.

14.

If thou wouldst keep vegetables fresh and green all the year, gather them in a dry day, and put them into an earthen vessel, which you must stop close, and set in a cold place: and they will, as says Glauberus[9], keep fresh a whole year.

15.

Do not expect to extract the essence of any vegetable unless by making use of the feces left after distillation: for if you take those feces, as for example of a nettle, and make a decoction thereof, and strain it and set it in the frost, it will be congealed and in it will appear a thousand leaves of nettles with their prickles, which when the decoction is again resolved by heat, vanish away, which shows that the essence of the vegetables lies in the salt thereof.

16.

In all your operations, diligently observe the processes which you read, and vary not a little from them; for sometimes a small mistake or neglect spoils the whole operation, and frustrates your expectation.

17.

Try not at first experiments of great cost, or great difficulty, for it will be a great discouragement to thee, and thou will be very apt to mistake.

18.

If any would enter upon the practice of Chemistry, let him apply himself to some expert Artist for to be instructed in the manual operations of things; for by this means he will learn more in two

---

[9] Johann Rudolph Glauber. His works are included in The R.A.M.S. Library of Alchemy. -pnw

months, than he can by his practice and study in seven years, as also avoid much pains and cost, and redeem much time which else of necessity he will lose.

19.

Enter not upon any operation, unless it be consistent with the possibility of nature, which therefore thou must endeavor as much as possible may be to understand well.

20.

Do not interpret all things you read according to the literal sense; for Philosophers when they wrote anything too excellent for the vulgar to know, expressed it enigmatically, that the Sons of Art only might understand it.

21.

In all thy operations propose a good end to thy self, as not to use any excellent experiment that thou shalt discover, to any ill end, but for public good.

22.

Understand well whether you shall prosper or not; before you begin anything pray to God, and you shall find all you desire in the Second book, made plain unto you.

The
Holy Guide,
Leading the way to
UNITE ART AND NATURE:

In which is made plain

ALL THINGS PAST, PRESENT, ΦιλοΥομοs

AND TO COME.

By John Heydon Gent. ΦιλοΥομοs

A servant of God and Secretary

of Nature.

Thus, have I declared unto you the descent of the secret power of Nature from God, even to this Earth. *The Harmony of the World*, Book 1. Chap. 1. page 9.

LONDON,

Printed by T.M. for the Author, 1662.

Book I. Chapter I.

Of God, Art and Nature.

1. Of God, of Man, of Creatures: 2. A divine pattern: 3. Frailty: 4. Happiness what? 5. A spirit that worketh all things: 6. Divine lights: 7. Plato's Crown: 8. The grounds of Knowledge: 9. Opinions: 10. Images of Heaven: 11. Single minds, Messengers and Angels: 12. Degrees of happiness: 13. Of three delights.

1.   God is our Holy Guide, therefore in all orderly speeches and matters of Learning, it first of all behooves a wise man to agree upon the thing in hand, what it is; And what is the bounds (or Definitions) of the same; it seems very needful in this discourse of the Rosie Crucian Medicines, to show first, their matter, in Nature and Art, their manner of working, all which we will here canonically and orderly make manifest, because it is a thing much in doubt and in question among the Learned.

Eyes that use to behold and view the reason and nature of things, may easily perceive by the outward shape and inward gifts of man, unlike and passing all other Wights (or living creatures) that he was made for some notable end and purpose above the rest, and so not for pleasure, honor, or enough of needful outward things, which they call Riches; nor yet for any other matters, which other sights void of wit and reason seek and follow; therefore a man ought not to make anything of that nature his end and happiness, unless he think it reason for the Master (and better workman) to learn of the Servant and worse; for what other pattern and end have we to follow? None at all; because we are the best Creatures in the world; than it is without the world, say you, and among the Blessed Minds, or the Æthereal Inhabitants, above and without all; neither yet have we found it, for they be our fellow-servants and subjects under one Almighty King.

2. Wherefore Eugenius Theodidactus says there remains nothing but God, and his happiness to be sought and set before our eyes; not with hope to overtake and reach it, that were madness; but with desire to attain so much thereof, as the proportion between him and us will suffer:

2.[10] Or if the unmeasurable and boundless, or infinite blessedness of God admit, no comparison, it was best, yea and by the example of Mr. Tho. Heydon, to make the bounds of our happiness in Long Life, Health and Youth, so much of the Service of God, as our whole power and nature will Hold and Carry; now if we knew that divine Pattern, and only gift of God, all were well: And this as almost all other truth, especially in case of life and manners, for the which this Book was chiefly written, by the witness and record of holy writ, and received to be known and proved: If that were not so strange and far off from this purpose, which is appointed as you see to run through the midst of Art, Nature, Reason, Philosophy and Physick.

3. Wherefore sit hence both in this and all other matters, Galen builds overmuch upon his own devices, not considering as some may object, that a man (especially a young man) may swerve, but we have assistance of the Rosie-Crucian Seraptical illuminated fraternity, and have besides a single judgement and manual experience in the Philosophers Pantarva, a double portion, of the spirits council, which

---

[10] The 1662 printed edition follows this section numbering, with two consecutive sections numbered "2." -pnw

said ἴσω ὅτι βελήαρδρας πάντα ἡδύτας, All other besides did not content us, because they were no more but men endued with ripe wits, and perhaps sound judgement in the course of kind (or Nature and Physick; Now I must look as near as I can to my own judgement, that it be still squared by the rule of truth and reason). And so, let us return to our purpose, long Life, Health, Youth, Riches, Wisdom and Virtue, are not to be found among those men that live like Hogs, always greedy after such things as beasts desire, and know no better than things auspicious to Swine.

4. Then to find this happiness and pleasure of heaven among men; To whom were it best to travel? Unto Poets think you? No; because they take their aim still at a vain mark; the peoples liking, as you may see by Mr. John Cleveland's Poems; for I will not draw of the dregs (when he says) If a man be rich, and have his health, with a contented mind, and honor, let him not care to be a God, nor for popular applause. This vain and worldly content is far from a Divine nature; Nor yet need we go to the lower or lesser houses of Physick, where as they be tainted and unfounded in other points of learning; so, in matter of manners they do not do well to place their content in honor, pleasure, or in such like outward things, no nor to set it in good life alone, and virtue.

5. Besides the opinion of Hermes, Tarthas, Apollonius, Phroates, and others: It is my thoughts, that that which is inferior or below, is as that which is superior or above, there being one universal matter and form of all things, differenced only by accidents, and particularly by that great mystery of rare faction and condensation, the inferior and superior, to work and accomplish the miracles of one thing, and to show the great variety and diversity of operations wrought by that

spirit that worketh all things, in all things; and as all things were from One, by the mediation of one God, having created all things in the beginning, which is the beginning of all things, and the wisdom of his Father; so all things sprung and took their Original from this one thing, by adoption, or sitting itself accordingly, in number, weight and measure; for wisdom builds her own house.

6. Plato and Pythagoras, for their matchless understanding in natural things, and Divine light in good order of life and manners, have been these many ages best accepted with the best, and followed in all things; therefore in this high point of manners, which we have touched, we will tell you the father of this one thing, or that which he uses instead of an Agent, and all the operations thereof, is the Sun, and the mother thereof, or which applies the place of a female and patient, is the Moon; the nurse thereof and her paps, all the influence of heat and moisture, of the Sulphur and Mercury of Natures; for the spirit of God moves not but upon the face of the waters; the earth, the wind, or air, is carried in its belly, as the sails in the chain, that tie the superior things to them that are below. This is the Father original, and Fountain of all perfection, and of all the secret and miraculous things done in the world, whose force is then perfect and complete. Now let us see what opinions others hold, and how near they come to Theodidactus his right line of truth.

7. To begin with Plato, the spring of this Philosophy, his Medicine and Happiness; he disputes in Philaebus, as near as I could gather, out of so large and scattered a speech, is nothing but Pleasure and Health in a Medicine. And yet this Divine man means not (lest you should marvel) with the herd of swine (though they were not the brothers of that foul opinion, but watered their gardens, as Tully says, with other men's springs) to set open all the gates of the fences, and to let in all

that comes; but only at a few narrow loops, to receive clean delight without all grief interlaced, and by name delight in colors, consent and some smells in health, wisdom and virtue.

And again he says in *Thaetus*, that justice and holiness, together with wisdom, make us like unto God; to let those two places serve for him, and to come to Pythagoras: as there are two sorts of men, one disposed to deal with others, which are called worldly men; and another quite contrarily bent to live alone, and to seek knowledge, which are called Philosophers; so he in his book appoints two several ends; for the first virtue (I mean adoring, and no idle virtue) garnished with outward helps, and gifts of body and fortune: for the next knowledge of the best things; and this he sets before that other, for many reasons vouched toward the end of his book; but especially, because God, whom we ought to follow, leadeth the same life.

8. These be the best grounds of Happiness and Pleasure that ever any Philosopher or Physician has said at any time; (for never a one has quite built it up;) let us see how they be squared: If the foul-fed Epicure may again be justly reproved, and reckoned as an impious person, whom never any heavenly thoughts touched for bringing in an idle God, neither ruling the world, nor regarding it; How can Aristotle seem wrongfully accused of impiety, & for the same banished out of the Academy, if there were no other proof against him (when he says) in that place, God leadeth no other than this beholding & and gazing life of his? Is it not an idle, and, as it were, a covetous life turned back upon itself, and estranged from all outward action applied and directed to others? Yea, and that in his own, and all other men's understandings; then to encounter him with his worthy Master Plato, if that were the best life, or the life of God, why did God make the World? He lived so before, if that had been the best life; but because he

was good, he would have others enjoy his goodness; and before he was busy in making, and is yet in ruling the world; and yet indeed it is no business, as we reckon it, that is no care and trouble, but an outward deed and action, clean contrary to the inward deed of a musing mind only shooting at his own good estate with his wisdom & knowledge.

9. But if he deny all this, as it's like he will, to increase the heap of sin, he grants no beginning; then what can be greater evidence then his own writings, one quite thwarting another, as cross as may be; for in another place he comes again, and says, that every man has so much happiness as he has Wisdom and Virtue, even by the witness of God himself, who is therefore happy, and not for outward goods; what can be more divinely spoken, and more cross to the former, foul and Godless opinion? nay, see the force of truth; he yields again according to the heavenly Master, that to forestall the place from the worse sort, good men ought to take office upon them, and to manage affairs of State: yea further, if they refuse (which if they be wise, they will, quote Zeno) *that they may rightly be compelled*; then; if this wise man has virtue in possession, an no doubt he has, he must as we see by his own confession, use it; and the same reason is of God himself in this great City of the World; but Plato by name, thinks those two so nearly allied and knit together, as he dare openly deny happiness to that Commonwealth where they be dislinked and stand asunder.

10. Then we see, that in the judgement of these two great philosophers and physicians, where they be best advised, and in deed and truth, the divine pattern of happiness, which we ought to strive unto, is no more, nor no less than that worthy couple of wisdom and virtue knit together in that bond of fellowship, which may never be parted asunder.

That Solomon desired, when God gave him his choice, and had him ask what he would have, and he would give it him, as you may read 1. Kings c. 1. He said, *Lord give thy servant an understanding heart, that he may judge between good and bad; and the speech pleased the lord, that Solomon had asked this thing; and god said unto him, because you have asked this thing, and has not asked long life, neither riches for thy self, nor the life of thine enemies, but has asked for thy self-understanding to discern judgement; behold, I have done according unto thy word. Lo I have given thee a wise understanding heart, so that there was none like thee before thee, neither after thee shall any arise like unto thee. And also, I have given thee that which you have not asked, riches and honor; and so, will god do to all those that mind wisdom and virtue.* In the first place, with an intent to do good to others. On the contrary, sad experience has witnessed even in our days, that many, whose whole aim was to be rich in this world, have been deprived of all, and forced to seek their bread in a strange land; so may others do yet for ought I know, if they be not all the wiser; they that swallow down riches, and not by right, shall vomit them up again; the Lord shall cast them out of their bellies, Job 20:15.

11.   But you may say, we have reared our Happiness, long life and Health aloft, and made it a fair and goodly work; but more fit for the dwelling of those clean and single minds (or spirits) above, which they call Messengers (or Angels) therefore is man buried here below in these earthly bodies, as we are scarce able to look up unto it: and therefore Pythagoras in his book, with good advice, often receives in enough of bodily and outward goods, to help the matter, (thought it not to be any other cause of joy, than the instrument is of Music:) and so Plato, we see, names his servants and helpers.

12.   Indeed, I grant, that this full and high pitch of happiness, etc. (I mean that measure above set) is free and easy to free and lively spirits;

but to us impossible, without outward means and helps, which, nevertheless, shall not be counted as any part of the frame of Health, needful to make up the whole, but, as it were, loose and hang-by steps and stairs leading up to it.

13. Then, if these be so needful as they be, it were of much need to lay them down, and keep just account, which those physicians do not, lest if there be two for one, happiness, etc. should halt; if again, too many, the idle parts might, infect, infect and mar the rest; as we may fear of Plato his first three delights, although they be not hurtful of themselves: Without more words, the just sum is thus: To obtain so much Happiness, etc. as our Nature is able to take and hold, the body had need be first willing and obedient, and then store of outward needful things to be at hand and ready; these every man knows; but for the body, that is obedient, when it is long lived, healthful, young, clear, and temperate; when all these helps flock together, we may be happy, if we will; if any want, we shall do what we can, as you shall bear hereafter.

Then let us marshal these things at last in order, and *The Holy Guide*, that compares Happiness to a Family, make that loving couple wisdom and virtue, as man and wife, and heads of the Household, the five proper eyes of the body like Children, and Riches as Servants. These again, if the chief of the house will suffer them to marry, will beget other two bond—children, to beautify the same house, honor and pleasure; but the wise and good Householder, will in no wise suffer it, lest his house be troubled with more than may be ruled; and, although true and right Honor and Pleasure will perforce follow, yet he shall not regard them, nor be minded towards them, as those grave men were towards Helen, and often use their saying, although they be such kind ones, yet let them go: and us follow our way to health and happiness, etc. See *The Harmony of the World*, etc.

**All Objections cast against the
Rosie Crucian Medicines Answered,
and the truth made manifest.**

Chapter II.

1. The way to Wisdom; 2. Hermes medicines; 3. Rules; 4. Possibilities and effects; 5. Faultless studies; 6. Approved reasons; 7. Opinions; 8. The stop-ship; 9. Secret truths; 10. Wonderous works; 11. Wisemen; 12. Alchemy; 13. Of the secret blast and motion of God; 14. Of Natures fault; 15. Divine truths; 16. Man's mind; 17. Of the life of God; 18. Raging Counsel; 19. Stingless Drones; 20. Dissention; the Emperor's folly.

1. Oh that we knew that health and happiness, we may when we will, go into the way where and how all men may be blessed: wherein I am quite bereaved of all helps from the Grecians, as men ever apt to speak & think well, rather than to do & perform any thing (though constancy & agreement, in their sayings) would have left blessedness as well as other good things in the power and reach of all men and I must fly for aide into Egypt, a people so far passing all other Nations, as it is better and nearer to God, to work and to do great wonderous things than to behold and look upon then.

2. For it is delivered to Ancient and true Record that one Hermes a king and law-giver of that Country, a man of rare and divine gift In knowledge above all that ever were, found out Medicines able to bring

all men to health & long life, etc. and left them behind him in writing to his people; & that they were after him, a long time by the wiser sort, closely wrought and used, until at last, they crept abroad and stole into ARABIA, when she flourished in Arms and Learning, and there got the name which they now commonly keep of Filius Solis Caelestis, Amicus Vitae, Pantarva, Ignis Vitae, Stella Vitae, Radix Vitae, Aqua Solis, Aqua Lune, Deliciae Vitae, Panacea, Succus Vitae, Medulla Vitae, Adjutrix Vitae, Salus Vitae, Sanguis Vitae, Aurum Potable; and indeed all these medicines are made of prepared Gold, etc. Now from thence in the same secret and dignified manner (for that is the wont of them, as becomes so deep secrets) they have traveled and spread themselves over all Nations; now and then opening and discovering themselves to a few of the better and wiser company.

3.   Then this is means to obtain blessedness, which I mean to take, and withal to prove it no pleasant dream and happy tale, if it were true as the common proverb goth of it; but as it is a Natural, Heroical, and almost a Divine deed, scarce to be reached or matched with any words, so I vow them a true and certain story, things often done, and again to be done as often, I am unfit, I grant, and unable to bear so great a burden, but that the desire I have both to defend the Truth from slander, and to do good to them that love it, makes it light and easy: and again, the hope upholds me, that if I chance to stumble or faint at any time, these will as gently and willingly lend their hand to stay me, or at least bear with the fall of misfortune.

Then for the common or wilder sort, which either for lack of good Nature, or rant of good Manners, use to wrangle about words, or twitch at things, I care not; and because I know them not, I pass them as unknown men; for neither was Hercules able, as they say, to match with many-headed hydra, nor yet with the awke and crooked crab.

4. Then to turn my speech, which way were it best to set forwards? not right and straight, the matter! No; because there is such crying out against the possibility of the good work which our Medicine promises; and that Awke for judgement of the matter has been the chief cause which has hitherto buried this Divine Art from the sight of good and learned men; I take it the best way of delivery, before I come to the point itself, to fetch about a little, and then show the possibility of those effects, and the way to work them by other or weaker means, as well as by Hermes Medicines. For although it be not so natural, in marching forward to move the least and weak part; yet I keep it right artificially, and then it shall agree with that good order of Art: First of all to put by a few of the light things laid against this blessed Science, because, albeit, they be gathered but by guess, besides all grounds of certainty; yet they have so wholly possessed the common people, yea, and some of the better and worse sort likewise, that without any further search or hearing of the matter, they have straight-way cast it off for false, and condemned it; for when as once sleep has taken the sort of the body, the senses yield, and can do nothing; so if wrong belief get once possession of the soul, reason is laid to rest, and cannot move again, before that must be loosened, put to flight and scattered.

5. First, say they, since there be seen in all places and times, so many hundreds with great pains, heed and cunning, to study this Art, and put the Receipts in practice: now if they were true and faultless, as others are, some should appear to hit the mark, and to gather the fruits of their travel, and to live as they do, of all men most miserable; or at least, because it is so ancient an Art, it would have been recorded in some public or private writing, besides their own, which be it bound with never so deep oaths (as it is) yet it is insufficient proof and witness In their own case.

6. These be the most capable reasons, and best approved among the people, wherewith they use to batter this exchanging sequence: but mark how light and weak they be, and easy to be wiped away; for how could the acts and deeds of these R.C. Philosophers & Physicians come into the writings and Records of men (to begin there with them) whose fame, nay, whose company they have ever shunned? and when their own Records, if they chance to light of anything that was not sown abroad, and published to the world, as is the life of worldlings; but left like most precious Jewels unto some friend of secret trust, which was counted as a Son adopted, upon condition to keep it still within the house and stock of Hermes, from the eyes and hands of the world and strangers, running evermore, like the wise Stars, a contrary race unto the world, that no marvel, though they be both, in like sort crossed by the world, and mis-called wanderers (or Planets,) when indeed and truth they go better. Now when they deem credit to be denied to the men's report & witness, it is a sign that either their own reports & witness is of light and little weight, whereby they judge of others; or else, that their thoughts are vain and phantastical, puffed up, I mean with that new kind of self-love and over-weening wisdom, to set up themselves, and pull down Authorities; of which sort it falls out most commonly in people, that while they strive to avoid the lake of superstition, they run headlong unawares down the river of impiety; for if such a wide breach and entry nay be suffered to be made into the credit and authority of the Writers, which are the life of Antiquity and light of Memory, great darkness and confusion will soon come in and overcast the world; yea, and so far forth at length, as naught shall be believed and judged true that is not seen; that even they which dwell in the main land, shall not grant a sea; a thing not only fond and childish among men, but also (ill be to me, if I speak not as I think)

wicked and godless amongst us Christians, whose whole Religion, as S. Augustine says, stands upon that ground.

7. Wherefore, if we must needs believe Records, yea, though they be sometimes lewd men, foolish and unlearned, as if they were as whole and harmless, as Xenocrates; but especially, although they had great cause to lie, and to speak more or less than the truth; who can in common reason refuse the solemn oaths of so good, and wise and learned men? For he is good for the love of Virtue itself; he that is wise, to avoid the shame of lying, will speak the truth: That shall I say Of Eugenius Theodidactus, that durst in times past own no other name, whose whole care and practice, drift & studies, now is nothing else but to find and set down the truth? But all is well & clear of all suspicion, if it may, be though those oaths and protestations to have sprung from himself, and others experienced in these undeniable truths, of more good will and desire to persuade the lovers of Wisdom and Virtue, than wrought out of fear or flattery, which may easily be judged in such men, as were all either then false protectors that cared not, or kings that needed not, as it is clear in all their eyes that are conversant in these kinds of studies. Wherefore such men as are so bold with our ground of reason to deny, and deny still all that comes, are, in my opinion, greatly to be looked unto; for although they, like Xerxes, pull not down Religion with hands openly, yet they are of another sort dangerous, that undermine it closely with wrong opinions. If our men avoid such plain untruths, as might be reproved by common sense and daily experience, as when Anexagoras said snow was black, and Xenophanes the Moon is inhabited, and full of hills and cities; and In cities of old, with some of late among the Stars (Sir Chr. Heydon Baron & Mr. John Heydon., and Mr. John Gadbury), but I speak not against Astrologers, but against such flattering liars that have gained their estates amongst silly foolish women, & ignorant people, that hold, that

the earth, the only moveable thing in the world, stands still, and such like ugly misshapen lies, wherewith Greece over-swarmed; then you had reason to use them with ill words and thoughts as you do: Now, although I was partly persuaded to be of the same opinion with those that hold the earths immobility: but being convinced, I relinquish my former opinion; for they maintain, that by a Heavenly Medicine they have great and wonderful changes turned all metals into Gold, Folly into Wisdom, Vice into Virtue, Weakness into Long Life, all Diseases into sound Health, and Age into Lustiness and Youth again; How can you disprove them? when did you see the contrary? you surely know the nature of the deeds and effects; for they require great knowledge; but the doing cause workmen, that you dislike is, their medicines you never saw, nor can imagine what it is, much less conceive the reason, strength & nature of it; Nay you see nothing, but grope and blunder in the dark, like blindfolded men at all things; else how could these exchanges have escaped, & been hid from you, in a world so full of all kinds of changes? I mean, you see great and admirable things (albeit you do not so take then, because you see them often) but you do not truly see them, that is, you perceive not the nature, cause and reason of them, and that makes you so childish to believe naught unseen, and count all things wondrous which are not common among you; much like that harmless and silly kind of people, of late discovered, which made miracles and wonders of many matters, that in other countries are ordinary and common, in so much as (to take one for all) they could not conceive how two men asunder could by letter certify one another, unless a spirit were wrapped up in the paper to make report, and tell the news; but if you and they could once by this Guide & Art, cut into the depth and nature of the great and marvelous works of kind and skill, which are common and daily among you; then, you would be ready and easy by comparison to receive almost anything unseen, and brought by report to you. Let me awake your wits a little;

you see daily, but not thoroughly, how the Moon by her Sympathy with the spirit of the water draws the Ocean after her, makes the ebbs and flowings thereof: it is likewise commonly known, that the loadstone in the roof of mahomet his church, draws up his iron tomb from the ground, and holds it ranging in the middle way; like as the miners in Germany, found their tools which they had left in such a Vault, hanging in the morning; which was accounted for a miracle, before such time as the cause, by the skillful, was seen and declared unto them. What should I say more of this Stone? it is not unknown that there are whole rocks thereof in India, at the Castle of the Adamant, erected by Jul. Caes. drawing ships that pass by loaded with iron unto them: & yet we see that this mighty Stone, in presence of the Diamond, the King of Stones, is put out of office, and can do nothing.

8. To come abroad, it has been often seen at Sea, that the little Stay-fish cleaving to the fore-ship, has stopped her full course.

9. I should now pass over to that other side of skill and craft, and call to mind many great and wonderous works there done and performed; the curious work of that Italian king, which held a clock besides a dial within it; these three common feats found out of late, passing all inventions of Antiquity, the gun, card, and printing, and many other dainty Devices of man's wit and cunning; if this short and narrow speech appointed would suffer any such out-ridings, let these few serve to awake you, and call your wits together: you see these things I say, and are never moved; but if you had never seen them, but heard the stories only reported, what would you have thought and said? and because no man judges so well of himself as of another; suppose a plain and harmless people, such as those Indians were, had from the beginning dwelt in a cave underground, let it be the Center if you will, and at the last one man more wise than the rest, had by

stealth crept out into the light: And by long travel & traffic with our people had seen and learned the course of nature of things which I have rehearsed unto you, and then returning home, had suddenly start up and begun to account the wonders which he had seen and learned: first, that he had found the earth hanging in middle of the air, and in like sort a bright and goodly cover compassing afar off the same; this cover beset and sprinkled with infinite lights and candles, and among the rest, one (to be short) of a foot in bigness to his sight, without all touching, or other means or instruments to be perceived, to hold and pull huge heaps of water after her, as she passed up and down continually, would they not shout and lift up their hands, and begin to suspect the man of infection with strange and travelling manners?

10.   But admit, when the noise were done, and all hushed, he went forward & told them of such a Church and Vault with other things, as well, and more strange than the earth (for that cannot be otherwise, unless heavy things flew up against Nature) hanged in the air alone, and of such hills, that as the Sun waters draws ships out of their courses, without any strength or means visible; furthermore, if he laid abroad the wonderful might of a little fish, like half a foot long, able to stay the main course of a ship under sail: do you not think with what sour countenances and reviling words, and reproaches, they would bait and drive him out of their company? But if the good and painful man burning with desire to reform the estate of this rude and deformed country, would not be stayed so, but spying a calmer time, durst come in presence, and step forth before them again, and say, that by his travel he had made such a ring as I speak of; such warlike Engines as should fall as fearful as thunder, and as hurtful as a canon fired at a fort, a mile off planted; with a kind of writing whereby four men might record as much in the same time as four thousand of the common Clerks; such a Card, wherewith a Countryman that never saw

the Sea, shall sit in the bottom of a Ship, and direct the course thereof throughout the world without missing; Is it not like they would apprehend him for a cousener[11], and adjudge him to punishment? then put the case you stood by and saw the matter, I appeal to your own experience, would you not think the Traveler worth pity and praise, and the people of reformation?

11. Well then, let us return to our purpose; there is a Nation of wise men dwelling in a soil as much more blessed (then yours) as yours is then theirs of the Deserts, that is, as they bide underground, and you upon the face of the roof: so these men inhabit the edge and the skirt of Heaven; they daily see and work many wondrous things, which you never saw nor made, because you never mounted so high to come among them; if any one chance to fly away from you to those heavenly places, & after like experience, to return & make the like reports, you give him the like rewards you give: (compare the rest) I say no more; but if God would give you leave and power to ascend those high places, I mean to those heavenly thoughts and studies, you might quickly, by view of deep causes, and divine secrets and comparison of one to another, not only believe the blessed Art, but also learn and perform the same, and cure all the diseased.

12. But they will not be rid so, but follow as fast again another way: that whereas so many have been, and are daily seen to wear away their lives in Alchemy, & to find nothing that good is, but contrary for the most part, to wit, untimely and unordinary death, sickness and age for long life, health and youth, and always smoke for golden Medicines, and folly for wisdom, and very near as often, bad and sad conditions for good and honest natures; (for by boiling themselves long in such

---

[11] Spelled exactly as in the 1662 printed edition. -pnw

deceitful stuff, as though they were burnt to the pots bottom, they carry most commonly, for ever after, an unsavory smack thereof;) it is plain sign the trade is vain, false and deceitful; this is the third charge they give unto us; let us see how to bear and withstand it. The most wise and great philosophers, and Rosie Crucian physicians, albeit they know God made mankind, for the happy life abovesaid, and that it was at first enjoyed, or else it had been made in vain, and that by corruption of ill custom (by his secret appointment) our kind is grown out of kind, and therefore may be restored, because It is a misleading, and no Intent of Nature; (which forecasting gave them occasion to seek the remedy) yet they thought It unlawful to teach these Medicines, set straight against the will of God, that all should be restored; for that he seemed on purpose to have sown good and bad, and great store of both together, in such sort as we see them, lest it all were alike, and in one state of happiness, the great Variety of business and stirring, and so the Society and Common-wealth among men should be clean taken away: like as the first sour striving seeds (whereof all things are made and sprung) were all alike, and one friend to another, all should be still and quiet, without succession, change and variety in the world, and so there should be no world; for God, when he cast his mind upon the building of the world, he went to make a beautiful and goodly work, meet for the Power, Wisdom and Pleasure of such a Builder, and therefore a stirring and changeable work, because there might be no cunning shown, no delight taken in one ever like or still thing; but light fighting for speed, is ever best in such a ground: let us away, and follow.

13.   Wherefore, by the example, and as it were by the secret blast and motion of God, after our men had found these restoratives, & used them for the time, & meant to leave them as becomes good men, to posterity, they took this way of counsel to lay them up safe in a strong

Castle, as it were in the which all, the broad gates and common easy entries should be fast shut up and barred, leaving one only little back door open, forefenced with a winding-mark, that the best sort, by wit, pains, and providence, might came into the appointed blessedness, the rest stand back forsaken; their Maze and plot is this: first, they hid themselves in low and untrodden places, to the end they might be free from the power of protectors, & etc. the eyes of the wicked world; and that they wrote their books with such a wary and well fenced skill, I mean, to overcast with dark and sullen shadows, and sly pretenses of likes & secret riddles drawn out of the midst of deep knowledge and secret learning, that it's impossible for any but the wise and well given, to approach or come near the matter.

14. And therefore it is, that when the godless and unlearned men, hovering over gain and honor, presume against Minerva's will to handle these words, when the things should rather be handled (for nothing is soft and gentle as speech, especially so thoroughly tempered) and yet all besides the secret meaning, thrust up in deep knowledge: then if these ways and phantasies they practice & set on work as fast (as their fingers itch) and miss as fast (as they must needs do) they say they followed our rules and precepts, and put our work in practice, and found them false, that were as if a cunning Archer and Huntsman had delivered as dark rules of shooting and hunting unto his Countrymen, and these by chance had fallen into the hands of another wild and untaught Nation, which simply missed by mistaking his drift and meaning, had made them ploughs to shoot in, and gored their Oxen to their game, and then missing of their purpose, cried out and blamed the Arts of shooting and hunting, & sought to blow away & abuse the man that taught them: would not a wise Judge hold and deem both these and them, and all other busybodies, that do use to mine and dig in other men's dealings, to be sent into their own trade

and business, wherefore they were made and fashioned? And to let the rest alone for the right owners? And for those of Hermes house, do not think they make claim, sue, and recover their own in open court, as others use (that were away in such a wicked world, to lose land, life and all together quickly) but in the secret sort, which falls out within the compass of your reproof. Neither would I have you follow too hard, and be so earnest upon the next reason, that albeit our men had cause to hide their works and practice, yet they would have shown the fruit and effect thereof advancing themselves, as others do, to Honor and Pleasure, and not have lived like the refuse of the world, in such mean plight and wretchedness; for that is the lightest of all other, though it seem greatest: if I list to rifle in the rolls of ancient Records, I could easily find and show you, that although the most part of people live in this harmless and safe estate, which I told you, yet some again were Kings, and men of great place and dignity (and yet I think by reminder, and not by purchase, so;) but I love not this kind of reasoning; let them that thirst, go to the fountain, and as I remember, that in the household of R. Crucian Riches are made but Servants, & not Masters & Rulers, because they be, for the most part, unruly and ambitious; and for that cause they have no liberty granted them, but are enjoyed to serve lowly their betters, and to look no further; so that if our men were happy, or at least lovers of the same, their riches ought to be employed in their own service, that is, to win Wisdom and Virtue, and not sent out to wait upon I know not what strangers, Honor and Pleasure; which, as they be strangers, yea, and dangerous strangers, lying open (as all high things do) to the blast of Envy, so most commonly they will not be ruled, no more than they which got them, and then rebelling against them which are their Lords and Rulers; do overthrow an happy estate.

15. Wherefore, what marvel is it if our men did this, when they did no more than wisdom requires, nor any more then all wise men have ever taught and followed? thinking and calling it a heavenly life, because it sunders the heavenly mind from the earthly body; not (as Pliny writes of Hermeticus) by sending the same out of the body to gather and bring hone news, but by an high contempt of earthly matters, and flying up to divine thoughts, not with the golden feathers of Euripides, but with the heavenly wings of Plato.

16. And therefore, this same divine man makes that mind alone the whole man, the body of a thing that is his, and belonging unto him, but unto his, that is the body; and, as I may term them, his man's men. And this thing also bias before him, did as well perform, when at the spoil of the City, having leave, he took not his carriage with him, and answered to the check of his friends, that he carried all his own things with him, which was nothing but a naked body.

17. Aristotle is of the same mind with Plato, as appears notably in his last *Book of Manners*, where he has laid down many sound reasons why this life is best, and so by wise men, is and ought to be taken; because it is, says he, the most quiet life, and fullest of true delight, and with all things needful best restored; for indeed it wants nothing; for what? as a mind is divine in respect of a body, so is the life of it, which is that we speak of, in regard of a civil and worldly life. And again, if our minds are ourselves, it were meet to lead our own life before strangers; but last of all, because God, our only pattern, leads no other life but this. I might be very large, if I list to seek about and traverse this matter: but here is enough to show the purpose and reason our men of Egypt had; It was in their choice to choose this kind of life which the world so despises: but how if I could bring them in bereaved of all choice and freewill, and driven by force of necessity to do the

same? would not that stop the widest mouths, think you, in all this lavish company? let us know first, that the mind of man being come from that high City of Heaven, desires of herself to live still that heavenly life, that is the blessed life above described; and if there be any let, as there is likely it is, in the weight and grossness of our body, over-weighing our minds down to the ground, and to all their own muddy matters, then that our men, after they have got this golden Stone, so famous in the world, do not, as they think, and would do, straight ways run to their Coffers; but first and chiefly gild their bodies with it; wherefore after that, by that mighty, fine and temperate Medicine, they have scoured out of all grossness and distemper of the body, the only lets to understanding and good manners, as we shall hear hereafter, and thereby leave the minds at large, and almost at her first freedom; she, and so they together laying aside, and, as it were, casting down all earthly matters, must return to their own former life again; so far, I mean, as the condition and state of man will suffer: and so put case you find your own dark and dusky eye—light, so soon taken with every foul, vain and worldly fancy, yet you must not judge these heavenly men thereby, but think the most sharp and clear sight of their understanding easily able to see the blemish and to avoid the Call of common love.

18.   Wherefore, to close up this point at last, with this happy craft of Hermes, for ought that they know, may be true and honorable; let the common and unlearned sort stay their judgement, and leave the trial sifting of any further matter unto the wise and learned, and there in all directions, if they have none of themselves, might learn better advise, before for the fault of some, they turn to any raging counsel, and bend the edge of Authority against all.

19. I grant, that as in all good Arts, so in this, because it is the most secret; there be some drones crept in among the friends: what then, as they are of another kind; or never begotten by Hermes, or any of his sons, so no reason they should slander the Name and House of Hermes, but bear the burden of their own faults; there may be sorted out and known from the holy stinged and profitable Bee: first, by their bigness in words and brags, and then (as follows lightly by the course of kind) by their stingless and unarmed weakness in all defense of learning; and thirdly, by their sloth and idleness; for although they never leave stirring, yet as Seneca says, *operose nihil agunt*,[12] they painfully do nothing, because all they do is to no purpose, all is fruitless and unprofitable. But Dioclesian lacked this discerning wisdom, and rashly ran upon all, and burnt the Book, much like that part of Lycurgus, who for the drunkenness of the people, cut down the Vines; had it not been better to have brought the springs of water nearer, and to have bridles, as Plato says, that made good with the sober?

Even so the Emperor might with better advice have tempered the heat of alchemy with the cooling Card of Discretion, and made it an Art lawful for a small number only, and with like charge to be practiced, which had been a Counsel worthy wise Princes, neither to let the hope of so great a Treasure go for a small loss, nor yet upon uncertain hopes, be it never so great, to lose a certain great thing, to wit, the life and goods of his Subjects well and orderly bestowed. Now let us join Art and Nature together, to know all things past, present, and to come; that Long Life, etc. may be with the more pleasure enjoyed; for after this methodically *Holy Guide*, Knowledge, the rest will be Imperfect: then Knowledge complete Happiness, Long Life perfects Knowledge;

---

[12] Laborious act. -pnw

Health comforts Long Life: Youth pleases Health; Riches rejoice Youth; Youth embraces Wisdom and Virtue, etc. which you shall find all in order.

**To The
Truly Noble by all Titles,
Sir Ralph Freeman,
Baronet, etc.**

**External, internal, and eternal happiness
be wished.**

The Rosie Crucians have a very excellent opinion (Most honored Sir) that we ought to labor in nothing more in this life, then that we degenerate not from the Excellency of the mind, by which we come nearest to God, and to put on the divine Nature: lest at any time our mind waxing dull by vain idleness; should decline to the frailty of our earthly body, and vices of the flesh. So, we should lose it, as it were cast down by the dark precipices of perverse Lusts.

Wherefore we ought so to order our mind, that it by itself, being mindful of its own dignity and excellency, should always both think, do, and operate something worthy of itself: But the knowledge of divine Science doth only and very powerfully perform this for us, when we by the remembrance of its Majesty, being always busied in divine Studies, do every moment contemplate divine things, by a sage and diligent inquisition, and by all the degrees of the Creatures ascending even to the Archtype himself, to draw from him the secret Practick, Theory of Art and Nature, according to the doctrine of the *Holy Guide*, which those that neglect, trusting only to natural and worldly things, are wont often to be confounded by divers errors and fallacies, and very oft to be deceived by evil spirits.

But the understanding of the *Holy Guide* purges the mind from errors, and renders it divine; giveth infallible power to our Rosie Crucian Guide, etc. drives far the deceits & obstacles of all evil Spirits, and together subjects them to our commands; yea, it compels good Angels, and all the powers of the world unto our service, viz. The virtue of our Art being drawn from the Archetype himself: To whom when we ascend, all Creatures necessarily obey us, and all the Quire of Heaven do follow us.

Seeing therefore (Learned Sir) you have a Divine and Immortal soul given you, which seeing the goodness of the Divine Providence, a well-disposed fate, and the bounty of Nature have in such manner gifted, that by the acuteness of your understanding & perfectness of senses, you are able to view, search, contemplate, discern and pierce through the pleasant Theatre of Natural things, the sublime house of the Heavens, and most difficult passages of Divine things.

I being bound to you by the band of these your great virtues am so far a debtor, as to communicate without Envy the true account of all opinions, these Rules, which we have read & learned, especially their precious Medicines & their greatest secrets of the Pantarva, etc. with their gift of healing, according to our complexion and capacity.

We present therefore now to you, a complete work in the *Holy Guide*, which we have perfected with diligent care, and very great labor and pains both of mind and body: and though it be rude and unpolished in respect of words, yet it is truly elaborate in respect of matter: wherefore I desire this one favor, that you would not expect the grace of an Oration, or the Elegancy of Speech in this Book, which we wrote long since and revised in our days of mourning, for the death of our fellow-Prisoner John Hewit, Doctor of Divinity, and others, who were spitefully thrust into Gaole with us, and many cruelly murdered by the Tyrant Cromwell, because they loved our Sovereign Lord the King.

And we expected to suffer for our Loyalty to His Sacred Majesty the King; but our Estates ransomed our Lives, etc.

Again, We have chosen the less Elegancy of Speech, abundance of matter succeeding in the place thereof; but seeing without doubt, many scoffing Sophisters will conspire against me, especially of those who boast themselves to be allied to God, and fully replenished with divinity; And the sect of self affectors, that will (unless some Judicious Patron be fixed to the Frontispiece, as the beams of the Sun to correct their saucy peering with blindness) not only disgorge their envy in words, but judge and condemn to the Fire the things, even before they have read or rightly understood anything of them, because these medicines agree not with their Bodies, nor such sweet Flowers with their nose. And also, by reason of that spark of hatred, long since conceived against me for my loving and serviceable endeavors to help the Royal Party to restore the King, and yet scarce containing itself under the Ashes. Therefore, Dear Sir, we further submit the Rules ascribed by me to the merits of your Virtue, and now made yours, to your censure, and commend it to your Protection, that if the base and perfidious Sophisters would defame it by the gross madness of their envy and malice, you would by the perspicacity of your discretion & candor of Judgement, happily protect and defend it and me.

Your most humble Servant
and true Honorer,
John Heydon.

The Holy Guide
LEADING THE WAY TO THE WONDER OF THE WORLD

Book II. Chapter I.

1. Of the wonderful Secrets of Numbers, 2. of their Signification. 3. How Moses showed so many signs by them. 4. How Joshua made the sun stand still by Numbers. 5. How by numbers Elijah called down fire from heaven upon his enemies. 6. How by these following Numbers the Rosie Crucians foretell all future things; 7. Command whole Nature, have power over devils, and angels, and do Miracles, & etc. 8. How by this number a river spoke to Pythagoras.

1.   I have observed, that the Numbers which are now vulgarly used amongst arithmeticians and calculators, have been in old time much more esteemed then they are now; the order of them is made after this manner, 1.2.3.4.5.6.7.8.9. to which is added a note of privation signed with the mark O, which although it signifies no Number, yet it makes others to signify, either tens, or hundreds, or thousands, as is well known to arithmeticians. The virtues and signification of these Numbers, the Hebrews are of opinion were delivered to Moses by God himself upon Mount Sinai, and then by degrees of succession without the Monuments of figures or letters was until the time of Esdras delivered to others by word of mouth only, as the Pythagorean opinions were formerly delivered by Archippus and Lysinus, who had Schools at Thebes in Greece, in which the Scholars keeping the precepts of their Masters in their memory, did use their wits and memory instead of Books.

2.   Moses delivers a double Science of this art: The one of Bresith, which they call cosmology, viz., explaining the power of things created, Natural and Celestial, and expounding the Secret of the Law and Bible by Philosophical reasons.

3.   Which truly upon this account differs nothing at all from Natural Magic, in which we believe King Solomon excelled; for it is written he was skilled in all things, even from the Cedar of Lebanon to the Hyssop that grows upon the wall.

4.   Also, in cattle, birds, creeping things and fishes: All which show he knew the Magical virtues of Nature and Numbers: The Rosie Crucians follow after this, as you may read in my book of Geomancy and Telesms, entitled, *The Temple of Wisdom*: and in my *Way to Bliss*, and *Rosie Crucian Physick*.

5.   They call the other Science thereof Mercara, which is concerning the more sublime contemplation of Divine and Angelic virtues, and of Sacred Numbers, being a certain Symbolical Divinity, in which Numbers are ideas of most profound things, and great Secrets. This is the Rosie Crucian *Infallible Axiomata*, which teaches of Angelical Virtues, Numbers, and Names in the Hebrew, also of the Conditions of Spirits and Souls in the Greek numbers and Names, which searches into the Mysteries of Divine Majesty as the Emanations thereof; and Sacred Names in Latin Numbers and Letters, which he that knows, may excel with wonderful Virtues, as that when he pleases, he say know all things past, present and to come; and command whole Nature, have power over devils and angels, and to do Miracles. By this they suppose that Moses did show so many signs, and turned the Rod into a serpent, and the Waters into Blood, and that he sent frogs, flies, lice, locusts, caterpillars, fire, with hail, Botches and Boils on the

Egyptians, and slew every first-born of man, and beast; and that he opened the Seas, and carried his thorow[13], and brought Fountains out of the Rocks, and Quails from Heaven, that he sent before his clouds and lightning by day, a pillar of fire by night, and called down from Heaven the voice of the Living God to the people, and did strike the haughty with fire, and those that murmured with the leprosy: and on the ill deserving brought sudden destruction, the earth gaping and swallowing them up.

6. Further, he fed the people with Heavenly Food, pacified Serpents, cured the envenomed, preserved the numerous multitudes from infirmity, and their Garments from wearing out, and made them Victors over their enemies. To conclude, by this Art of Numbers and Letters, Joshua commanded the sun to stand still; Elijah called down fire from heaven upon his enemies, restored a dead child to life, Daniel stopped the mouths of the lions; the three children sang songs in the fiery Oven: Moreover, by this idea of Letters and Numbers, the incredulous Jews affirm; that even Christ did so many Miracles. The Rosie Crucians very well know the angels and spirits that govern these Numbers and therefore deliver Charms against devils, and their bonds, and the manner of Conjurations; for against Diseases, they heard a Brother make a Spirit cry out, Οἴμοι τῶν Τριπόδων

7. Pythagoras was not only initiated into the Mosaical Art of numbers, but arrived also to the power of working miracles, as his going over a River with his Companions testifies that he speaking 80. & π in a Table to the River, the River answering him again with an audible & clear voice, Χαῖρε Πυθαγόρα, Salve Pythagora; that he

---

[13] "thorow" is the exact spelling used in the printed 1662 edition. -pnw

showed his thigh to Abaris the Priest, and that he affirmed that it glittered like Gold, and thence pronounced that he was Apollo; that he was known to converse with his friends at Metapontium and Tauromenium; (the one a Town in Italy, the other in Sicily, and many days journey distance) in one and the same day. This makes good my Apologue at the beginning of A New Method of Rosie Crucian physick, and the way to bliss.

8.   Porphyrius and Iamblicus report very strange things of him, which I willingly omit: I shall only add his Predictions of Earthquakes, or rather, because that may seem more natural, his present shaking of Places in Cities, his silencing of violent Winds and Tempests; his calming the rage of the Seas and Rivers, etc., which skill Empedocles, Epimenides, Cathartes, and Abaris having got from him, they grew so famous, that Empedocles was surnamed Alexanemus, Epimenides, Cathartes, and Abaris, Aethrobates from the power they had in suppressing of storms and winds, in freeing of Cities from the Plague, and in walking aloft in the Aire: which skill enabled Pythagoras to visit his friends after that manner at Metapontium and Tauromenium, in one and the same day.

Chapter II.

1. Of the Power 2.3. and Virtues 4.5. of Hebrew, 6, 7. Greek and 8. Latin Letters, when the numbers are attributed to them.

1. The Pythagoreans say that the very elements of letters have certain divine Numbers, by which collected from proper names of things, you may draw conjectures concerning Secret things to come.

2. And there is an uneven Number of Vowels of imposed names, which did betoken Lameness, or want of Eyes, and such like misfortunes, if they be assigned to the right side parts: but an even number to them of the left: And by the Number of Letters you may find out the ruling Planets of any one that is borne, and whether the Husband or Wife shall dye first, and know the prosperous or unhappy events of the rest of our works,

3. The Latin, Greek, and Hebrew Letters deputed to each Number, shall show you, being divided into three Classes, whereof the first is of units, the second of tens, the third of hundreds, and seeing in the Roman Alphabet there are wanting four to make up the Number of twenty-seven Characters, their places are supplied with I. and U. simple Consonants, although the Germans for who the Aspirate use a double W, the true Italians and French in their Vulgar speech put G joined with U, instead thereof writing thus, Vuilhelmus, and Guilhelmus.

John Heydon

1	2	3	4	5	6	7	8	9	10	20	30	40	50
A	B	C	D	E	F	G	H	J	K	L	M	N	O

60	70	80	90	100	200	300	400	500	600	700
P	Q	R	S	T	U	X	Y	Z	I	V

800	900
Hi	Hu

1	2	3	4	5	6	7	8	9	10	11	12	13	14
α	β	γ	δ	ε	ζ	η	θ	ι	κ	λ	μ	ν	ξ

15	16	17	18	19	20	21	22	23	24
ο	π	ρ	σ	τ	υ	φ	χ	ψ	ω

Capitals.

1	2	3	4	5	6	7	8	9	10	20	30	40	50
A	B	Γ	Δ	E	ϛ	Z	H	Θ	I	K	Λ	M	N

60	70	80	90	100	200	300	400	500	600	700
Ξ	O	Π	ϟ	P	Σ	T	Υ	Φ	X	Ψ

800	900
Ω	ϡ

Now the Classes of the Hebrew Numbers are these.[14]

Now if you desire to know the Ruling Planet of any that is born, compute his name, and of both his Parents, through each Letter according to the Country he was born in, and the Number above written, and divide the sum of the whole being gathered together by 9, subtracting it as often as you can; and if there remain a unity, or 4, both signify the Sun; if 2 or 7, both signify the Moon, but three Jupiter; five Mercury; six Venus; eight Saturn; nine Mars. And the reasons thereof I have showed you in my Book of Geomancy and Telesmes, entitled, *The Temple of Wisdom*.

In like manner, if you desire to know the ascendant of any one that is born, compute his name, and of his Mother and Father, and divide the whole collected together by 12, if there remain 1, it signifies the Lion; if Juno 2, Aquarius; if 3, Capricorn; if 4, Sagittarius; 5, Cancer; if Venus 6, Taurus; if Palladium 7, Aries; if Vulcans 8, Libra; if Mars his 9, Scorpio; if 10, Virgo; if 11, Pisces; if Phoebus 12, they represent Geminos.

---

[14] These and many other symbols are shown in the Gematria section of "Alchemical Symbols," third edition, in The R.A.M.S. Library of Alchemy Volume 21. -pnw

6. And now let no man wonder that by the Numbers and Letters all things may be known, seeing the Pythagorean Philosophers and Rosie Crucians testify the same: in those number, lie certain hidden mysteries, found out by few; for the most High created all things by Numbers, Measure and Weight, from whence the truth of Letters and Names had its original, which were not instituted casually, but by a certain Rule, although unknown to us.

7. Hence Saint John in the Revelation says, Let him which has understanding compute the Number of the name of the Beast, which is the Number of a man; yet these are not to be understood of those names, which a disagreeing difference of Nations, and divers Rites of Nations, according to the causes of places, or education have put upon men, but those which were inspired into every one at his birth, by the Heavens, with the conjunction of the Stars.

8. Moreover, Tucer, Rabanus, and R. Lully have dedicated to the Elements and Deities of Heaven, sacred numbers; for to the Aire they have designed the number eight, and to Fire five, to Earth six, to Water twelve. Besides unity is ascribed to the Sun, in which God put his Tabernacle; and that this also is of Jupiter, doth the Causative power of his Ideal and intellectual Species testify, who is the Head and the Father of the Gods, as Unity is the beginning and Parent of Numbers, 1, engraved in Brass, they say bringeth a Spirit, in the shape of a black man standing, and clothed in a White Garment, girdled about, of a great body, with reddish eyes, and great strength, and he appears like a man angry, and he giveth Boldness, Fortitude, & makes a man lofty.

Chapter III.

The Number of Happiness.

1, 2. The Pythagorick names or nature of a Monad or Unit. 3, 4 applied to the first day's work: 5, 6, What are the upper waters: 7, 8, And that Souls that descend εἰσγεγεσιν, are the Naiades or Water Nymphs, in Porphyrius: 9. That matter of itself is unmovable: R. Bechai his Notation very happily explained in my 11. Temple of Wisdom: 12. Of the Number One, and the signification, and what Angel rules it.

1.   I admire the goodness of God towards his Creatures, how fit the Number is to the Nature of every day's work: And so I conclude, that God ordered it so on purpose, and that in all probability Pythagoras was acquainted with his Axiomata, and that was the reason the Pythagorians made such a deal of do with Numbers, as you shall find in Order, putting other conceits upon them than any other arithmeticians do and that therefore if such Theorems as the Pythagoreans held, be found suitable and compliable with Moses his Text, it is a shrewd presumption that these are the right Rosie Crucian Axiomata thereof.

2.   Philo makes this first day spent in the Creation of immaterial and spiritual beings, of the intellectual world, taking it in a large sense for the *mundus vitae*, the world of life and forms: And the Pythagoreans call an εἰδος Form, and Ζωη Life. They call Ζηνὸς πύργος or the Tower of Jupiter, giving also the same name to a Point or Center; by which they understand the vital Formality or Center of things: the

rationes seminales: and they call an unite also λόγος σπερματίτης, which is seminal form; But a very short and sufficient account of Philo's pronouncing that spiritual substances are the first day's work, is, That as a Unit is indivisible, you cannot make two of one of them, as you may make of one piece of corporeal Matter two by actual division or severing them one piece from another: wherefore what was truly and properly created the first day, was immaterial, indivisible, and independent of the matter, from the highest Angel to the meanest seminal form.

3. And for the potentiality of the outward Creation, since it is not so properly any real being, it can breed no difficulty; but whatever it is, it is referable fitly enough to incorporeal things, it being no object of sense, but of intellect, and being also impassible and indiminishable, and so in a sort indivisible; the power of God being indiminishable, and it being an adequate consequence of his power; wherefore this potentiality being ever one, it is rightly referred to the first day. And in respect of this the Pythagoreans call a Unit as well as the BINARY, as also ἀλαμπία & σκοτωδία, which names plainly glance at the dark potentiality of things, set out by Moses in the first days Creation.

Νυκτός δ' αὖτ' αἶπνε τε και ἡμέρα ἐξεγένοντο.
       Plato.
But of the Night, both day & sky were born.

4. God Created now Corporeal matter (as before the world of Life) out of nothing, which universal matter may well be called for extension is very proper to corporeal matter; Castellio translates it *liquidum*, and this universal Matter is most what fluid still, all over the World, but at first it was fluid universally.

5. But here it may be, you will enquire, how this corporeal Matter shall be conceived to be betwixt the waters above, and this underneath; for what can be the waters above? Maimonides requires continued Analogy in the hidden sense of Scripture; as you may see in his Preface to his *Moreh Nevochim*: But I need not fly to that general refuge; for me thinks that the seminal forms that descend through the matter, and so reach the possibility of the parts of the outward Creation, and make them spring up into Art, are not unlike the drops of rain that descend through the heavens or air, and make the earth fruitful; Besides, the seminal forms of things be round, and contracted at first, but spread when they bring any part of the possibility of the outward Creation into Art, as drops of rain spread when they are fallen to the ground, so that the Analogy is palpable enough, though it may seem too elaborate, and curious. We may add to all this, concerning the Naiades, or Water Nymphs, that the Ancients understood by them, Τας εις γένεσιν κατιούσας ψυχας ποίνως ἁπάσας (i.e.) all manner of souls that descend into the matter and generation, and this is the number, by which it is said, they raise the dead to life; wherefore the watery powers may be here indigitated by the name of the upper waters.

6. The frequent Complaints that the noble Spirit in Pythagoreas and Plato make against the incumbrances and disadvantages of the body, make the *Holy Guide* very true and probable; and it is something like our Divines fancying Sheol to be Created this day.

7. This is consonant to Plato's School, who make the matter unmovable of itself, which is most reasonable; for if it were of its own Nature moveable; nothing for a moment would hold together, but dissolve itself into infinitely little particles; whence it is manifest, that

there must be something besides the matter, either to bind it or to move it; so that the Creation of immaterial Beings was by 1. and is in that respect also necessary.

8. For this Agitation of the matter brought it to my fancy in the second principle of the Rosie Crucian physick, which is the true Æther, or rather שמים for it is as liquid as water, and yet has in it the fiery principle of fire, which is the first Element, and made by the number; as the heavens were, and called שמים because they are מים+אש fire and water; for the round particles, like water (though they be not of the same figure) slake the fierceness of the first principle, which is the purest fire; and yet this fire in some measure always lies within the Triangular intervals of the round particle, as my Book above named declares at large.

9. And this Number 1. is called a number of concord, of piety, of friendship, which is so knit that it cannot be cut unto parts; for unity doth most simply go through every number, and is the common measure, Fountain, and original of all numbers, contains every number joined together in itself entirely, the beginning of every multitude, always the same, and unchangeable; whence also being multiplied into itself, produces nothing but itself; it is as I told you above indivisible, void of all parts; but if it seem at any time to be divided, it is not cut, but indeed multiplied into unities: yet none of these unities is greater or lesser then the whole unity, as a part is less than the whole, it is not therefore multiplied into parts, but into itself. Therefore, it is named Cupid, because it is made alone, and will always bewail itself, and beyond itself it has nothing, but being void of all haughtiness, or coupling, turns its proper heat into itself: It is therefore the 1. beginning and end of all things; and all things which are, desire that

one, because all things proceeded from one; and that all things may be the same, it is necessary that they partake of that one: And as all things proceed of 1. in many things, so all things endeavor to return to that one 1., from which they proceeded; it is necessary that they should put off multitude.

10.   One therefore is referred to the high God, who seeing he is one, and innumerable, yet creates innumerable things of himself, and contains them within himself; there is therefore one God, one world of the one God, one sun of the one world; also, one phenix in the world, one king amongst bees, one Leader amongst Flocks of cattle: 1. Ruler amongst herds of beasts, and Cranes follow 1. and many other Animals honor Unity; amongst the members of the body, there is one principal, by which all the rest are guided, whether it be the head, or as some will, the heart: there is one element overcoming and penetrating all things: viz. fire. There is one thing created of God the subject of all wondering which is on earth, or in heaven; it is actually animal, vegetable, and mineral, everywhere found, known by few, called by none by its proper name, but covered with Figures and Riddles, without which neither alchemy, nor Natural magic can attain to their complete end or perfection; from 1. man, Adam all men proceed, from that one all became Mortal; from that one Jesus Christ, they are regenerated.

11.   And as St. Paul says, one Lord, one faith, one baptism, one God, and father of all, one mediator betwixt God and man, one most high creator, who is over all, by all and in us all; for there is one father, God, from whence all, and we in him, one Lord Jesus Christ by whom all, and we by him, one God Holy Ghost, into whom all and we into him; and in the exemplary world, I. Divine Essence, the fountain of all virtues & power, whose name is expressed with one most simple

Letter I. God; And in the intellectual world there is I. Supreme Intelligence, the first Creature, the Fountain of Lives, the soul of the world: And in the Celestial world, there is one King of Stars, Fountain of Life, the Sun: And in the Elemental world, there is I. Subject and instrument of all virtues, natural, and supernatural, and that is, the philosopher's stone: And in the lesser world, there is I. first living, and last dying, and that is the heart, and in the infernal world, there is one Prince of Rebellion of Angels, and darkness, and that is Lucifer. By this number and Letters of the Hebrew, it is said Moses showed so many signs in Egypt. This number signifies England, and the King thereof.

12. They say if at 1 of the clock under a fortunate Horoscope you cast One, and Agiel in a piece of Gold, Agiel the angel that rules that number will immediately come, and personally attend you and fulfill your desires; by this number Plato was born, and the number 45. educated him, this Number Telesmatically engraved in Gold will easily make you understand the first book, viz. happiness and its effects.

The Holy Guide

Chapter IV.

THIS NUMBER UNITES ARTS AND NATURE.

1,2,3,4, That Universal Matter is the Second days Creation, 5,6,7,8. fully made good by the Names and Properties of the Number two; 9,10,11, its virtues.

1. How fitly doth the Number 2. agree with the nature of the work of this day, which Is the Creation of Corporeal Matter, and the Pythagoreans, call the number 2. ὕλη matter, and Simplicius speaking of the

PYTHAGOREANS, Εἰκότως ἓν μὲν τὸ εἶδος ἔλεγον, ὡς ὁρίζον ὅπερ ἂν καταλάβῃ καὶ περατοῦν, δύο δὲ τὴν ὕλην ὡς ἀόριστον, καὶ ὄγκου καὶ διαιρέσεως αἰτίαν.

They might well (says he) call 1. Form, as defining and terminating to certain shape and property whatever it takes hold of, and 2. they might well call matter, it being indeterminate, and the cause of bigness and divisibility, and they have very copiously heaped upon the number 2. such appellations as are most proper to Corporeal matter. As Ἀσχημά-τιστος, Ἀόριστος, Ἄπηρος, unfigured, undetermined, unlimited, for such is matter itself till form take hold of it. It is called also REA from the fluidity of the Matter Ἀερία, Ἀγερία, because it affords substance to the Heavens and Νεῖκος, Μοῖρα, Θάνατος, (i.e.) contention, fate, and death, for these are the consequences of the souls being joined with Corporeal matter Κίνεσις, γένεσις, Διαίρεσις. Motion, Generation, and Division, which are properties plainly appertaining to bodies; they call the number 2. also Ὑπομονή it is the Ὑποκείμενον, the subject that endures and undergoes all the charges and alterations the Active forms put upon it; wherefore it is plain the Pythagoreans understood Corporeal matter by the number 2. which no man can deny but that it is a very fit Symbol of division that eminent property of matter.

2. But I might cast in a further reason of the שמים being Created the second day: for the Celestial matter does consist of two plainly distinguishable parts, viz. The first Element and the second, or the *Materia Subtilissima*, and the round particles, as I said before.

3. And 2 is called number of Science and Memory, and of Light, and the number of Man, who is called another world, and the lesser world; it is also called the number of charity and mutual love, of marriage & society: The first number is of 2. because it Is the first multitude, it can be measured by no number besides unites alone, the common measure of all numbers. it is compounded, but more properly not compounded, the number 3 is called the number uncompounded.

4. But the number 2 is the first branch of unites, and the first procreation: Hence it is called generation, and Juno and an Imaginable corporation, the proof of the first motion, the first form of parity, the number of the first equality, extremity, and distance betwixt, and therefore the peculiar equity, and the proper Art thereof, because it consists of 2 equally poised; it is a number of Conjunction and profit of increase, as it is said by the Lord, two shall be one flesh, and Solomon says, it is better that two be together than one; for they have a benefit by their mutual society; if one shall fall he shall be supported by the other; woe to him that is alone, because when he falls he has not another to help him. And if two sleep together, they shall warm one

the other; how shall one be hot alone? and if any prevail against him, two resist him.

5.   And it is called a number of wedlock and sex; for there are two Sexes, masculine and feminine, and two Doves bring forth two Eggs; out of the first of which is hatched the Male, out of the second the Female; 2. is called middle, that is capable, that is good and bad partaking; and beginnings of Division of multitude and destruction, and signifies Matter; 2. is also sometimes the number of discord and confusion, of misfortune and uncleanness, whence Hierom and Jovianus says, that therefore it was not spoken In the second day of the Creation, and God said, that it was good, because the number 2 is evil.

6.   Hence also it was, that God commanded all unclean Animals should go into the Ark by Couples, because, as I said, the number 2 is a number of uncleanness, and is most unhappy in their conjuration and invocations of spirits and souls of the dead, especially any of those that are under the Angels deputed to Saturn or Mars, for these 2 are accounted by geomancers and astrologers unfortunate, it is also reported that the number 2 doth cause apparitions of fiery ghosts and fearful goblins, and bring mischief of evil Spirits to them that travel by night; Pythagoras says the unity is God and a good intellect, and that Duality is a Devil, and an evil intellect, In which is a Material multitude; wherefore the Pythagoreans say, that 2 is not a number, but a certain confusion of Unites; and Eusebius says, that the Pythagoreans called Unity Apollo, and 2 strife and boldness.

7.   And 3 justice, which is the highest perfection, and is not without many Mysteries. Hence there were two Tables of the Law in Sinai, two cherubims looking to the propitiatory, in Moses, two Olives dropping

Oil, in Zachariah, two natures in Christ, Divine and Human: Hence Moses saw two appearances of God, viz. his face and back parts.

8.   By the number 2 also they say, if it be engraved in Copper, It will bring to you a genius that is good for to procure the love of women; sometimes print it in lapis lazulus, and sometimes in virgin wax, and write the names of the man & woman in virgin parchment: to which appears a naked maid having a looking-glass in her hand, and a chain tied about her neck, and nigh her a handsome young man, holding her with his left hand by the chain, and with his right hand be will be playing with her hair, and smiling on her, and these are sent by one of those angels of the number.

9.   Also 2 Testaments, 2 Commands of love, 2 first dignities, 2 first people, 2 kinds of Spirits, good and bad, 2 intellectual Creatures, an angel and soul, 2 great Lights, 2 Solstitia, 2 Equinoctials, 2 Poles, 2 Elements, producing a living Soul, viz. Earth and Water. By this number 2 it is said Elijah called down fire from heaven upon his enemies. And the name of God in the Exemplary world is expressed with two Letters,

יה Jah אל Eli. And there are two intelligible substances in the intellectual world, viz. an Angel, and the Soul; and two Lights in the Celestial worlds, the Sun, and the Moon; and two principal seats of the Soul in the lesser world, viz. the Heart, and the Brain; and there are two chiefs of the Devils in the infernal world, viz. Beemoth and Leviathan, also two things Christ threatens to the damned, viz., weeping and gnashing of teeth.

10.   The number 2 is said to signify a thing lost, and here they enquire whether a man shall be rich or poor.

11. This number is commonly made upon Brass, that which is red or Copper, at the hour of 2, and Jejajel is the Angel that rules it, and 325. by that number was this book made.

Chapter V.

The Number of Long Life.

1. The Nature of the Third days' work 2. set off by the Number 3. That the most learned do agree that the Creation was perfected at once, The Notation of מכב strangely agreed with the Notorious conclusion of the *Temple of Wisdom* of the signification of the Number 3:

1. In this third day was the waters commanded into one place; the Earth adorned with all manner of plants, Paradise and all the pleasure and plenty of it created, wherein the Serpent beguiled Eve, etc. What can therefore be more likely than that the Pythagoreans use their Numbers as certain remembrancers of the particular passages of this History of the Creation? when as they call their Number 3. Τει Των ἀ θαλατω ϑιος (i.e.) Triton and Lord of the Sea, which is in reference to Gods commanding the water into one place, and making thereof a Sea, they call also the ternary, Κέρας αμαλθέας & οφιων, the former intimates the plenty of

Paradise, the latter relates to the Serpent there; but now besides this, we shall find the ternary very significant of the nature of this day's work; for first, the earth consists of the 3 Elements in my Book, entitled, *The Temple of Wisdom*: (for the truth of that Book will force itself in here whether I will or no:) And indeed I had no thoughts of this, when I wrote that; and then again, there are three grand parts of this third Element necessary to take an Earth habitable, the dry land, the Sea, (whence are springs and rivers) and the air: And lastly, there are in vegetables, which is the main work of this day, three eminent Properties, according to my Cousin Heydon's Philosophy, viz. nutrition, accretion, generation; and also if you consider their duration, there be three Cardinal points of it, Ortus, Arme, Interitus, you may call in also that minerals, as the Arabians call them, which belong to this day as well as Plants, that both Plants and they, and in general all terrestrial bodies have the three Chemical principles in them, salt, sulphur and mercury.

2.   As the matter of the Universe came out in the second day, so the contorting of this Matter into Suns and Planets is contained in this fourth day: The Earth herself not excepted, though it is said she was made in the first day, and as she is nurse of Plants, said to be uncovered in the third, yet as she is a receptacle of Light, and shines with borrowed rays like the Moon and other Plants[15], she may well be referred to the fourth days Creation.

3.   Nor will this at all seem bold or harsh, if we consider that the learned have already agreed, that all the whole Creation was made at once, As for example, the most rational of all the Jewish Doctors, R.

---

[15] "Plants" exactly matches the spelling used in the 1662 printed edition, but the author may have meant "planets". -pnw

Moses Egyptius, Philo Judeus, Abraham Judeus, Procopius, Gareus, Cardinal Cajetane, Saint Augustine and the Schools of Hillel and Samai; so that leisurely order of days is thus quite taken away, & all the scruple that may arise from that hypothesis.

4.   Wherefore I say the number 3 is an uncompounded number, a holy number, a number of perfection, a most powerful number; for there are three persons in God, there are three Theological virtues in Religion: hence it is that this number conduces to the ceremonies of God and Religion, that by the solemnity of which prayers and sacrifices are thrice repeated; and the Pythagoreans use it in their sanctifications and purifications, and it is most fit in bindings or ligations.

5.   And in Johannes De Spagnet it was the custom in every Medicine to spit with three deprecations, and hence to be cured. The number of 3 is perfected with 3 Argumentations, long, broad and deep, beyond which there is no progression of dimension whence the first number is called square, Hence it is said, that to be a body that has 3 measures, and to a square number nothing can be added; wherefore Cardanus in the beginning of his speech concerning heaven, calls it as it were a law, according to which all things are disposed; for Corporal and Spiritual things consist of three things, viz. beginning, middle and end, by three the world is perfected Harmonie, necessity and order, (i.e.) concurrence of causes, which many call fate, and the execution of them to the fruit of increase, and a due distribution of the increase; the whole measure of time is concluded in 3, viz. past, present and to come: All magnitude is continued in 3. Line, Superficies and body: every day consists of 3 intervals, length, breadth, thickness: Harmonious Music contains 3 consents in time, Diapason, Hemiolion, Diatessaron: there are also 3 kinds of Souls, vegetative, sensitive, and

intellectual. And God orders the world by number, weight, and measure; as the number 3 is deputed to the Ideal forms thereof, as the number 2 is to the creating matter, and unity of God the maker of it: Rosie Crucians doe constitute 3 Princes of the world, Oxomasis, Milris, Axamcis, (i.e.) God, the Mind, and the Spirit; by the 3 square or solid the 3 number of 9 of things produced are distributed, viz. of the super celestial in nine, orders of Intelligences; of Celestial into 9 Orbs; of inferior into 9 kinds of generable and corruptible things: Lastly, in this ternal Orb, viz. 27. all musical proportions are included, as Plato and Proclus do at large discourse.

6.   And the number of 3 has in it a Harmony of 5, the grace of the first voice, also intelligence; there are 3 hierarchies of angelical spirits; there are 3 powers of intellectual Creatures, memory, mind and will: there are three orders of the blessed, viz. of martyrs, confessors and innocents: there are three quaternions of Celestial signs, viz. fixt, moveable and common, and also of Houses, viz., angels, succedents, cadents. There are also 3 faces and heads in every sign, and 3 Lords of each triple-city; there are 3 fortunes amongst the Planets, 3 graces amongst the Goddesses, 3 Ladies of destiny amongst the infernal crew, 3 Judges, 3 Furies, 3 headed Cerberus; you read also of three double Hecats.

7.   Three months of the Virgin Diana, three persons in the super-substantial Divinity, three times, of Nature, Law and Grace; three Theological Virtues, faith, hope and charity; Jonas was three days in the Whales belly, and 3 days was Christ in the grave.

8.   In the Original world there are three Persons in the Trinity, viz. The Father, the Son, the Holy Ghost; and there are three Hierarchies of Angels in the Intellectual world, viz. Supreme, Middle, and Lowest;

and three degrees of the blessed, Innocents, Martyrs, Confessors. And there are three degrees of Elements, viz. Simple, Compound, thrice Compounded; in the lesser world there are three parts answering the three-fold world, viz. the head, in which the intellect grows answering to the intellectual world, the breast where is the heart the seat of life, answering to the Celestial world; the belly, where the generation is, and the genital members answering the Elemental world; and in the infernal world there are three Furies, viz. Alecto, Magaera, Tesiphone; three infernal Judges, Minos, Aarus, Rhadamantus; three degrees of the Damned, wicked, apostates, infidels.

9. The chaos is self in every first analysis is also three—fold, the Sapphire of the chaos is likewise three—fold. And here is six parts, which is the Pythagorean Senarius, or numerus Conjugal: In these six the influx of the Metaphysics, called unity, is sole Monarch, and makes up the seventh number, or Sabbath, in which at last by the assistance of God the body shall rest; again, every one of these parts is twofold, and these Duplicities are contrarieties. Here you have twelve, six and six in a desperate division, and the unity of peace amongst them: these Duplicities consist of contrary Natures; one part is good, one bad, one corrupt, one incorrupt; one rational, one irrational; these bad, corrupt, irrational seeds, are the Tares and Sequels of the Curse. This is the *Holy Guide*.

10. *Septem partibus*, says Zoroastes, *insunt duo tern aria, & in medio stat unum duodecim stant in bello, tres amici, tres inimici: tres viri vivificant, tres etiam occidunt, & deus rex fidelis ex sua sanctitatis atrio dominatur omnibus, unus super tres, & tres super septen, & septen super duodecim, & sunt omnes stipati, alius cum alio.*

11. By this number 3 in a Telesme of Tin Jophiel, carried Philip to Azotus.

12. The number 3 engraved in Quicksilver fixed according to Art, will bring to you an Angel in the form of a handsome young man bearded, having in his left hand a rod & a Serpent about it, and in his right sometimes he holds a Dart; and he they say confers knowledge, eloquence, diligence in merchandizing and gain by Sea; this makes a man fortunate in gaming, and to win. This number thus engraved, makes men understand the way to Long Life.

Chapter VI.

The Number of Nature and Health.

1. Of the signification of the Number 4. 2,3,4,5, how the Corporeal world was universally erected into form and Motion on the fourth day, 6. is most notably confirmed by the Titles and property of the number 4. 7,8 The infallible Rosie Crucian or Pythagorick Oath, 9, wherein they swore, 10, 11, by him that taught them the Mystery of the Tetractus. 12. that the Tetractus was a Symbol of the whole Art, that lay couched in numbers and letters: 13,14. The mystery of the Number 4.

1. The Earth, as one of the Primary Planets, was created the fourth day, and I translate תבליסם Planets, primary because of ת Emphatically, and Planets because the very Notation of their name

implies their Nature, for כוכב is plainly from כו, or burning, and כב extinction, nouns made from כבית & תבבית as ת and את from תית and אכת, according to unexceptionable Analogy. And the Earth, as also the rest of the Planets, their Nature is such, as if they had been once burning and shining Suns; but their light and heat being extinguished, they afterwards became Opaced Planets; this conclusion seems here plainly to be contained in the Rosie Crucians and Moses, but is at large demonstrated in my new method of Rosie Crucian physick.

2. Nor is this Notation of כוכב enervated by Alleging that the word is ordinarily used to signify fixed Stars, as well as the Planets; for I do not deny, but that in a Vulgar Notion it may be compatible to them also. For the fixed Stars according to the imagination of the rude people, may be said to be lighted up, and extinguished, so often as they appear and disappear; for they measure all by obvious sense and fancy, and may well look upon them as so many Candles, set up by Divine Providence in the night, but by day frugally put out for wasting: and I remember Theodidactus in his περί προνοίας, has so glibly swallowed down the Notion, that he uses it as a special Argument of Providence, that they can burn thus with their heads downwards, and not presently swell out and be extinguished, as our ordinary Candles are; wherefore the word כוכב may very well be attributed to all the Stars, as well fixed as Planets, but to the fixed only upon vulgar seeming grounds, to the Planets upon true and Natural; and we may be sure that is that which Eugenius Theodidactus, The Rosie Crucian would aim at, and lay stress upon, in the Book M. Wherefore in brief ת Emphatical in תבוכבם contains a

double Emphasis, intimating those true כבבום or Planets, and then the most eminent amongst those truly so learned. Nor is it at all strange, that so abstruse conclusions of Philosophy should be lodged in this Numeral and Literal Text; for as I have elsewhere intimated, Moses has been aforehand with Rosie Crucians, the ancient patriarchs having had will, and by reason of their long lives, leisure enough to invent as curious and subtle theorems in Philosophy, as ever any of their posterity could hit upon, besides what they might have hid by tradition from Adam: and if we find the Earth a planet, it must be acknowledged forthwith that it runs about the Sun, which is pure Rosie Crucian, and a shrewd presumption that they were taught that mystery by Moses: but that the Earth is a Planet, besides the Notation we have already insisted upon, the necessity of being created in this fourth day amongst the other Planets, is a further Argument, for there is no mention of its Creation in any day also, according to this *Holy Guide*.

The Hebrew is על תארץ. And I have made bold to interpret it not of this one individual Earth, but of the whole Species: and therefore I render it the world at large, תארם, is not an individual man, but mankind in general.

3. This fourth days Creation is the contrivance of Matter into Suns and Planets, or into Suns, Moons and Earths; for the Ætherial Vortices were then set a going, and the Corporeal world had got into a useful order and shape. And the ordering and framing of the Corporeal world, may very well be said to be transacted into the Number 4. four being the first body in Numbers, and therefore preferred before all the virtues, and the foundation and root of all Numbers is four; whence also all foundations, as well in Artificial things as Natural and Divine,

are four square, as I shall show you; and it signifies solidity, which also is demonstrated by a four square figure, and in an equilateral pyramid, which figure also is a right Symbol of Light, the rays entering the eye in a pyramidal form, and Lights now are set up in all the vast Region of the Ætherial Matter which is heaven, The Pythagoreans also call this Number σῶμα & κόσμος, body and the world, intimating the Creation of the Corporeal world therein, and further signifying in what excellent proportion and harmony the world was made. See Cornelius Agrippa.

4. They call this Number 4. Armonia and ὑργία & Βακχασμὸν ἀνεγείριον Harmony, Urania, and the stirrer up of Divine fury and ecstasy, insinuating that all things are so sweetly and fittingly ordered in the world, that the several Motions thereof are as a comely Dame, or ravishing Music, are able to carry away a contemplative Soul into Rapture and ecstasy upon a clearer view, and attentive animadversion of the order and economy of the Universe; and the Rosie Crucians in the head of a Catalogue of the most famous Lawgivers, do much Pythagorize, in the expression of Moses: they say that this Number 4. contains the most perfect proportions in Musical Symphonies; viz. Diatessaron, Diapente, Diapason, and Disdiapason,

Τῆς μὲν γὰρ διὰ τεττόρων ὁ λόγος ἐπίτριτος &

etc. For the proportion of Diatessabon is as four to three, of Diapente as three to two, of Diapason, as two to one, or four to two, of Disdiapason as four to one. I might cast in also the consideration of that Divine Nemesis, which God has placed in the frame and Nature of the universal Creation, as he is a distributer to every one according to his works, from whence himself is also called nemesis by Plato, τὸ τῆς ἀξίας διανεμητικός Because he everywhere distributes what is due to every one; this is in ordinary Language, Justice; and both Philo and Plotinus out of the Pythagoreans, affirm that the Number four is a Symbol of justice, all which makes towards what I drive at, that the whole Creation is concerned in this Number four, which is called the fourth day. And for further Eviction we may yet add, that as all Numbers are contained in four virtually, (By all Numbers is meant ten, for when we come to ten, we go back again) so the root and foundation of all the Corporeal Creation is laid in this fourth days' work, wherein Suns, Earth, and Moons, and the ever whirling Vortices; for as Philo observes, Pythagorean like, ten, (which they call also Κόσμος, οὐρανός, & παντέλεια the World, Heaven, and all perfectness) is made by the scattering of the parts of four thus, one, two, three, four; put these together now and they are ten. Παντέλεια, τὸ πᾶν, the Universe; this was such a secret amongst Pythagoras his Disciples, that it was a solemn Oath with them, to swear by him, that delivered to them the mystery of the Tetractis Tetrad or Number four.

I WITH PURE MIND BY THE NUMBER FOUR DO SWEAR THAT'S HOLY, AND THE FOUNTAIN OF NATURE ETERNAL, PARENT OF THE MIND, ETC.

5. Thus, they swore by Pythagoras, as is conceived, who taught them this Mysterious Tradition, had it not (thank you) been a right worshipful mystery, and worthy of the solemnity of Religion and of an Oath, to understand that one; two, three, four, make ten, and that ten is all which rude mankind told first upon their fingers, and arithmeticians discover it by calling them Digits at this very day. There

is no likelihood that so wise a man as Pythagoras was, should lay any stress upon such trifles, or that his Scholars should be such fools as to be taken with them; but it is well known, that the Pythagoreans held the motion of the Earth about the Sun, which is plainly employed, according to the *Holy Guide* of this fourth days' work. So much of his secret got out to common knowledge and fame as I conceived, that the choicest and most precious treasure of knowledge being laid open in the R.C. *Infallible Guide* of the fourth day; from thence it was that so much solemnity and Religion was put upon that Number, which he called his *Tetractis*, which seems to have been of two kinds, the one the single number of four, the other thirty six made of the four first masculine numbers. AM the four first feminine, viz. 1,3,5,7. and 2,4,6,8. wherein you see that the former and more simple Tetractis is still included and made use of; for four here takes place again in the assignment of the masculine and feminine numbers; whence I further conceive, that under the number of this more complex Tetrad, which contains also the other in it, be taught his Disciples the Mystery of the Creation, opening to them the nature of all things, as well spiritual as corporeal,

Ὁ γὰρ ἄρτιος ἀριθμὸς τὸ θῆλυ τινὸν ἔχει καὶ παθητινὸν, ὁ δὲ περιττὸς τὸ ἄτμειτον καὶ ἀπαθὲς καὶ διατήσιον. Διὸ ὁ μὲν διπλοῦς ὀνομάζεται ὁ ἄρρην,

as Plato writes, for even number carries along with it divisibility, but an odd-number indivisibility, impassibility, and activity, wherefore that is called feminine, this masculine.

6. Wherefore the putting together of the four first masculine numbers to the four first feminine, is the joining of the active and passive principles together, matching the parts of the matter, with congruous forms from the world of Life, so that I conceive the Tetractis was a Symbol of the whole System of Pythagoras philosophy: which is the

very same with the mosaical or Rosie Crucian *Infallible Axiomata*: and the root of this Tetractis is six, which signifies the six days work.

7.   And Fowl and Fish were made by the number four; for there is affinity betwixt them, because Fowl frequent the water in their kind, and the Elements themselves of Air and water are very like one another; besides, the fins of fishes and wings of Birds, the Feathers of the one, and Scales of the other are very analogical they are both also destitute of ureters, Dugs and Milk, and are Ouiparous, further their Motions are mainly alike. The Fishes as it were flying in the water, and the Fowls swimming in the Air, according to that of the Poet concerning Dadalus, when he had made himself wings.

Insultum per iter gelidas enavit ad arctos.

Cast in this also, that as some Fowls dive and swim under water, so some Fishes fly above the water in the air for a considerable space, till their Fins begin to be something stiff and dry.

8.   The number four is the first four square plain, which consists of two proportions, whereof the first is of one to two, and of two to one, the later of two to four, and it proceeds by a double procession and proportion, viz. of one to one, and of two to two, beginning at one and ending at four, which proportions differ in this, that according to arithmetic they are unequal to one the other, but according to geometry are equal; therefore a four square is ascribed to God the Father, and also contains the mystery of the whole Trinity. For by its single proportion, viz. by the first of one to one, the unity of the paternal substance is signified, from which proceeds one Son equal to him; by the next expression also simple, viz. of two to two, is signified by the second expression the Holy Ghost from both; that the Son be

equal to the Father by the first expression: and the Holy Ghost be equal to both, by the second expression.

Hence the super-excellent, and great name of the Divine Trinity of God is written with four letters, viz. JOD, HE, and VAU, HE; where it is the aspiration, HE, signifies the proceeding of the Spirit from both: for HE being duplicated terminates both Syllables and the whole name, but is pronounced JEOVA, as some will have it, whence that JOVIS of the Heathen, which the Ancients did picture with four ears, whence the number four is the Fountain and Head of the whole Divinity, and the Pythagoreans call it the perpetual Fountain nature; for there are four degrees in nature, viz. to be, to live, to be sensible, to understand; there are four motions In nature, viz. Ascendant, Descendant, going Forward, Circular.

9. There are four Angles in the Heaven, viz. Rising, Middle, Falling of the Heaven, and the Bottom of it; there are four Elements under Heaven, viz. Fire, Air, Water, Earth; according to these there are four triplicities In Heaven. There are four first qualities under the Heaven, viz. cold, heat, dryness, and moisture; for these are the four humours, blood, flegme, choler, melancholy; there are four parts of a year, Spring, Summer, Autumn, and Winter; also, the wind is divided into Eastern, Western, Northern, and Southern; there are also four Rivers of Paradise; viz. Pyson, Gibon, Hiddekel, and Parath; and so many infernals, viz. Phlegeton, Cocitus, Styx, Acheron.

10. And the number four makes up all knowledge; first it fills up every simple progress of numbers with four terms, viz. with one, two, three, and four, constituting the number ten; it fills up every difference of numbers, the one even, and containing the first odd in it, it has the grace of the fourth voice. Also it contains the instrument of four

strings, and a Pythagorean Diagram, whereby are found out first of all Musical tunes; and all Harmony of Music for double, treble, four times double, one and a half, one and a third part, a concord of all, a double concord of all, of five of four, and all consonance is limited within the bounds of the number four; It doth also contain the whole of mathematics in four terms, viz. point, line, superficies, and profundity: it comprehends all nature In four terms, viz. substance, quality, quantity, and motion; also all natural Philosophy, in which are the seminary virtues of nature, the Natural springing, the growing Form, and the compositum: Also metaphysics is comprehended in four bounds, viz. being, essence, virtue, and action; Moral Philosophy is comprehended with four virtues, viz. Prudence, Justice, Fortitude, Temperance. It has also the power of Justice, for Times or Terms in the year at Westminster Hall; all manner of Cases are Tried and Suits in Law, to the content of the people. Hence a fourfold Law of Providence from God; fatal from the soul of the world, of Nature from Heaven, of Prudence from man; Of this you may be better satisfied in my Book entitled, *The Idea of the Law*, etc.

11. There are also four Judiciary powers in all things being, viz. the Intellect, Discipline, Opinion and Sense: It has also great power in calling of James, of Spirits and Angels, and in Predictions, and in all Mysteries; hence the Rosie Crucians do ratify the Number 4 with an Oath, as if it were the number that God had fixed on them, to confirm their Faith, as appears in these Verses;

Ου μα τον ημετερα ψυχα παραδοντα τς τραντον Πασαν αεννασ ριζαν εχειο υσεως.

i.e.
By him that did disclose
The Tetrads Mystery:

> Where Natures found that overflows,
> And hidden root doth lie.

Now there are four Gospels received from 4. Evangelists throughout the whole Church; the Hebrews received the chief name of God written with four Letters: Also, the Egyptians, Arabians, Persians, Magitians, Mahumetans, Grecians, Tuscians, Latins, write the name of God with only four Letters, viz. thus, Thet, Alla, Sire, Orsi, Abdi, θεός, Esar, Deus. In the Original world, from whence the Law was received, The Name of God is written with four Letters יהוה. In the intellectual world whence the fatal Law was received, there are four Triplicities, or intelligible Hierarchies, Seraphim, Cherubim, Thrones; Dominations, Powers, Virtues; Principalities, Archangels, Angels; Innocents, Martyrs, Confessors: And there are four Angels, ruling over the four Corners of the world, viz. Michael, Raphael, Gabriel, Uriel; four Rulers of the Elements, Seraph, Cherub, Tharsis, Ariel; four consecrated Animals, the Lion, the Eagle, Man, a Calf; four Triplicities of the Tribes of Israel; Dan, Asser, Nephtalim; Judah, Issachar, Zabulun; Manasseh, Benjamine, Eperaim, Reuben, Simeon, Gad; four Triplicities of Apostles, Matthias, Peter, Jacob the elder; Simon, Bartholomew; Matthew, John, Philip, James the younger; Thaddeus, Andrew, Thomas; four Evangelists, Matthew, Mark, Luke and John.

12. The Celestial World is constituted by the Law of Nature: four Triplicities of Signs, Aries, Leo, Sagittarius; Gemini, Libra, Aquarius; Cancer, Scorpius, Pisces; Taurus, Virgo, Capricornus; four Elementary Stars and Planets, Mars and the Sun, Jupiter and Venus, Saturn and Mercury, the fixed Stars and the Moon.

Four qualities of the Celestial Elements, Light, Diaphanousness, Agility, Solidity, where Generation and Corruption is according to the Elemental Law; there is Fire, Aire, Water and Earth; and four qualities, Heat, Moisture, Cold, Dryness; and four kinds of mixed bodies, Animals, Plants, Metals, Stones; four kinds of Animals, Walking, Flying, Swimming, Creeping; and four things answer the Elements in Plants, viz. Seeds, Flowers, Leaves, Roots; so in Metals, Gold, Iron, Copper and Tin, Quicksilver, Lead and Silver; and in Stones there are four answer the Elements, bright and burning, light and transparent, clear and congealed, heavy and dark.

13.   In the Law of Prudence, Man, there are four Elements, the Mind, the Spirit, the Soul, the Body; four powers of the Soul, the Intellect, Reason, Phantasy, Sense; four Judiciary powers, Faith, Science, Opinion, Experience; four Moral Virtues, Justice, Temperance, Prudence, Fortitude; four Senses relating to the Elements, Sight, Hearing, Taste and Smell, Touch; four Elements in the body, Spirit, Flesh, Rumors, Bones; a fourfold Spirit, Animal, Vital, Generative, Natural; four Humours, Choler, Blood, Phlegm, Melancholy; four manners of complexions, Violence, Nimbleness, Dullness, Slowness.

By the Law of Punishment in the Infernal World, there are four Princes of Spirits that rule the four evil Angels of the world, Oriens, Paynion, Egin, Amaimon. This number cast in a Telesme of silver by Art, brings to you the form of a Virgin, adorned with fine clothes, with a Crown on her: This number giveth acuteness of wit, and the love of men.

14.   The number 4 they say, Telesmatically engraved in silver with the name of the Angel of the day and hour, and the parties name, and the name of his genius, according to the Letters in the Figure of the world, all gathered together, will produce the society of an Angel, who

The Holy Guide

appears like a man sitting in a chair, holding a balance in his hands; and they burned Nutmegs and Frankincense before him; and he made the number fortunate and happy to the bearer by Merchandizing: this number according to Art engraved, makes a man healthful in all Countries.

Chapter VII.

The Number of Youth.

1. Of the signification of the Number 5; 2. And what Angels may be called by it; 3. And how they may appear.

1. The Number 5 is of no small force; for it consists of the first even, and the first odd, as of a female and a male; for an odd number is the male, and the even the female; Rosie Crucians call that the Father, and this the Mother: Therefore the number 5 is of no small perfection, or virtue, which proceeds from the Mixtion of these numbers; it is also the just middle of the universal number, viz. 10. for if you divide the number 10. there will be 9 and one, or 8 and 2, or 7 and 3, or 6 and 4. and every collection makes the number 10. and the exact middle is always the number 5. and its equidistant; and therefore it is a number of Mirth and good fellowship; it is called by the Pythagoreans the number of Wedlock, as also of Justice, because it divides the number 10 even; there is also in it an emblem of Generation; as an Eagle engendering with an Eagle brings forth an Eagle; and a Dolphin engendering with a Dolphin, etc. wherefore the Pythagoreans this number Cytherea, that is Venus, and Τάμος, Marriage; and in Birds

it is evident that they choose their Mates: That the Image of God consists in this, rather than in the dominion over the Creature, I take to be the right sense, and more Philosophical, the other more Political.

It is a wonder, says Grotius, to see how the explication of the Rabbins upon this Fra:[16] and those passages in Plato's *Symposion* agree one with another; which, notwithstanding whatsoever proceeded, I make no question, says he, but they are false and vain; and I must confess I am fully of the same opinion: but this strange agreement Aristophanes his Narration, in the forenamed Symposium, and the Comments of the Rabbins, upon the Rosie Crucian M. is no small argument that Plato had some knowledge of Moses, which may well add the greater authority and credit to this our *Holy Guide*; but it was the wisdom of Plato to own the *Holy Guide* himself; by such unwarrantable Fancies as might rise from the Numbers and the Text, to cast upon such a ridiculous shallow Companion as Aristophanes, it was good enough for him to utter in that club of wits, that Philosophic Symposium of Plato.

2.   And there be five senses in Man, Sight, Hearing, Smelling, Feeling; Tasting, five powers in the Soul, vegetative, sensitive, concupiscible, irascible, rational: 5 fingers on the hand, five wandering Planets in the Heavens, according to which there are five-fold terms in every sign; in Elements there are five kinds of mixt bodies, viz. Stones, Metals, Plants, Plant-animals, and so many kinds of Animals, as men, four-footed beasts, creeping, swimming, flying. And there are five kinds, by which all things are made of God, viz. Essence, the same, another, sense, motion; the Swallow brings forth but five young, which she feeds with equity, beginning with the eldest, and so the rest according to their age: also this number has great power in Expiations; for in

---

[16] This exactly matches the spelling in the 1662 printed edition. -pnw

holy geomancy and telesmatical Figures, it drives away Devils, in natural things it expels poison; it is called the number of fortunateness and favor; and it is the seal of the Holy Ghost, and a bond that binds all things, and the number God loves; it is the number of the cross, yea eminent with the principal wounds of Christ, whereof he vouchsafed to keep the Scars in his glorified body.

The Philosophers did dedicate it as sacred to Mercury, esteeming the virtue of it to be much more excellent then the number four, by how much a living thing is more excellent than a thing without life; for this five in a Figure of Geomancy, Noah found favor with God, end was preserved in the Flood of waters that overflowed Palastina, Apamia, and the Country of the East.

3. In the virtue of this Number, Abraham being a hundred years old, begat a Son of Sarah, being ninety years old, and a barren woman, and past child-bearing, and grew up to be a great people. Hence in time of grace, the name of Divine Omnipotence is called upon with five Letters; for in time of Nature the Name of God was called upon with three Letters: In the time of the Law the ineffable Name of God was expressed with four Letters. And in the exemplary world, The Name of God is expressed with five Letters, Eloim; and the name of Christ with five Letters, Ihesu: In the intellectual world there are five intelligible substances, viz. Spirits of the first Hierarchy called Gods, or the Sons of God: Spirits of the second Hierarchy, called Intelligences; Spirits of the third Hierarchy, called Angels, or Messengers, which are sent, souls of Celestial bodies, heroes or blessed Souls: In the Celestial world, there are five wandering Stars, Lords of the Terms, Saturn, Jupiter, Mars, Venus, Mercury.

There are in the Elementary world five kinds of corruptible things, Water, Air, Fire, Earth, a mixed body, and five kinds of mixt bodies, Animal, Plant, Metal, Stone, Plant Animal; And in the Infernal world there are five Corporeal Torments, deadly bitterness, horrible howling, terrible darkness, unquenchable heat, and piercing stink. The Number five engraved Telesmatically in Gold, with the Numbers of Letters of names, of Angels, Planets, and days, they say will bring to you a woman Crowned with the gesture of one dancing in a Chariot, drawn with four horses; and a flame of Lightning attends her: This number makes a man fortunate in Physick, Chemistry, Astrology, Geomancy, and happy in sporting with woman; they say it makes Ladies beautiful, and beloved of whom they please, etc.

Chapter VIII.

The Number of Riches.

1. Of the Signification of the Number six; 2. And why the whole Creation was comprehended within it. 3. And how a Spirit carried away a Quaker. 4, 5, 6, Of the virtue of the Number.

1.   The Senarius or number six has a double reference, the one to this particular days' work; the other to the whole Creation. For the particular days' work, it is the Creation of sundry sorts of Land Animals, divided into male and female; and the number six is made up of male and female; for two into three is six; hence the Pythagoreans call this number Τάμος Matrimony, adding more, that they did it in reference to the Creation of the world, set down by Moses: This

number also in the same sort that the number five, is a fit Emblem of Procreation; for six into six makes thirty six. Here is something also that respects man, particularly the choicest result of this six days labor. The number of the Brutish Nature was five. But here is an unite superadded in man; reason reaches out to the knowledge of a God, and one added to five makes six.

But now for the reference that six bears to the whole Creation, that the Pythagoreans and Rosie Crucians, did conceive it was significant thereof, appears by the Titles they have given it, for they call it

Διάρθρωσις, Ἀμμω Ἀμμων Κόσμου Παντός

The articulate and complete reformation of the Universe, the Anvil, and the world. I suppose they call it the Anvil from that indefatigable shaping out of new forms and figures upon the matter of the Universe, by the virtue of the active Principles that ever busies itself everywhere; but how the senary should Emblematize the world, you shall understand thus; the world is itself complete, filled and perfected by its own parts; so is the Senarius, which has no denominated part of a six, three and two, viz. one, two, three, which put together make six; a perfect number, is that which is equal to its parts; wherefore this number sets out the perfection of the world, and you know that God in the close of all saw, that all that he made was very good; then again the world is Ἀρρήνη Θῆλυ mas. & femina, that is, it consists of an Active and Passive principle, the one brought down into the other, from the world of Life; and the Senary is made by drawing of the first masculine Number into the first feminine; for three into two is six.

2. Thus you see continuedly that the property of the number sets off the nature of the work of every day, according to those mysteries that the Pythagoreans have observed in them; and besides this, that the numbers have ordinarily got names answerable to each day's work, which as I have often intimated, is a very high probability, that the Rosie Crucians had an *Infallible Guide,* referring to Moses his text; six is a number of perfection, because It is the most perfect in nature, in the whole course of numbers, from one to ten, and it alone is so perfect, that in the Collection of its parts it results the Bane, nothing wanting nor abounding; for if the parts thereof, viz. the middle, the third and ninth part, which are three, two, one, be gathered together, they perfectly fill, up the whole body of six.

Now this perfection all the other numbers want; hence by the Rosie Crucians it is said to be altogether applied to generation and marriage, and is called the balance of the world, for the world is made of the number six; neither doth it abound or is defective; hence that is, because the world was finished by God the sixth day; for the sixth day God saw all the things which he had made, and behold they were exceedingly good, and the evening and morning were the sixth natural day.

Therefore the Heaven, and the Earth, and all the Hosts thereof were finished; it is also called the number of man, because the sixth day man was Created; and it is also the number of our redemption; whence there is a great affinity betwixt the number six and the Crosse, Labor and servitude; hence it is in the Law commanded, that in six days the work is to be done, six days Manna is to be gathered, six years the ground was to be sown, and that the Hebrew servant should serve his master six years; six days the Glory of the Lord appeared upon Mount Sinai, covering it with a cloud; the Cherubim had six wings.

3. And there are six Circles in the Firmament, Artic, Antarctic, two Tropics, Equinoctial and Ecliptic; six wandering Planets, ♄, ♃, ♂, ♀, ☿, ☽, running through the Latitude of the Zodiac, on both sides the Ecliptic; there are six substantial qualities in the Elements, viz. sharpness, thinness, motion, and the contrary to them, dullness, thickness, rest; there are six differences of position, upwards, downwards, before, behind, on the right side, on the left side. There are six natural offices, without which nothing can be, viz. Magnitude, Color, Figure, interval, Standing, Motion. Also a solid figure of a four square thing has six Superficies; there are six Tones of all Harmony, viz. five Tones and two half Tones, which make one Tone, which is the sixth: and the Name of God in the exemplary world are written with six Letters אלהים and אלבבדר; and six orders of Angels or Messengers in the intelligible world will not obey the call of Inferiors, because they are not sent: and there are six degrees of men in the lesser, the Intellect, Memory, Sense, Motion, Life, Essence. And six Devils in the infernal world, which are the Authors of all Calamity, Arteus, Magalesius, Ormenus, Licus, Nicon, Minion. The signs to distinguish, whether the Spirit be good or evil, are the same by which we distinguish whether a man or a tree be good or evil; namely Actions and fruits; and wicked men cannot converse with Angels:

As for Example, a Quaker or a Presbyterian contrived a number in Silver, and could often converse with an Angel, as they called it, at Fell Fens Furnace in Lancashire, for there they lived, although not devout before God; thus, they dissembled, as their manner is, with God and the world, and freely and frequently courted this familiar Spirit. But there are lying Spirits; so it happened the 15 day of June, 1660 being at dinner with his wife and four children, there knocked one at the door, he thought not of his familiar Devil I suppose, but opened the door, at which entered a man in black clothes, and after salutation, said, he must go with him; and this the Quaker was afraid; and one of his sons run out to call neighbors; the rest cried, and the woman also; so one of the children said, O Jesus, mother this man has feet like a Cow; and the woman casting her eyes upon him, blessed her, and said, sweet Husband forsake these foolish fancies of Quaking; immediately as the word went out of her mouth, for all his striving, this monster carried him away, and top of the house also, to the astonishment of all their neighbors, who by this time were come to save the man, but he was gone before, and never heard of after.

4. This may forewarn ungodly Saints the meddling with the Sacred things written in this Book; for the Devil may appear to you like an Angel of Light; wherefore you are commanded in Scripture to judge of the Spirits by their Doctrine, and not of the Doctrine by the Spirits; for miracles, our Savior has forbidden us to rule our faith by them, MAT. 24., 24. and Saint Paul says, Golat. 8. though an angel from heaven preach to you otherwise, etc. let him be accursed; wherefore it is plain, that we are not to judge whether the Doctrine be true or not by the Spirit, but whether the Spirit be good or no by the Doctrine: so likewise, Job. 41. Believe not every spirit: for false prophets are gone out into the world, V. 2. Hereby shall ye know the spirit of God, V. 3. Every spirit that confesses not that Jesus Christ is come in the flesh, is not of God; and this is the spirit of antichrist, V. 15. Whosoever confesses that Jesus Christ is the son of God, in him dwelleth God, and he in God.

5. The knowledge therefore we have of good and evil Spirits, cometh not by vision of an Angel that may teach it, nor by a miracle

that may seem to confirm it: but by conformity of doctrine with this Article and Fundamental point of Christian Faith, which also Saint PAUL says is the sole Foundation, That Jesus Christ is come in the flesh, 1 Cor. 3. 11.

6. For wisdom and knowledge in Philosophy, the Law and Divinity, they engraved the name of the Angel of the day Planet and hour, with the man's name at length, and the number attributed unto each Letter, and he it is said receives virtue from an Angel that appears like a man riding on a Peacock, having Eagles feet, and on his head a Crest, in his right hand holds fire, and in his left a cock.

And now you know how to try a Spirit; but some deny the appearing of Spirits, and that there are any good or bad; but we shall prove that there are, and the difference of Spirits in the third Book; to which we add some Rules to cure those that are possessed and diseased. One main design to make men happy by Knowledge, long Life, Health, Youth, Riches, Wisdom, and Virtue, and how to alter, change, cure and amend all diseases in young or old, with the art of repairing Rosie Crucian Medicines, and times to administer them, and their virtues and uses.

Chapter IX.

The Number of Virtue.

1. 2. 3. 14. 5. 6. 7. 8. 9. 10. 11. 12. 13. 14. 15. 16. 17. 18. 19. 20. 21. The Signification of the Number 7. 22. 23. 24. 25. 26. And that 7. is a fit Symbol of the Sabbath or Rest of God.

1. The Hebdomad or Septenary is a fit Symbol of God, as he is considered having finished these six days Creation; for then, as this *Holy Guide* intimates, he creates nothing further, and therefore his condition is then very fitly set put by the number 7. All numbers within the decade are cast into the Ranks, as Plato observes, οἱ μὲν γεννῶσιν ὁ γεννώμενοι δὲ γεννῶνται μὲν, ὁ γεννῶσι δε. οἱ δὲ ἀμές τερα, χỳ γεννῶσι ᾗ γεννῶνται. (i.e.) some beget, but are not begotten, others are begotten, but do not beget; the last both beget and are begotten; the number 7 is only excepted; that is neither begotten, nor begets any number, which is a perfect Emblem of God, celebrating this Sabbath; for he now creates nothing of a new, as himself is uncreatable, so that the creating and infusing of souls, as occasion shall offer, is quite contrary to this *Holy Guide*. And 7 is of various and manifest power; for it consists of 1 and 6, or of 2 and 5, or of 3 and 4, and it has a unity as it were the coupling together of two threes, hence it is called a number of Marriage, and the Astrologers and Geomancers are resolved by the seventh House, whether the Querent shall marry the party desired; 7 is called (7 is called the Occult Intelligence) the Vehiculum of man's life, which it doth not receive from its part so, as it perfects by its proper right, of its

whole; for it contains body and soul; for the body Consists of four Elements, and is endowed with four qualities.

2.   Also, the Number 3 respects the soul, by reason of the 3-fold power, viz. Imaginative, irascible and concupiscible. The number 7 relates to the generation of men, and it causes man to be received, formed, brought forth, nourished, live, and indeed altogether to subsist; for when the genital seed is received in the womb of the woman, if it remain there 7 hours after the effusion of it, it is certain that it will abide there for good: Then the first seven days it is coagulated, and is fit to receive the shape of a man; then it produces Infants called mature perfection, which are called Infants of the seventh month, because they are the seventh month.

3.   After the birth, the seventh hour tries whether it will live or not: for that which will bear the breath of the air after that hour, is conceived will live; after seven days it casts the reliques of the Navil; after twice seven days its sight begins to move after the light; after 21 days it turns its eyes and whole face freely; after seven months it breeds teeth; after fourteen months it sits without fear of falling; after 21 months it begins to speak; after 28 months it stands strongly and walks; after 35 months it begins to refrain sucking its nurse; after seven years its first teeth fall, and new are bred, fitter for harder meat, and its speech is perfected; after fourteen years boys wax ripe, and then is a beginning of Generation; at 21 years they grow to be men in stature, and begin to be hairy, and become able and strong for Generation; at 28 they begin to burnish and cease to grow taller; in the 35 year they attain to the perfection of their strength; at 42 they keep their strength, at 49 years of age they attain to their utmost discretion and wisdom, and the perfect age of man.

But when they come to the tenth seven year, where the number seven is taken for a complete number, then they come to the common term of life; the Prophet saying our age is seventy years; the utmost height of man's body is seven foot.

4. There are also seven degrees in the body, which complete the dimension of its altitude from the bottom to the top, viz. Marrow, Bone, Vein, Artery, Flesh, Skin. There are seven which by the Greeks are called black members, the Tongue, the Heart, the Lungs, the Liver, the Spleen and two Kidneys: there also seven principal parts of the body, the Head, the Breast, the Hands, the Feet, and the privy Members: it is manifest concerning breath, and meat, that without drawing of the breath, the life doth endure above seven hours, and they that are starved in prison by blood-thirsty Creditors, or otherwise with famine, live not above seven days. The veins and arteries are moved by seven.

5. All judgements in diseases are made with greater manifestation upon the seventh day, being called critical or judicial; also of seven portions, God creates the soul, the soul receives the body by seven degrees; all differences of voices proceed to the seventh degree; after which there is the same resolution: Again, there are seven modulations of voices, Ditonus, semi-Ditonus, Diatessaron, Diapente, with a Tone Diapente, with a half Tone, and Diapason.

6. There is also in Celestials a most potent power of the number 7 for seeing there are four corners of Heaven diametrically looking one towards the other; which indeed is accounted a most full and powerful Aspect, and Consists of the number seven, for it is made from the seventh sign, and makes a cross the most powerful figure of all; but this you must not be ignorant of, that the number seven has a great

communion with the Cross; by the same Radiation and number the Solstice is distant from Winter, and the Equinox from the Summer; all which are done by seven signs; there are also 7 circles in the Heavens according to the Axel-tree; there are seven Stars about the Artic pole, greater and lesser, called Charles Waine: also seven Stars called the Pleiades, and seven Planets, and the Moon dispenses the influence Spiritual and Natural, gives the light of seven to us; for in 28 she runs her appointed course; which number of days the number seven, with its seven terms, viz. from one to seven, doth make and fill up as much as the several numbers, by adding to the Antecedents, and makes four times seven days, in which the Moon runs through, and about the Longitude of the Zodiac by measuring, and measuring, and measuring again; with the like seven of days it dispenses its light, by changing it.

7. For the first seven days unto the middle as it were of the divided world, it increases; the second seven days it fills its whole Orb with light; the third by discretion, it again is contracted into a divided Orb; but after 28 days it is renewed with the last diminution of its light; and by the same seven of days it disposes flux and influx of waters; for in the first seven of the increase of the Moon it is by little lessened; in the second by degrees increased; but the third is like to the first, and the fourth doth the same as the second.

8. And also seven is applied, to which ascending from the lower in the seventh Planet, which betokens rest, to which the seventh day is ascribed, which signifies the seven thousand, wherein (as Saint John witnessed) the Dragon, which is the Devil, being bound, men shall be quiet, and lead a peaceable life: Moreover, the Rosie Crucians call seven the number of Virginity, because the first is that which is neither generated, or generates, neither can it be divided into two equal parts, so as to be generated of another number repeated, or being doubled, to

bring forth another number of itself, which is contained within the bounds of the number ten, which is manifestly the first bound of the numbers; and therefore they dedicated the number seven to Pallas.

9. It has also in Religion most potent signs of its esteem, and it is called the number of an Oath; hence amongst the Hebrews to swear, is called Septenary, to protest by seven. Abraham, when he made a Covenant with Abimelech, appointed seven Ewe-lambs for a Testimony; it is called the number of blessedness, or of Rest, viz. in soul and body. The seventh day the Creator rested from his work, as I showed you above; wherefore this day was by Moses called the Sabbath, or the day of Rest: hence it was that Christ rested the seventh day in the Grave.

10. Besides, it is most convenient in Purification, whence Apuleius says, And I put myself forthwith into the Bath of the Sea to be purified, and put my head seven times under the Waves, and the leprous person is cleansed, being sprinkled seven times with the blood of a Sparrow; and Elijah said to a leprous person, go and wash thyself seven times in Jorden, and thy flesh shall be made whole: And he washed seven times and was cleansed; seven is a number of Repentance and Remission, and seven years Repentance was ordained for sin, according to the opinion of the Wise-man, saying, And upon every sinner seven-fold.

11. And also, the seventh year there were granted Remissions of all debts and trespasses, in full testimony of Love and fulfilling of the Law. And Christ with seven Petitions finished his speech of our satisfaction; and at the end of seven years every Apprentice doth challenge liberty to himself; seven Is suitable to Divine praises; and David said, seven times a day do I praise thee, because of thy righteous

judgements; it is moreover called the number of Revenge, as Cain shall be revenged seven-fold.

12. The number 7 engraved by Art in Silver, will produce to your light a man leaning on a staff, having a bird on his head, and a flourishing tree before him; and this is made for travelers against weariness; the number seven is Christal by Art engraved, brings to you a woman cornuted, riding on a Bull, with seven heads, holding in her right hand a Dart, and in her left a Looking-glass; they will come to you in white and green, and grant you the love of any woman you shall nominate.

13. And seven is described to the Holy Ghost, who is seven-fold, according to his gifts, viz. the Spirit of Wisdom and Understanding, the Spirit of Council and Strength, the Spirit of Knowledge and Holiness, and the Spirit of the Fear of the Lord, which is the 7 eyes of God; there are seven Messengers or Angels that wait the will of God, and seven Lamps burn before the Throne of God, & seven Golden Candlesticks, and in the middle was one like the Son of God; and he had in his right hand seven Stars; there are seven Angels in the presence of God, that stand before him, and seven Spirits before the Throne; and there are seven Trumpets, I mean seven Angels, that stood before the Throne of God. A Lamb had seven horns and seven eyes, and a book was opened with seven seals.

And when the seven seals were opened, there was made silence in Heaven; by seven Julius Caesar did bind and repel evil spirits, as you read in Lucan.

> I will now call you up by a true name,
> The Stygian dogs; I in the light Supreme
> Will leave and follow you; also, through grave,

> From all the urns in death I will you save.
> The number Seven unto the Gods will show,
> To whom I address the self in other hew
> Thou wast wont with wan form and without grace,
> And thee forbid to change Erebus his face.

14. And also of all clean beasts, seven were brought into the Ark, and of Fowls seven; and after seven days the Lord rained upon the earth, and upon the seventh day the Fountains of the deep were broken up, and the waters covered the earth; and Abraham gave Abimelech seven Ewe-lambs; and Jacob served seven years for Leah, and seven more for Rachel; and seven days the people bewailed the death of Jacob.

15. And you read of seven Kine[17], and seven Ears of Corn, seventy years of plenty, and seven years of scarcity; and in numbers seven Calves were offered on the seventh day; and Balaam erected seven Alters; seven days Mary the sister of Aaron went forth leprous out of the Camp; and in Joshua seven Priests carried the Ark of the Covenant before the Host; seven days they went round the Cities; and seven Trumpets were carried by the seven Priests; and the seventh day, the seven Priests sounded their Trumpets.

16. And Abessa reigned in Israel seven years; Samson kept his Nuptial seven days, and the seventh day he put forth a riddle to his Wife; he was bound with seven new Cords, and seven Widths; and seven Locks of his head were shaven off; seven years the Children of Israel were oppressed by the King of Maden.

---

[17] Cows. -pnw

17. And Elias prayed seven times, and at the seventh time, behold a little cloud: seven days the Children of Israel pitched over against the Assyrians, and the seventh day they joined Battel; and seven times the Child needed that was raised by Elisha: seven Gifts of the Holy Ghost: seven Petitions in the Lords Prayers: seven words of Christ upon the cross: seven words of the Virgin Mary.

18. Moreover, this Number has much power, as in natural, so in sacred, spiritual, ceremonial, and mysterious: seven hours were Adam and Eve in paradise; and there were seven men foretold by an Angel, before they were born, viz. Ismael, Isaac, Sampson, Jeremiah, John Baptist, James the Brother of our Lord, & our Savior Jesus Christ: And in the Original world, they write the Name of God with seven letters, ARARITA. And in the Intelligible world there are seven Angels that stand in the presence of God, Zaphiel, Zadkil, Camael, Raphael, Hanael, Michael, Gabriel. In the Elementary World, there are seven Birds, that are used in calling or Angels, viz. the Lapwing, the Eagle, the Vulture, the Swan, the Dove, the Stork, the Owl: and seven Fishes, viz. the Sea-Cat, the Mullet, Thimallus, the Sea-Calf, the Pike, the Dolphin, the Thurlefish; and seven Animals, the Goat, the Cat, the Ape, the Mole, the Hart, the Wolf, the Lion: and seven Metals, Lead, Tin, Iron, Gold, Copper, Quicksilver, Silver: and seven Stones; the Onyx, the Saphira, the Diamond, the Carbuncle, the Emerald, the Achates, the Crystal.

19. This number seven engraved in Silver will bring Gabriel to you, and he will bring you a Genius suitable to yourself in nature, number and name, and he will appear like a man clothed in comely apparel, or like a man and woman sitting at a Table playing, and this giveth mirth, riches, and the love of woman.

20. Again, the Number seven is of two kinds, the one is, ἡ ἐντὸς δεκάδος ἑδδομας. The other, ἡ ὁντος. The Septenary within the Decade is nearly seven unites; the other is a seventh number, beginning at an unite, and holding on in a continual Geometrical proportion till you have gone through seven proportional Terms; for the seventh Term, there is this Septenary of the second kind, whose Nature Plato fully expresses in these words, Ἀιθι γὰς ὁ ὑπὸ μονάδος συντιθέμενος ἐν διπλασίοις ἢ τριλασίοις ἢ συνόλως ἀναλογέσιν, ἑβδόμος ἀριθμὸς κύβος τε, καὶ τετράγωνος ἐστιν ἀμφοέσα περιέχων, τῆς τε ἀσωματο καὶ σωματῖνης οσίας. Τῆς μὲν ἀσώματο κατὰ τὴν ἐπίπεδον ἂν ἀποτελῦοι τετράγωνοι, τῆς δὲ σωματικῆς κατὰ τὴν ἑτέραν ἂν ἀποτελῦοι κύβοι, i.e.

21. For always beginning from an unite, and holding on in double or triple, or what proportion you will, the seventh number of this rank is both square and cube, comprehending both kinds, as well the Corporeal as incorporeal substance; the incorporeal according to the Superficies which the squares exhibit, but the corporeal according to the solid Dimensions which are set out by the Cubes.

22. As for example, 64 or 7. 2. 9. these are numbers that arise after this manner: each of them is a seventh from an unite, the one arising

from double proportion, the other from Triple; and if the proportion were quadruple, quintuple, or any else, there is the same reason. Some other seventh number would arise, which would prove of the same Nature with these, they would prove both Cubes and squares, that is, Corporeal and Incorporeal; for such is 64. either by multiplying eight into eight, and so it is a square, or else by multiplying four cubically, for four times four times four is again sixty-four, but then it is a Cube. And so 7. 9. is made either by squaring of 27. or Cubically multiplying of nine, for either way will 729. be made: and so is both Cube and square corporeal and incorporeal: whereby is intimated, that the world shall be reduced in the seventh day to a seer spiritual consistency, to an incorporeal condition; but there shall be a cohabitation of the spirit with the flesh, in a Mystical or Moral sense, and that God will pitch his Tent amongst us. Then shall be settled everlasting righteousness, and rooted in the Earth, so long as mankind shall inhabit upon the face thereof.

23.   And this truth of the Reign of Righteousness in this seven thousand years is still more clearly set out to us in the Septenary within ten, *Τῇ ἐντὸς δέα δ☉ ἑβδομάδι*, as Plato calls it, the naked number seven. For the parts it consists of are three and four, which put together make seven. And these parts be the sides of the first orthogonian Numbers; the very sides that include the right angle thereof. And the orthogonian what a foundation it is of trigonometry, and of measuring the Altitudes, Latitudes, and Longitudes of things, everybody knows that knows anything at all of the Mathematics: and this prefigures the uprightness of the holy Generation, who will stand and walk, *κατ' ὀρθάς*, including neither this way, nor that way, but they will approve themselves of an upright and sincere heart; and by the Spirit of righteousness will these Saints be enabled to find out

that depth, and breadth, and height of the wisdom and goodness of God, as somewhere the Apostle himself phrased it.

24. But then again, lastly, this three and four comprehended also the Conjunction of the corporeal and incorporeal Nature; three being the first superficies. And four the first body; and in the seven thousand years there will be so great union betwixt God and man, that man shall not die, but partake of his Spirit. And the Inhabitants of the Ætherial Region will openly converse with these of the Terrestrial, and such frequent conversation, and ordinary visits of our cordial Friends of that other world, will take away all the toil of life, and fear of death amongst men, they being very cheerful and pleasant here in the body, and being well assured, no pain shall afflict them, when they please to go out of it; for Heaven and Earth by this number shall shake hands together, or become as one house; and to die shall be accounted but to ascend into a higher Room.

25. And though this dispensation for the present be but very sparingly set afoot, because of my youth; yet I suppose the more ancient and Learned may have a glimpse at it; concerning whom accomplished posterity may happily utter something answerable to that of our Saviors concerning Abraham, who tasted of Christianity before Christ himself was come in the flesh: Abraham saw my day and rejoiced at it. And without all question, that plenitude of happiness that has been reserved for future times, the presage, prediction and representation of it has in all Ages been a very great Joy and Triumph to all holy men, Rosie Crucians and Prophets; Adam, Seth, Enos, Cainan, Mahalaleel, Jared, they died, not enjoying the Riches of Gods Goodness in their bodies; but Enoch, who was the seventh from Adam, was by seven taken up alive into Heaven by Gabriel: and seems to enjoy that great BLISS in the body, I have showed you, in my Book called *The Way to*

Bliss: The World then in the seventh Chiliad will be assumed up into God, snatched up by his Spirit, enacted by his power: The Jerusalem that comes down from Heaven, will then in a most glorious and eminent manner flourish upon earth; God will, as I said, by seven, pitch his Tabernacle amongst us: and for God to be in us and with us, is as much as for us to be lifted up into God.

26. By seven Cornelius Agrippa being at Paris writ to a Friend of his at the Court, the signification of a saying in Jeremiah: viz. our crown is fallen, woe unto us, because we have sinned; which (says he) I wish might not be so truly applied to you; for truly that verse, the Numeral Letters being gathered together, M C V L expresses the year M D X X I V, wherein, according to the account, your king was taken at Papia.

27. The number seven engraved in a Jasper stone, with the parties Name, and the Numbers attributed to it, the Angel of the Planet, and day of the week; and this they say is good to obtain the favor of a Lady in Marriage, and the Love of Virgins; it makes the bearer fortunate also in Gaming, as Tables, Cards, Dice, Horse-racing, Bowling, Shooting, Cock-fighting, etc. And by this Number they say will appear a little Maid, clothed in Long white Garments, with her hair spread abroad, holding flowers in her right hand, and she gives virtue to this number, that the bearer of it shall go at his pleasure invisible: if it be engraved with his name and Genius, etc. in a Diamond.

By this Number they give Physick to renew Youth, etc. but some do not believe there is any such thing as Spirits; but we shall prove their Existence and Apparition in the third Book, etc.

Chapter X.

The Number of Wisdom.

Of the signification of the Number 8, and what Angels say be called by it, and how they appear.

1. I Am not angry at the Fanatic dispositions of man, that rail against my Writings, viz. Francis Osborn, Author of *Advice to a Son*, and Elias Ashmole, Esq; that made public my imperfect Copy; and some others I pity, being the worm-eaten memorials of defaced Histories & etc.

2. These ignorantly wonder what those officious spirits should be I so much talk of in my book, called the *Familiar Spirit*, that so willingly appear, and after my directions, offer themselves to consociate with a man.

3. O my enemies! Whom I pity, more than despise, I shall tell you lovingly, as I speak to the capacities of people; They are Angels uncapable of incorporation into human bodies and souls of the deceased, that have affinity with mortality and human frailty; and these will not appear to you at all times, but every first, third, fifth, seventh, or ninth year, they will come, and forever accompany you in a fortunate telesme.

4. And eight will bring to you a Genius you shall know in its proper place; And Orpheus was used to beseech Divine Justice by light,

and he usually swore by eight Deities, viz. Fire, Water, Earth, the Heaven, Moon, Sun, Planets, the Night: and Virgil speaking of Magic, says;

> I many times with eight have Maeris spy'd,
> Chang'd to a Wolf, and in the Woods to bide:
> From Sepulchers would souls departed charm,
> And corn bear standing from another's Farm.

5. The Rosie Crucians call eight the number of Justice and Fulness: first, because it is first of all divided into numbers equally even, viz. into four, and that division is by the same reason made into two times two, viz. two times two twice; and by reason of this equality of Division, it took to itself the name of Justice; but the other received the name, viz. of Fulness, by reason of the Contexture of the corporeal solidity, since the first makes a solid body. There are eight only visible Spheres of the Heavens; also, by it the property of corporeal matter is signified, which Tyrius comprehended in eight of the Sea-songs: this is also called the Covenant, or Circumcision, which was commanded to be done by the Jews the 8th. day.

6. By eight it is said, a Gentleman in the Kings Army at Edge-Hill battle showed the effects of Spirits, viz. the appearance of armed men fighting and encountering one against another in the sky, and hundreds saw these things besides himself.

7. And through eight Tsagarith showed the Citizens of Jerusalem for forty days, or five times eight together, horseman running in the air in cloth of Gold, and armed with Lances, like a band of soldiers, and troops of horsemen in array, encountering and running one against

another, with shaking of shields, and multitudes of pikes, and drawing of swords, and casting of darts, and glittering of golden ornaments, and harness of all sorts: and thus he predicted the great slaughter of no less than 80000 made by Antiochus: this is also recorded in the second of Maccabees, Ch. 5.

8.   And in eight was seen a Sea-fight over Sydmouth in Devonshire, by Caleb Perkinson, a Captain of a Ship and his company; and he predicted a Sea-fight, which happened betwixt us and the Hollanders.

9.   Eight is an evil number, and always signifies diseases, or death; and in the old Law there were eight Ornament of the Priest, viz. a Breastplate, a Coat, a Girdle, a Mytre, a Robe, an Ephod, a Girdle of an Ephod, a Golden Plate: hither belongs the number to Eternity, and the end of the World, because it follows the number seven, which is the mystery of time.

10.   By eight Hammel caused a Spirit, called Eckerken, to appear always in the shape of a hand, and haunt those that troubled him, pulling down travelers off their horses, and overturning carriages.

11.   They say this number was dedicated to Dionysius, because he was born the 8th. month; in everlasting memory whereof, Naxos the Island was dedicated to him, which obtained this prerogative, that only the women of Naxos should safely bring forth in the eighth month, and their children should live; whereas the children of the eighth month in other Nations die, and their mothers then bringing forth are in manifest danger.

12. By eight artificially charactered in a proper subject, Apollonius Tyaneus told the Ephesians of an old man that was a Specter, and how the walking Plague was by his means in the City here and there; whereupon they stoned the shape, and after a while they uncovered the heap; whereupon appeared the shape of a great black dog, as big as any Lion: Yet I know some able Artists in England, that will not believe but this is Melancholy, and fraud of a Priest: But the learned Grotius, a man far from all levity and vain credulity, is so secure of the truth of Tyaneus his Miracles, and Telesmatical Numbers, that he does not stick to term him imprudent that has the face to deny them.

13. In the Infernal world there are 8 rewards of the damned, Prison, Death, Judgement, the wrath of God, Drunkenness, Indignation, Tribulation, Anguish; and eight kinds of blessed men in the lesser world, the Poor in spirit, the Merciful, the Meek, the Mourners, they that hunger and thirst after Righteousness, the Peacemakers, they which are persecuted for Righteousness sake: eight particular qualities in the Elementary world, the heat of the Fire, the moistness of the Air, the coldness of the Water, the dryness of the Earth, the dryness of the Fire, the heat of the Air, the coldness of the Earth, the moistness of the Water; and eight rewards of the Blessed; Inheritance, Incorruption, Power, Victory, the Vision of God, Grace, a Kingdom, Joy; and the Name of God in the Original word is written sometimes with eight letters, Eloha

Vadaath ולדעתאלה   Iehovae Vedaath הלעת יהוה

and these Questions are resolved by eight.

14. The number eight telesmatically engraved, as others were in an onyx stone, they say renders a man powerful in good and evil, so that he shall be feared of all; and whosoever carries it, they give him the power of charming diseases, and he shall terrify men by his looks

when he is angry: this makes also a man bold and fortunate in Wars: by this number appears a Spirit like a man riding upon a Lion, having in his right hand a naked sword: and by this number they preserve goods from stealing, and cause Thieves to bring again what they have taken away; this helps all diseases in the secrets and bowels: by this number they know when the party will die that is sick.

Chapter XI

The Number of Changing Bodies

Of the signification of the number nine; how that by nine Julius Caesar called up Spirits, and did what he pleases: how Galleron by nine went invisible, and had the society of a familiar Genius.

1.   And nine is a powerful number in all things: Julius Caesar made this number in Gold telesmatically in the hour by Art, and carried with him in his march to the River Rubicon, which divides Gallia Citerior from Italy, and upon the Banks side his army saw appear at his command, a Genius, like a proper man, playing on a Reed; the strangeness of his actions, as well as the pleasantness of his music, had drawn several of the Shepherds unto him, as also many of the Soldiers; amongst whom were some Trumpeters, which this Triton (if I may so call him) or Sea-god, well observing, nimbly snatches away one of the Trumpets out of their hands, leaps forthwith into the River, and sounding to horse, with that strength and violence, that he seemed to rend the heavens, and made the air ring again with the mighty forcibleness of the blast: in this manner he passed over to the other side of the River: whereupon Caeser taking the omen, leaves off all further

dispute with himself, carries over his army, enters Italy, secure of success, from the so manifest tokens of the favor of the gods. The number nine is dedicated to the Muses, by the help of the order of the celestial spheres, and divine Spirits.

2. And there are nine Spheres; and according to these there are nine Muses, viz. Caliope, Urania, Polymnia, Terpsichore, Clio, Melpohome, Erato, Euterpe, and Thalia; which mine Muses indeed are appropriated to the nine Spheres, so that the first resembles the supreme sphere, which they call *Primum Mobile*; and so descending by degrees, according to the written order, unto the last, which resembles the Sphere of the Moon.

3. So the Name of God in the Original Word, is sometimes written with nine letters, הוהיצבאוה JEHOVAH SABOATH, ךלהעחוהי Ιεόυα ξιδχευν אלהיםגבלרי F△οIMTIBOP; and there are in the intelligible World nine Quire of Angels, Seraphim, Cherubim, Thrones, Dominations, Powers, Virtues, Principalities, Archangels, Angels; and nine Angels ruling the Heavens, Metattron, Ophaniel, Zaphkiel, Zadktel, Camael, Raphael, Haniel, Michael, Gabriel. In the Celestial World Calliope is appropriate to the Primum Mobile; Urania to the Starry Heaven, Polymnia to Saturne; Terpsichore to Jupiters; Clio to Mars; Melpomene to the Sun, Erato to Venus, Euterpe to Mercury, Thalia to the Moon: and they engrave nine upon a Sapphire, Emerald, Carbuncle, Beryl, Onyx, Chrysolite, Jasper, or Topaz: but properly and most effectually to be resolved of their Questions, or to obtain their desires, they Telesmatically in an hour engrave it in Sardis or Silver; and this will make a man (they say) go invisible, as Caleron, Alexanders brother-in-

law sometime did, when he lay with his brothers Concubine as often as himself: This number obtains the love of women.

4. And the ninth hour our Savior breathed out his Spirit; and in nine days the Ancients buried their dead; and nine years Minos received Laws from Jupiter, as you may read in my Book, called *The Idea of the Law*: and nine was most especially taken notice of by Homer, when Laws were to be given, or answers were to be given, or the sward was like to rage; the astrologers and geomancers also observe the number nine in the ages of men, no otherwise then they do of seven, which they call a climacterically year; which are eminent for some remarkable change: yet sometimes it signifies imperfectness and incompleteness, because it doth not attain to the perfection of the number ten, but is less by one, without which it is deficient, as Austin interprets it of the ten Lepers; neither is the longitude of nine Cubits of Og, King of Basan, who is a type of the Devil, without a Mystery: and there are nine senses inward and outward together in man viz. memory, cogitative, imaginative, commonsense, rearing, seeing, smelling, tasting, touching: and nine orders of Devils in Sheol, viz. false spirits of lying, vessels of iniquity, avengers of wickedness, jugglers, or lylians, airy powers, furies, sowing mischief, sifters, or tryers, tempters, or ensnarers.

> I aim at the Celestial Glory;
> Below the Moon all's Transitory.

5. The number nine, the number of the Planet, day of the week, Angel of the day, and hour engraved in Gold, will (they say) bring down to you an Angel like a King crowned, sitting in a Chair, having a Raven in his bosom, and under his feet a Globe; be wears Saffron—colored clothes, & he, they report, gives virtue to the number & makes

the bearer of it invincible & honorable, & helps to bring their business to a good end, and to drive away vain Dreams. This number prevails against Fevers and the Plague, and they made it in a Balanitis—stone, or a RUBIE: This number causes long life and health: by this Number Plato so ordered himself, that he could by it cause Nature to end his days at his pleasure, and by departing upon the same 81 years after his Birth, to fulfill of purpose nine times nine, the most perfect Number.

Chapter XII.

The Number of Medicines.

Of the signification of the number ten, how by this Number ten, Socrates in a Monitory Vision had a Swan in his lap, and of Plato's birth and education.

1.   The Number ten is a Number of honor and preferment, and Pythagoras had honor by it. And Plato by this Number had the knowledge of the more sacred Mysteries of God, and the state of the soul of man in this world, and that other deservedly got to himself the title of Divine, ὁ Θεῖος Πλάτων

2.   But as for Miracles, I know none he did, though something highly miraculous happened, if that Fame at Athens was true, that Steusippus, Clearchus, and Anaxilides report to have been, concerning his Birth, which is, that Aristo, his reputed Father, when he would forcibly have had to do with Peritione, she being indeed exceeding fair and beautiful, fell short of his purpose, and surcease from his attempt,

that he saw Apollo in a Vision, and so abstained from meddling with his Wife, till she brought forth her son Aristocles, who after was called Plato.

3. But that is far more credible, which is reported concerning the commending of him to his Tutor Socrates, who the day before he came, dreamed that he had a young Swan in his lap, which putting forth apace of a sudden, flew up into the Air, and sung very sweetly; wherefore the next day, when Plato was brought to him by his Father, τῦ τν, ἐι τἐιεῖν ἐιν τόν ορνιν, he presently said, this is the Bird, and so willingly received him for his Pupil.

4. But his acquaintance with the *Holy Guide*, as it is more credible in itself, so I have also better proof: As Aristobulus the Jew in Clemens Alexandrianus Saint Ambrose, Hermippus in Josephus against Appions: And lastly, Numennius the Platonist, who ingeniously confesses,

Τι γαρ ὅτι Πλατων αΜωσης Ἀττικιζων

That is Plato, but Moses in Greek, as I have elsewhere in my Book called, *The Idea of the Law*, alleged amongst Lawgivers; the Number ten is called, every Number, Complete, signifying the full course of life; for beyond that we cannot number within itself, or explain them by itself, and its own by multiplying them.

Wherefore it is accounted to be of a manifold Religion and Power, and is applied to the purging of souls: Hence the Ancients called Ceremonies Denary, because they that were to be expected, and to offer Sacrifices were to abstain from some certain things for ten days, whence amongst the Egyptians, it was the custom for him that would sacrifice to Jao (i.e.) Jar, to fast ten days before; which Apuleius testifies of himself, saying, it was commanded that I should for the space of ten days refrain all meat, and be fasting.

5. There are ten Sanguine parts of Man, the Menstrue, the Sperm, the Plasmatick spirit, the Masse, the Humours, the Organic body, Vegetative part, the Sensitive part, Reason, and the Mind: There are also ten simple Integral parts constituting a man, the Bone, the Cartilage, Nerve, Fiber, Ligament, Artery, Vein, Membrane, Flesh, Skin. There are also ten parts of which a Man consists intrinsically, the Spirit, the Brain, the Lungs, the Heart, the Liver, the Gall, the Spleen, the Kidneys, the Testicles, the Matrix.

6. There are ten Curtains in the Temple, ten strings in the Psaltery, ten musical instruments which with Psalms were sung, the names whereof were Neza, on which their Odes were sung: Nablum the same as Organs, Mizmor on which the Psalms, Sirs on which the Canticles, Tebila on which Orations, Berach on which Benedictions, Ralel on which Praise, Rodaia on which thanks, Afre on which the felicity or bliss and happiness of any one, Hallelujah on which the praises of God only, and Contemplations: there were also ten figures of Psalms, *a. ω. †* etc.

7. And there were ten Singers, viz. Adam, Abraham, Melchisedick, Moses, Asaph, David, Solomon, and the three sons of Chorab and the name of God in the original world is written with ten Letters, אבצ אנא אלהים Ελοιμζαʼβαοθ: There are ten Commandments, and the tenth day after the Ascension of Christ, the Holy Ghost came down: There are ten Names of God, אלהא Ἐστὲ יהוה ιοδ

Ιεova יהוה'אלהים Ιεova Ελοιμ אל El ימיבר
אלה Ελοιμ μιCop אלאה Ελoá היאכצית-
יהן Ιεova ζαβοαθ היצמים אל Ελοιμ
ζαβαοθ שדי ζαδαι אדיאי Αδoναι μελεu.

8.   And the Number ten cast in Gold was it, in which Jacob wrestling with the Angel all might overcame, and at the rising of the Sun was blessed, and called by the name of Israel. By, and in this Number, Joshua overcame thirty-one Kings, and by ten in a Carbuncle David overcame Goliath, and the Philistines; and in ten Daniel escaped the danger of the Lions.

9.   There are ten Messengers that carry the souls down from God through the Heavens, Spheres, Stars, and Planets to the bodies of men, and these be their names, Ketber, Hechmach, Binar, Hlsed, Geburar, Tiphereth, Nezah, Jod, Hesod; Malchuth; this Number is as circular as unity, because being heaped together, returns into an unity, from whence it had its beginning, and it is the end and perfection of all Numbers, and the beginning of tens as the Number ten flows into unity back again, from whence it proceeded; so everything that is flowing, is returned back to that from which it had the beginning of its Flux, so water returns to the Sea from whence it had its beginning; the body returns to the earth from whence it was taken; time returns to eternity from whence it flowed; the spirit shall return to God that gave it, and lastly, every creature returns to nothing, from whence it was created, neither is it supported, but by the Word of God, in Whom all things are hid.

10.   By the Number ten Polomides an Abbot of Malego in Spain could tell almost at any distance how the affairs of the world went,

what consultations or transactions there were in all the Nations of Christendom, from whence he got to himself the reputation of a very holy man, and a Rosie Crucian; but other things came to pass, no less strange and miraculous, as that at the celebrating of the holy Eucharist; the Priest should always want one of his round Wafers, which was secretly conveyed to this Priest or Abbot, by the administration of Angels, he receiving into his mouth, eat it in the view of the people, to their great astonishment, and high reverence of the Saint.

11. At the elevation of the Host, Polomides being near at hand, but yet a wall betwixt, that the wall was conceived to open, and to exhibit Polomides to the view of them in the Chapel. And thus when he pleases he would partake of the Consecrated bread; when this Abbot came into the Chapel himself, upon some special day, that he would set off the solemnity of the day by some notable and conspicuous Miracle; for he would sometimes be lifted up above the ground three or four Cubits high, other sometimes bearing the Image of Christ in his Arms, weeping savourly, he would make his hair to increase to the length and largeness, that it would come to his heels and cover him all over, and the Image of Christ in his arms, which anon notwithstanding would shrink up again to its usual size: And after this he called an Angel by ten in Gold Telesmatical engraved, to show the true Religion to him; and the Angel bid him turn Protestant, in the best sense of the Church of England; and afterwards all his life he preached, and was created Bishop in England, and preached to the Protestants in France, and known well by the name of the Bishop of Spalatta.

12. And all things with the Number ten, and by the Number ten make a round, taking their beginning from God, and ending in him: God therefore the first unity, or one thing, before he communicated himself to inferiors, diffused himself into the first of Numbers, viz. the

Number three, then into the Number ten, as into ten Idea's and measures of making all Numbers, and all things, which the HEBREWS call ten Attributes and blessed Souls, viz. Haloeh, Hakades, Ophanim, Abalim, Hasmallim, Seraphim, Melachim, Elohim, Ben Elohim, Cherubim Issim; and ten ruling Angels, Metratton, Jophiel, Zaphkiel, Zadkiel, Camael, Haviel, Michael, Gabriel, the soul of Messiah: Ten Spirits of the World that rule the Sphere, Reschith Kagallalim, Masloth, Sabbathi, Zedek, Madim, Schemes, Noga, Chocab, Levanah, Holom, Jesodoth: ten consecrated Animals, viz. a Dove, a Lizard, a Dragon, an Eagle, a Horse, Lion, Man, Genitals, Bull, a Lamb: They account ten Divine names, for which there cannot be a further Number.

13. Hence all tens have some divine things in them, and in the Law are required of God as his own, together with the first fruits as the original of things, and beginning of Numbers, and every tenth is the end given to him, who is the beginning and end of all things: and ten Orders of the Damned in Hell, viz. false gods, lying spirits, William Lilly The King of Sweden's Juggler, etc. Vessels of Iniquity, Tempters or Ensnarers, Sisters of Triers, Revengers of wickedness, Furies the Seminaries of Evil, Aery Powers, wicked souls bearing Rule.

> All spirits were created pure at first,
> But by their self-will after were accurse.

14. The Number ten, and the Letters and Numbers of Angels, etc. engraved in a Carbuncle, renders a man flee from diseases, and makes him live long, fresh, and beautiful, this helps all diseases in the Kidneys, it causes the party that bears it to live secure from Thieves, for no Thief can enter his house in the night, nor have power to carry away a Sheep or Horse, etc. if once slept in the ground, he shall not come out again until the party that owns the ground pleases: and this

The Holy Guide

Number you must character or cast in Copper, and lay it under the gate in the earth, and in the ground on the East side of the place you would have guarded, be it House, Garden, or an Orchard, etc. And it's said the Angel that gives virtue to this, appears like an old man leaning on a staff, having in his hand a Sword, and he seems in black Clothes. By this Number they know when to begin any work in this book.

Chapter XIII.

The Number of preparations of Gold.

Of the signification of the number 11, how by it we know the bodies of Devils, and their natural ὀφιωγεύς in Pherecydes Syrus;

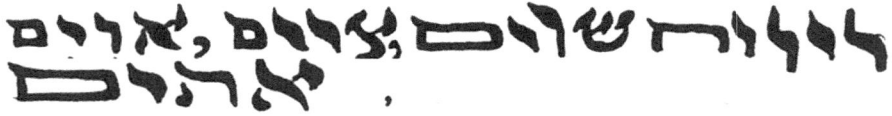

, names of Spirits haunting fields and desolate places: of Theophilus Fulwood, who had the continual society of a Guardian Genius: of Plotinus and Olympius.

1. By 11 Theophilus Fulwood, the Rosie Crucian, knew the Demones Metallici, and Guardian Genii, who told him, That the bodies of Spirits were cold; & indeed it stands to very good reason, that the bodies of Spirits being nothing but coagulated airs, should be cold, as well as coagulated water, which is snow and Ice, and that it should be a more keen and piercing cold; it consisting of more, subtill particles than those of water, and therefore more fit to insinuate, and more accurately and stingingly to effect and touch the nerves. Hence we

may also discover the folly of the opinion that makes the very essence of spirits to be fire; for how unfit that would be to coagulate the air, is plain at first; for it would rather melt and dissolve those consistencies, then constringe them, and freeze them in a manner; but it is rather manifest, that the essence of Spirits is a substance specifically distinct from all corporeal matter whatsoever; but my intents is not to philosophize concerning the nature of Spirits.

2. Pherecydes Syrus the Rosie Crucian, and Master of Pythagoras, by eleven knew the fauni and sylvani, and onocentauri, δαιμόνων γενωχαθυλὸν κỳ σκοτεινόν τῇ ἐνιφανεία A kind of spirits that frequent the woods, and are of a dark Color; they cause a noise and stir in those desolate places, and therefore he forewarns his scholars to beware of their acquaintance,

> HAEC LOCA CAPRIPEDES SATYROS, NYMPHASQUE TENERE.
> FINITIMI FINGUNT, & FAUNOS ESSE LOQUUNTUR;
> QUORUM NOCTIVAGO STREPITU LUD QUEJOCANTI
> AFFIRMANT VULGO TACITURNA SILENTIA RUMPI.
> i.e.
> These are the places where the Nymphs do won;
> The Fauna and Satyrs with their cloven feet,
> Whose noise, and shouts, and laughters loud do run,
> Through the still mire, and wake the silent night.

3. But the Jews understand by 11, and לילית a She-devil, an enemy to women in child-bed; whence it is, that they write on the walls of the room where the woman lies in לילית

אדם חוה עוץ Adam, Eve, out of doors Lilith. And our Savior Christ in the 12th. of Mat. 43. ver. plainly allows of this Doctrine, that evil spirits have their haunts in the fields and deserts, which Grotius observes to be the opinion of the Jews, and that שדים Daemons have their names for that reason, from ת שד the field; for if it were from שוד, it would be rather שדם Skiddim then שודם Shedhim, as Grammatical Analogy requires.

4. The number 11, as it exceeds the number ten, which is the number of the Commandments; so, it falls short of the number twelve, which is of Grace and Perfection, therefore it is called the number of sins, and the penitent. Hence in the Tabernacle there were commanded to be made 11 coats of hair, which is the habit of those that are penitent, and lament for their sins; whence this number has no Communion with Divine or Celestial things, nor any, attraction attending to things above; neither has it any reward; but yet sometimes it receives a gracious favor from God, as he which was called the eleventh hour to the Vineyard of the Lord, received the same reward as those who had born the burthen and heat of the day; and I never knew but one spirit that ever appeared by this number that was a good spirit; and that Theophilus Fulwood had, as he gathered from certain Monitory dreams and visions, although other spirits would speak to him, this would not, but yet he was forewarned as well of several dangers as vices; that this spirit discovered himself to him after he had for a whole year together earnestly prayed to God to send a good Angel to him; and he engraved 11 in silver for it, to be the guide and

governor of his life and actions, that he might not be deluded by evil spirits.

5.   Adding also, that before and after prayer he used to spend two or three hours in meditation and reading the Scriptures, diligently enquiring with himself, what Religion amongst those so many Controverted in the world, might be best; beseeching God that he would be pleased to direct him to it; and that he did not allow of their way, that at all adventures pray to God to confirm them in that opinion they have already preconceived, be it right or wrong.

6.   That while he was thus busy with himself, he light upon the Book of Common-Prayer, and in it he found a paper, in which was written, episcopacy meliorated, is the best religion; and that a good and holy man can offer up no greater, nor more acceptable sacrifice to God, then the obligation of himself, his soul; and under it was D. G. BISHOP of C. and therefore following the Bishops Council, that he offered his soul to God.

7.   And that after that, amongst many other Divine Dreams and Visions, he once in his sleep seemed to hear the voice of God saying to him, I will save thy soul; I am he that before appeared unto thee: Afterwards, that the Spirit every day would knock at the door about three or four a:clock in the morning, though be rising and opening the door could see no body, but that the Spirit persisted in this course, and unless he did rise, would thus rouse him up.

8.   This trouble and boisterousness made him begin to conceit that was some evil spirit that thus haunted him; and therefore, he daily prayed earnestly to God, that he would be pleased to send a good

Angel to him, and often also sung Psalms, having most of them by heart.

9. Wherefore the Spirit afterwards knocked more gently at the door, and one day discovered himself to him waking, which was the first time that he was assured by his senses that it was he; for he often touched and stirred a drinking-glass that stood in his chamber, which did not a little amaze him.

10. Two days after, when he entertained a Gentleman of the Kings, a friend of his, at supper with him, that this friend of his was much abashed while he heard the Spirit thumping on the bench hard by him, and was stricken with fear; but he bid him be of good courage, there was no hurt toward; and the better to assure him of it, told him the whole truth of the matter.

11. Wherefore from that time, says Eugenius Theodactus, he did affirm, that this Spirit was always with him, and by some sensible sign did ever advertise him with things, as by striking his right ear, if he did any ways amiss; if otherwise, his left; if anybody came to circumvent him, that his right ear was struck; but his left ear if a good man & to good ends accosted him; if he was about to eat or drink anything that would hurt him, or intended or purposed to himself to do anything that would prove ill, that he was prohibited by a sign; or if he delayed to follow his business; that he was quickened by a sign followed him.

12. When he began to praise God in Psalms, and to declare his marvelous acts, that he was presently raised and strengthened with a spiritual and supernatural power.

13. That he daily begged of God, that he would teach him his Will, his Law, and his truth: and that he set one day apart in the week for reading the Scripture and Meditation, with singing of Psalms, and that he did not stir out of his house all that day; but that in his ordinary conversation he was sufficiently merry, and of a cheerful mind, and he cited that saying for it, *vidi facies sanctorum latas*; but in his conversing with others, if he had talked vainly or indiscreetly, or had some days together neglected his Devotions, that he was forthwith admonished thereof by a dream, that he was also admonished to rise betimes in the morning, and that about four of the clock, a voice would come to him while he was asleep, saying; Who gets up first to pray?

14. He told Eugenius also, how he was often admonished to give alms, & that the more charity he bestowed, the more prosperous he was; and that on a time when his enemies sought after his life, and knew that he was to go by water, that his Father in a Dream brought two horses to him, the one a white, the other a bay; and that therefore he bid his servant hire him two horses, and though he told him nothing of the colors, that yet he brought him a white one and a bay one.

15. At another time, when he was in very great danger, and was newly gone to bed, he said, that the Spirit would not let him alone till he had raised him again; wherefore he watched and prayed all that night; the next day after he escaped the hands of his persecutors in a wonderful manner; which being done, in his next sleep he heard a voice saying, now sing, *qui sedet in latibulo altissimi*.

16. Eugenius asked him why he would not speak to the Spirit for the gaining of the plainer and more familiar converse with; he answered, that he once attempted it, but the Spirit took away the

Number and Plate, and struck it against the door with that vehemence, as if he had intended to have beat it down, whereby he gathered his dislike of the matter.

17. But though the Spirit would not talk with him, as those that appear by other Numbers, yet he could make use of his judgement in the reading of his books, and moderating his studies; for if he took an ill book into his hands, and fell a reading, the Spirit would strike it, that he might lay it down, and would also sundry times, be the books what they would, hinder him from reading and writing overmuch, that his mind might rest, and silently meditate with itself: he added, also, that very often, while he was awake, a small, subtle, inarticulate sound would come into his ears.

18. Eugenius further enquiring, whether he ever did see the shape and form of the Spirit; he told him, that while he was awake, he never did see anything but a certain light, very bright and clear, and of a round compass and figure; but that once being in great jeopardy of his life, and having heartily prayed to God that he would be pleased to provide for his safety; about break of day, amidst his slumbering's and waking's, he espied on his bed where he lay, a young boy clad in a white garment, tinctured somewhat with a touch of purple, and of a visage admirable lovely and beautiful to behold.

19. And this was the first Rosie Crucian that ever I saw, being about seven years since; but being now one of the fraternities, I asked him of souls and spirits: and what numbers were fittest to be engraved for a good Genius, and how to go invisible, and in several shapes.

20. Now, says he, by 11 a good Angel will come and make you invisible, and transform you into any of these shapes, a Boy, a Lamb, a

Dove, a beam of light; and the Spirit gets into the body, and by his subtle substance more operative and searching than any Æther, or lightning, melts the yielding compages of the body to such a consistency, and so much of it as is fit for his purpose, and makes it pliable to your imagination; and then it is as easy for him to work it into what shape he pleases, as it is to work the air into such forms and figures as he ordinarily doth; nor is it any more difficulty for an Angel to mollify what's hard, then it is to harden what is soft and fluid as the air.

21. And he that has this power, you can allow him that which is lesser, viz. to instruct men how they shall for a time forsake their bodies, and come in again: for can it be a hard thing for him that can thus melt and take in pieces the particles of the body, to have the skill and power to loosen the soul, a substance really distinct from the body, and separable from it, which at last is done by the easy course of nature at final dissolution of soul and body, which we call death; but no course of nature ever transforms the body of man into the shape of a Lamb, or a Dove; so that this is more hard and different from the course of nature then the other; I, you'll say the greatness and incredulity of the Miracle is this, that there should be an actual separation of soul and body, and yet no death: But this is not at all strange, if we consider that death is properly a disjunction of the soul from the body, by reason of the bodies unfitness any longer to entertain the soul, because of diseases or age.

22. But this is not such a miracle, nor is the body properly dead, though the soul be out of it; for the life of the body is nothing else but that fitness to be actuated by the soul, the conservation whereof is helped by aurum potable, and numbers engraved in Gold, Silver, precious Stones, and in Metals, which keeps out the cold, keeps in the heat

and spirits, that the frame and temper of the body may continue in fit case to entertain the soul again at her return; so the vital stem of the carcass being not spent, the pristine operations of life are presently again kindled, as a Torch new blown out, and yet reeking, suddenly catches fire from the flame of another, though at some distance, the light gliding down along the smoke.

23. Wherefore the flying in the air, walking in Ladies chambers invisibly, and bringing of messages from one lover to another, and discovering secrets, etc. it is easy; for they be then really out of the bodies: And Socrates laying in the field for quietness sake, being far from the noise of his brawling wife Zantippe, fell asleep, and being asleep, Euripides espied a thing come out of his mouth very lovely to behold, of a whitish Color, little, but made like a Cony running in the grass, and at last coming to a Brook side, very busily attempting to get over, but not being able, one of the standers by made a bridge for it of his sword, which it passed over by, and came back again with the use of the same passage, and then entered into Socrates his mouth, and they saw it no more afterwards; when he waked, he told how he dreamed he had gone over an iron bridge, and other particulars answerable to what Euripides and his fellows had seen beforehand; all those that transform themselves into Lambs, Doves, Bryes, or little Birds, or Conies, have their understandings unchanged, they have the mind and memory of a man as before.

24. Mistake me not; all that can do these miracles, are not Rosie Crucians; for many of the Witches and Sorcerers in Egypt could do miracles as well as Moses, who was taught of God as these Rosie Crucians are.

25.   To persuade you to the truth of numbers, when consecrated to God with Divine names, and engraved upon consecrated subjects, and what wonderful virtues they have in natural and supernatural things, I shall amplify and prove by Plotinus; for that which Porphyrius records of him, falls little short of a miracle, by the number 11 as being able by it engraved, as his enemy Olympius confessed, to retort that Magic upon him, which he practiced against Plotinus, and that sedately sitting amongst his friends, he would tell them; Now Olympius his body was gathered like a purse, and his limbs beat one against another. But your Witches, Sorcerers, Conjurers, and Enchanters are not able to stand before Rosie Crucians, no more than Iannes and Iambres could stand before Moses, who did really those things, and abundance more than the other could imitate by delusions, sleight and Legerdemain; and this proves the truth of Angels, fallacies of Devils: the one makes a happy man, the other makes him miserable. The Devil promised our Savior more than he could perform; but God performed more to Abraham then he promised: And to come again to Plotinus, although he was not instructed by the Jewish Priests and Prophets, yet he was a familiar friend of that hearty and devout Christian, and learned Father of the Church, Origen, whose authority I would also cast in together with the whole consent of the learned amongst the Jews; for there is nothing strange in the Metaphysical part of this *Holy Guide*, but what they had constantly affirmed to be true; but the unmannerly superstition of many is such, that they will give more heed to an accustomed opinion, which they have either taken up of themselves, or has been conveyed unto them by the confidence of some private Theologer, then to the authority of either Fathers, Churches, workers of Miracles, or what is best of all, the most solid reasons that can be propounded; which if they were capable of, they could not take any offence at the admittance of the Rosie Crucian Philosophy into this present *Holy Guide*; but the principles and most

notorious conclusions thereof, offering themselves so freely, and unaffectedly, and so aptly and fittingly taking their place in the Text, that I know not how, with Judgement and conscience, to keep them out.

26.   In an elected hour they engrave 11 in cast metal, and the numbers, Angels and Letters belonging to it; and this makes the bearer to gain in his trade, cures all diseases in the legs, viz. the Gout, etc. And to this appears an Angel like a beautiful man, that makes a man prosperous by Sea.

By this number they know times when to give Medicines, and how Devils offer themselves; by this number you shall know an Angel from a Devil, as you shall see in the third book all in order.

Chapter XIV.

The Number of Knowledge, of Dissolving Gold, & etc.

Of the signification of the Number twelve, of its natural virtue: twelve Magical Aphorisms of Janbosher: Of Angels, and their nature and dignity: What these Guardian Genii may be; whether one or more of them be allotted to every man, or to some none; what may be the reason of Spirits so seldom appearing: And whether they have any settled shape or no: What their manner is of assisting men in either Devotion, or Prophesy, or Love: Whether every man's complexion is capable of the society of a good Genius: And lastly, whether it be lawful to pray to God to send such a Genius or Angel to one or no, that in the Number and Name we desire at the engraving.

1. The Number twelve is Divine, and that whereby the Celestials are measured: It is also the Number of the signs in the Zodiac, over which there are twelve Angels as chief, supported by the Irrigation of the great Name of God: In twelve years Jupiter perfects his course, and the Moon runs through twelve signs in twenty-eight days or thereabouts. There are twelve chief joints in man's body, in hands, elbows, shoulders, thighs, knees, and vertebrae of the feet: there is also a great power of the Number twelve in divine Mysteries: God chose twelve Families, and set over them twelve Princes: So many stones were placed in the midst of Jordan, and God commanded that so many should be set on the breast of the Priest: twelve Lions did bear the brazen Sea that Solomon made: there are so many Fountains in Helim; and so many Spies sent to the Land of Promise; and so many Apostles of Christ set over twelve Tribes; and twelve thousand people chosen; the Queen of Heaven crowned with twelve Stars; and twelve angels are set over the twelve Gates of the City; and twelve stones of the Heavenly Jerusalem. In inferior things, many breeding things proceed from this Number: so, the Cony being most fruitful brings forth twelve times in the year; and the Camel is so many months in breeding, and the Peacock brings forth twelve Eggs, & there are 12 months in the year, as Virgil sings.

> How the Sun doth rule with twelve Zodiac Signs,
> The Orb that's measured roundabout with Lines,
> It doth the Heavens starry way make known,
> And strange Eclipses of the Sun and Moon;
> Arcturus also, and the Stars of rain,
> The seven Stars likewise, and Charles his Wain:
> Why Winters Sun makes toward the West so fast;
> What makes the Nights so long ere they be past?

2. And there are twelve Magical Aphorismes.

1. Ante emnia punctum extitit: non τὸ ἄτμον, aut Mathematicum, sed diffusivum, Monas erat explicite: implicite Myrias, Lux erat & mox principium, & finis principii, omnia, & nihil, est, & non.

2. Commovit se monas in Diade: & per triadem egressae sunt facies Luminis secundi.

3. Exivit ignis simplex, increatus: & sub Aquis indust se tegumento ignis multiplicis, creati.

4. Respexit ad fontem superiorem: & inferiorem deducto typo, triplici vultu sigillavit.

5. Creavit unum unitas: & in tria distinxit; trinitas eat & Quatenarius, nexus & medium reductionis.

6. Ex visibilibus primum effulsit Aqua: Faemina incumbentis ignis & figurabilium gravida mater.

7. Porosa erat interius, & corticilus varia, cujus venter bubuit Caelos convolutos, & Astra indiscreata.

8. Separatus Artifex divisit hanc in Amplas regiones, & à parente, faetu, disparuit Mater.

9. Peperit tamen Mater filios Lucidos, influentes in terram Chai.

10. Hi generant Matrem in noviseimis: cujus fons cantat in Luco Miraculoso,

11. Sapientiae condus est hic: esto qui potet, promus.

12. Pater est totius Creati: & cx Filio Creato vivam Pilii Analysin, Pater generatur, hahes summum Generantis Circuli Mysterium: Filii Filius est, qui Filii Pater fuit.

3. The Name of God among angels is spoke with twelve letters אב קיורלחקרש, Father, Son, Holy Ghost.

4. There are twelve Messengers that bear the Commands of God, and have influence on the nine Orders of Angels, and Quire of Blessed Souls, and ten Sephiroth into the Angels of the Celestial Spheres, and Angels of the Planets, and Planets themselves; and into the Angels of the Signs, whose names are these, Malchidiel, Asmodiel, Ambriel, Verchiel, Hanaliel, Zuriel, Barbiel, Adnarciel, Hanael, Gabriel, Barchiel; and these have influence upon the twelve Signs of the Zodiac, Aries, Taurus, Gemini, Cancer, Leo, Virgo, Libra, Scorpio, Sagittarius, Capricornus, Aquarius, Pisces: and from these Angels after this Order doth man receive a good genius, according to the Number of his Name, engraved in the Metal, or in one of these twelve Stones, a sardonius, a carnoel, a topaze, a calcedony, a jasper, emrald, the beril, an amethist, the hyacinth, a chrisoprasus, a christal, a saphir.

5. And there are twelve Tribes, twelve Prophets, twelve Apostles, twelve Months; twelve Plants, Sang-Upright, Vervain, Bending-Vervain, Comfry, Lady-Seal, Calamyn, Scorpion-Grass, Mugwort, Pimpernel-Dock, Dragonwort, Aristolochy: and twelve principal

Members, the head, the neck, the arms, the breast, the heart, the belly, the kidneys, the genitals, the hams, the knees, the legs, the feet: By the Number twelve Spirits appear that resolve all manner of Questions, as Janboshar, Adams Tutor says, and you may find it recorded in the Indian Books, written by Isagarith a hundred years before Adam. Now they say in old time the Months were called and reckoned for years; but this I leave to more curious pens, and pass on to my design: And if these things practiced be found true, and answer the expectation of the Reader, let him then say faithfully his thoughts of Art, neither scandalizing, nor smutting it with disgraceful words.

6. Now it cannot but amuse a man's mind to think what these Officious spirits should be, that so willingly by Numbers sometimes offer themselves to consociate with a man; whether they may be Angels uncapable of incorporation into human bodies, which vulgarly is conceived: or whether the souls of the deceased, they having more affinity with mortality and human frailty then the other, and so more sensible of our necessities, and infirmities, having once felt themselves, and separate souls are in a condition not unlike the Angels themselves.

7. But there are Angels in Heaven, that are set over man as Guardians, and their names you find by these Axiomata in the Numbers and Letters, as these willingly come to us. Now we are to inquire, whether every man has his Guardian Genius or no: that Witches have many, such as they are, their own confessions testify: The Pythagoreans were of opinion that every man has two genii, a good one, and a bad one, which Mahomet has taken into his Religion, adding also that they sit on men's shoulders with table-books in their hands, and that the one writes down all the good, and the other all the evil a man doth: But such expressions as these I look upon as Symbolical rather than Natural.

8.   And I think it more reasonable, that a man changing the frame of his mind, changes his genius withal, or rather unless a man be very sincere and single-hearted, that he is left to common providence; as well as if he be not desperately wicked, or deplorably miserable, scarce any particular evil spirit interposes, or offers himself a perpetual assistant in his affairs and fortunes.

9.   But extreme poverty, irksome old age, want of friends, the contempt, injury, and hard-heartedness of evil neighbors working upon a soul low sunk into the body, and wholly devoid of divine life, doth sometimes kindle so sharp, so eager, and so piercing a desire of Satisfaction and revenge, that the shrieks of men while they are a murthering, the howling of a Woolf in the fields in the night, or the squeaking and roaring of tortured beasts, do not so certainly call to them those of their own kind, as this powerful Magic of pensive and complaining soul in the bitterness of its affliction, attracts the aid of these our officious spirits; so that it is most probable, that they that are the quickest to hang Witches, are the first that made them, and have no more goodness nor true piety, then these they so willingly prosecute, but are as wicked as they, though with better luck or more discretion, offending no further than the Law will permit them; and therefore they severely starve the poor helpless man; though with a great deal of clamor of Justice, they will revenge the death of their Hog or Cow.

10.   And now it were worth our disquisition, why spirits so seldom nowadays appear, especially those that are good; whether it be not the wickedness of the present Age, as I have already hinted, or the general prejudice men have against all spirits that appear, that they must be straight-ways Devils, or the frailty of human nature, that is not usually able to bear the appearance of spirits, no more than other Animals are;

for into what Agonies Horses and Dogs are cast upon their approach, is in every ones mouth, and is a good circumstance to distinguish a real Apparition from our own imaginations.

11.   Or lastly, whether it be not the condition of spirits themselves, who, it may be, without some violence done to their own nature, cannot become visible, it being happily, as troublesome a thing to them to keep themselves in one steady visible consistency in the Air, as it is for men that dive to hold their breath in the water.

12.   Now although Spirits appear upon Numbers and Names engraved upon Metals, Minerals, or precious Stones, it may deserve our search, whether spirits have any settled form or shape: Angels are commonly pictured, like good plump boys, which is no wonder the boldness of the same Artist, not sticking to picture God Almighty in the shape of an old man: In both as it pleases the Painter.

13.   But this story seems rather to favor their opinion, that say, that Angels and separate souls have no settled form, but what they please to give themselves upon occasion by the power of their own fancy. Ficinus, as I remember, somewhere calls them aerial stars. And the good genii seem to me to be as the benign eyes of God running to and fro in the world, with love and pity beholding the innocent endeavors of harmless and single-hearted men, ever ready to do them good, and to help them.

14.   What I speak here of the condition of the soul out of the body, I think is easily applicable to other genii or Spirits; and this I conceive of separate souls and spirits.

Like to a light fast locked in Lanthorn' dark

Whereby by night our wary steps we guide
In shabby streets, and dirty channels mark
Some weaker rays from the black top doe glide,
And flusher streams perhaps through th' horny side;
But we've past the peril of the way,
Arriv'd at home, and laid that case aside,
The naked light how clearly doth it ray,
And spread its joyful beams bright at Summers day!
Even so the soul in this contracted state
Confin'd to these strait instruments of sense
More dull and narrowly do operate;
At this hole bears, the sight must ray from thence,
Here tastes, there smells, but when she's gone from hence,
And round about has perfect cognizance;
What e're in her horizon doth appear,
She is one Orb of sense, all eye, all Airy ear.

15. Now you know by the virtue of Name and Numbers how spirits appear; let us inquire how these good genii become serviceable to men, for either heightening their Devotions, or enabling them to Prophesy, etc. whether it can be by any other way then by descending into their bodies, and possessing the heart and brain: For the Euchites, who affected the gift of Prophesy by familiarity with evil spirits, did utterly obliterate in their souls the πατειχά σόμβελα, the principles of goodness and honesty (as you may see in Posellus, περι ενιργιας δαιμόνων that the evil spirits might come into their bodies, whom these sparks of virtue, as they said would drive away, but those being extinguished, they could come in and possess them, and enable them to prophesy.

16. And that the Imps of Witches do sometimes enter their own bodies, as well as theirs to whom they send them, is plain in the story of Witches in Trismegistus.

17. It is also the opinion of R. Lully, that these spirits get into the veins and Arteries both, of men and beasts.

18. Wherefore concerning the holy Rosie Crucians, it may be conceived reasonable, that the good genii insinuate themselves into their very bodies, as well as the bad into the bodies of the wicked; and that residing in the brain, and figuring of it by this or that object, as we ourselves figure it, when we think the external senses being laid asleep, those figurations would easily be represented to the common sense; and that memory in the Rosie Crucian E. T. recovering them when he awaked, they could not but seem to him as other dreams did, saving that they were better, they ever signifying something of importance unto him.

19. But these Raptures of Devotion by day, might by the spirits kindling a purer kind of love flame in his heart, as well as by fortifying and raising his imagination, and how far a man shall be carried beyond himself by this redoubled soul in him, none I think, can well conceive unless they had the experience of it.

20. And if this be their manner of communion, it may be enquired by this Number, whether all men be capable of consociation with these good genii. Cardan somewhere intimates that their approaches are reprehensible by certain sweet smells, where the mind doth not stink with pride and hypocrisy, have some natural advantage for the gaining their society. But if there be any peculiar complexion or natural condition required, it will prove less hopeful for everyone to

obtain their acquaintance; yet Regeneration comes to its due pitch: though it cannot be without much pain and anguish, may well rectify all uncleanness of nature; so that no singular good and sincere man can reasonably despair of their familiarity. For he that is so highly in favor with the King, it is no wonder he is taken notice of by his Courtiers.

21.   Some question these Numbers, and the virtues I attribute to them when engraved, whether God assists us or not, and whether it be lawful to pray to God for such a good Genius or Angel: But the examples of Enoch, Moses, Joshua, Elijah, Jeremiah, Ezekiel, Daniel, St. John Baptist, And St. John the Divine, with many others, as Rector Of Troy, Alexander, Julius Caesar, Judas Maccabeus, King Arthur Of England, Charlemagne, or Charles the Great, HUON Of BORDEAUX, GODFREY of Bulloyn, and thousands more I could name, seems a sufficient warrant.

22.   But I conceive faith and desire ought to be full sail to make such voyages prosperous, and our end and purpose pure and sincere; but if pride, conceitedness, or affectation of some peculiar privilege above other mortals, spur a man up to so bold an enterprise, his devotions will no more move either God, or the good genii, then the whining voice of a counterfeit will stir the affection of the discreetly charitable. Nay this high presumption may invite some real friends to put a worse jest upon him, then was put upon that tattered Rogue Guzman, by those Mock-spirits, for his so impudently pretending kindred, and so boldly intruding himself into the knowledge and acquaintance of the Gentry end Nobility of Genoa.

But the safest Magic is the sincere consecrating a man's soul to God, and the aspiring to nothing but so profound a pitch of humility, as not to be conscious to ourselves of being at all touched with the praise and applause of men, and to such a free end universal sense of charity, as to be delighted with the welfare of another as much as our own; they that solely have their eyes upon these, by Numbers and Names, will find coming in whatever their heart can desire; but they that put forth their hand to catch at high things, as they fancy, and neglect these, prove at last but a plague to themselves, and a laughing stock to the world.

In a convenient season they engraved the Number twelve in a white and clear stone, with the Letters of the name of the party, and the Genius, Angel, and Planet, etc. And this increases felicity, honor, and confers benevolence and prosperity, and freedom from enemies; and this Number cures all diseases in the feet: to this they say appears a Genius, whose figure is a man, having the head of a Lion, or a Ram, and Eagles feet, and he seems to be in Blue, and a flame of light attends him.

By this Number they know whether the Medicine will prosper or not.

Chapter XV.

Of what Angels appear by the virtue and power of Numbers above twelve.

1. In thirteen for the Agreement of Married Couples, and for the dissolving of the Charms against Copulation, they added the Numbers of their Wanes together, and divided them by mine, and the remainder

was engraved with thirteen upon a plate of Beril and Zedeck, and then a Genii would appear like a man and woman in white embracing.

2. Now the Numbers that are above twelve, you see are endowed with many and various effects and virtues, whereof you must understand by their originals and parts, as they are made of a various gathering together of simple Numbers, or manner of multiplication; sometimes as their significations arise from the lessening, or exceeding of another going afore, especially more perfect, so they contain of themselves the signs of certain Divine Mysteries, so you see the third Number above ten shows the Mystery of Christs appearing to the Gentiles, for the thirteenth day after his Birth a Star was a guide to the Magicians.

3. The fourteenth day doth typify Christ, who the fourteenth day of the first Month was sacrificed for us: upon which day the Children of Israel were commanded by the Lord to celebrate the Passover, 14. Matthew, doth so carefully observe, that he passed over some Generations, that he might everywhere observe this Number in the Generations of Christ. To cure the sick, they made this Number in Gold, and then an Angel would appear like the head of a Lion, and they would make a perfume of Amber.

4. The fifteenth number is a token of spiritual Ascensions, therefore the Song of Degrees is applied to that in fifteen Psalms. And fifteen years were added to the life of Hezekiah; and the fifteenth day of the seventh month was observed and kept holy: This Number they engraved with the man's name in Virgin Wax and Mastic, and then would appear a King crowned, before whom they would burn lignum aloes, and he would reconcile him with his King whom he offended.

5. The Number sixteen the Pythagoreans, Porphirians and Platonists call the Number of Felicity. It also comprehends all the Prophets of the Old Testament, and the Apostles and Evangelists of the New. They engraved this in a silver Ring, whose table was square, and then the Genius would appear in the shape of a woman well clothed, sitting in a chair, to whom they would burn musk, camphire, and calamus aromaticus. They affirmed, that she giveth happy fortune, and every good thing.

6. The Number seventeen is called of R. Lully, a Number of Victory; by it engraved with the Letters and Numbers of his Name, added together in red wax, appeared a Genius like a Soldier sitting on a Horse, holding a Pistol cocked in his right hand ready to fire; and they burnt red Earth and Storax before him: And this enabled Julius Caesar to come into this Kingdom of England.

7. By the number eighteen, Israel served Eglon King of Moab: your name and number engraved in Iron, they say will preserve you against Thieves and Robbers, for a Genius in the form of an Ape will attend.

8. By the Number nineteen engraved in Copper, appearing a woman holding her hands upon her face, and they burnt liquid Storax before it, that might facilitate birth, and provoke the Menstrues.

9. By the Number twenty, Jacob served, and Israel was sold; and amongst creatures that have many feet, there is none that have above twenty feet, and they say that this number engraved in Tin, with the Number of the Hunters name, will bring you a Genius like Sagittary, half a man, and half a horse, and before this they burnt a Wolfs head, and it made them prosperous in hunting.

10. By twenty-one, with the Number of the Kings name, for the destruction of his enemies, and to overcome Kingdoms, they engraved it in Gold, and finely wrought it, and then appeared a Genius in the image of a man, with a double countenance before and behind, and before this they burnt brimstone and Jet.

11. Twenty-two signifies the fulness of wisdom, and so many are the Characters of the Hebrew Letters, and so many books doth the Old Testament contain: by this Number engraved in silver, a little Virgin appears, and is reported to increase the light of the eyes, to assemble Spirits, to raise Winds, to reveal secret and hidden things.

12. Twenty-three, engraved with the man's name, and the Numbers of it, in a sapphire, makes appear the Genius of a man willing to make himself merry with Musical Instruments, and he makes a man honored before Kings and Princes, and helps the pain of the teeth, he bestows the favor of men and Aerial spirits.

13. Now I shall say nothing of twenty four, because it is evil, and giveth a Genius of a wicked man, whose name was Cain, and the name of any Spirit you may find by the number and name of the man, for what remains of Addition, and Division, tells you the number and the name of the spirits; I have told you of all men's names, what Angels rule them as you heard before; these numbers are said to be good and Prosperous, viz. 1, 2, 3, 4, 7, 9, 11, 13, 14; very good, 16, 17, 19, 20, 22, 23, 10, 26, 27; indifferent good, 5,6, 8, 12, 15, 18, 21; very ill, 24, 25, 28, 29, 30. Worst of all.

Chapter XVI.

Of Kings, Lords, or other People that fight, or go to Law One against another, which shall have the Victory.

1. Now we have showed you the power, virtue, end signification of numbers; we shall next teach you the Use of them: And first, you must know the proper names of them which would fight or go to Law one against the Other: and according to the letters and numbers in the second Chapter of this book: Join unto each letter of the said names the number that is attributed to it, and sum the said numbers together each man by himself, and divide the sum of each mane name by nine, and Judge by these Rules following; and if it fortune, that in dividing the whole by nine, there remain nothing; then the last number of nine must be it, you must add to this name.

2. And if the names be both one, the Numbers will be the same, as John against John; and you must remember to write the names in the Nominative Case singular.

3. And if one to one remain, then in combat, he that is of the lesser stature shall overcome the other, because the lesser loves Clamors, Seditions, Rebellion, Deceit, Strife, Debate, and is Captain of ill company, that strive to overrun and kill men, and by that means is feared; some men say, the younger shall overcome the elder; but I observe not that rule: the greater is a mighty man, strong and cruel, proud, and gives to fight; but yet he shall be hurt in the head; and the lesser shall have, the choice, of weapon, and overcome the other; in Law the lesser shall obtain the suit.

4.	Two to two, the greater shall have the Choice of weapon, and shall have the victory with long weapon, because he is noble and handsome, and of good reputation, and loves good company: The lesser is a man of good nature, and well beloved; but yet he loves to kiss in a corner, and therefore he shall be hurt in the face, and on the arm: In Law the greater shall Obtain this suit; and this trouble is or will be about women.

5.	Three against three; here the lesser shall choose and overcome with short weapon, because he is Princely, and full of spirit; but the greater is a poor soldier, that has nothing but his sword; he has been hurt in the arm, and is servant to them that have likewise been hurt, and have lost some of their limbs, and shall now be hurt in the stomach: In Law the lesser being witty shall obtain his suit.

6.	Four to four; now the greater shall have the choice of weapons, and shall have the victory with long weapon; he is lofty in his deeds, and takes pleasures in Arms, being very handsome and amiable in complexion, full of words, contentious: In Law-suit the greater shall obtain the inheritance of his father or mother, or the goods of the other that is in controversy, with him; is one of wisdom, beauty and policy, and well beloved, yet by deceit and treason would he beguile this friends, but he fails.

7.	Five to five; then the lesser stature shall choose the weapon, and overcome, with short weapon; yet the party is listless; and weak in generation, however honest, and therefore, the greater shall be hurt on the side, and on the head, and shall surely dye, because he loves Unjust quarrelling in the Law, the lesser shall in two Terms obtain his suit.

8. Six to six; again, the taller shall overcome, but the lower shall choose his weapon, the other is an ingenious man, full and active of body, a lover of good clothes, Guns, Crossbows, Horses and Harness.

9. Seven to seven; and again, the lesser shall vanquish with the choice of weapon, which is short; she is a great Lady, angry, and a fighter, and seeks nothing but strife and quarrels, a favorer of Rectors, and men of War for her defense, and to be maintained by them in her controversies; and the greater shall be shot in the arm, and hurt in the head and stomach.

10. Eight to eight; the greater shall overcome with long weapon, & the lesser shall have the choice of weapon, the greater is a very fawning deceitful Knave, full faced and bodied, of a brown hair, much given to Witchcraft, charms and Enchantments; a great embracer of women, and therefore shall hurt the lesser in the belly, side and knee: In Law the greater shall obtain his suit; and there is like to be murther, for the lesser is a good man of countenance and condition, and loves good clothes, but high spirited; and so there is like to be blows given, with more loss to him that shall win the suit then it is worth.

11. Nine to nine; here the lesser shall have the choice of weapon, and beat the other with short weapon; he is a man very noble in his actions, aims and high things, with a little pride; the other is one has great power to do evil, applying himself unto nothing but revenge to murder and slaughter, and to rob and deceive: A Phanatick Anabaptist in the fear of God will cut your throat, he shall be for all his cunning hurt on the knee, and on the side: In suit the lesser shall obtain without trouble, and they go to Law for Heritages, or Woman's apparel found; the lesser shall be content to take part rather than trouble.

12.   1 to 2, the 2 shall have the choice of weapon, and overcome one with short weapon, and he shall be hurt in the head for all his gallantry, and dye thereof. In Law one shall win, and have more favor in his suit then he looks for; and this suit is brought for Gowns, Garments, and women's money.

13.   One to three; one is a man will choose the long weapon and beat the other; three shall be hurt on the arm, and on the stomach: In Law one shall obtain the suit, the declaration is upon bond or debts.

14.   One to four; here four shall chase the field and day of battle and overcome his enemy with long weapon; and one shall be hurt at the heart: In Law four shall by deceit obtain his suit, which is about Succession or inheritance.

15.   One against five; The first shall choose the field and day of battle, and overcome five with short weapon: In Law one shall obtain his suit, which is some gift of a Lord or Knight.

16.   One to six; here six shall beat one, and hurt him in the belly and head with a long weapon, and imprison him. And in Law six shall obtain their suit, which is brought about with money or merchandisable wares; and at last they shall be friends.

17.   One to seven; in this business one shall beat seven with a short weapon, and hurt him on the side, although he had the choice: In Law one shall obtain his suit about women, or marriage goods; and the suit shall be long.

18.   One against eight; here one shall be hurt on the side, and in the genitals; with long weapon, because he is poor and malicious: he will

be long angry, he is hurt or blemished in his face, one of his eyes is out, one of his members is cut away, and he is an old man: and eight also is poor, perplexed, proud and sorrowful, having one of her members cut away: in Law eight shall obtain his suit, which is about goods and household-stuff, and such things as are unmovable, as houses and the appurtenances, etc.

19.   One to nine; one shall have the choice of weapon, and beat nine, and hurt him on the side: In Law one shall win, and the suit is about honor, preferment, or some preeminence.

20.   Two to three; here three shall be beat with short weapon, and two shall be hurt on the arm: In Law 3 shall obtain his suit by the help of the Parson of the Parish, because it is amongst kindred about heritages.

21.   Two to four; now two shall have the better in fighting, and hurt four on the stomach, and on the arm: In Law two shall win by the subtilty of his Attorney.

22.   Two against five; here five shall in fighting beat two, and hurt him in the flank and breast: in Law five shall obtain the suit, which is about women and women's clothes.

23.   Two to six; here two shall overcome and beat six with long weapon, and hurt him in the body, and on his shoulders; In Law two shall obtain his suit without any great trouble; two is a man delighting in strife and contention, and unjustly he practices, to take away the goods of others; six is a fine fellow, well made, and of good countenance, proud, and therefore they will agree, and the suit is about Mer-

chandise, or money lent; for it is better to please a knave then an honest man.

24.   Two to seven; now seven will beat and hurt two on the side, although two shall have the choice of weapon, yet he shall be overcome: In Law seven shall obtain his suit by delays; two keeps company with martial men, or with little men, that have their eyes sunk in their heads, and a small beard, ready to do a mischief; wherefore seven will easily be content to be quiet.

25.   Two to eight; here two shall have the choice of weapon, and beat eight, and hurt him on the stones and bottom of his belly: In Law two shall obtain his suit, and he is a man sad and pensive, punishing his body by extreme melancholies, and he loves to dig in the earth, for to find treasure: and therefore two shall be assisted by men of small stature, counterfeit, and as it were monstrous: eight is a man very happy, and taketh nothing in hand, but it shall turn to his praise; but here it happens eight shall have good right, but that the subtilty of two is of such force, that eight shall lose: and this suit is for moveable goods.

26.   Two to nine; now nine shall beat and hurt two at the heart with short weapon: In Law nine shall obtain his suit, which is about gifts or goods of the dead.

27.   Three to four; by short weapon, four shall beat three, and hurt him in the head and arms: In Law four shall obtain his suit, which is about his Fathers goods, and his kinsfolk would beguile him.

28. Three to five; by short weapon three shall hurt five on the side or shoulder, at last friends: In Law three shall obtain his suit, and after they shall be friends.

29. Three to six; with long weapon six shall hurt three in the belly: In Law six shall obtain his suit, which is for merchandise.

30. Three to seven; here three shall beat seven, and hurt him in the leg and arm: In Law the suit shall be long, yet at last obtained by three.

31. Three against eight; by long weapon three shall be hurt on the body and entrails by eight: In Law eight shall obtain his suit, which is about the apparel, dowry, or things of woman.

32. Three to nine; now with short weapons three shall hurt nine in the head, whereof be shall die: In Law three shall obtain his suit by the help of some Lords of the Kings Court.

33. Four to five; Here five shall beat four and kill him: In Law five shall obtain his suit, because he is an honest man in heart, and the suit is about goods given by the Prince.

34. Four to six; with long weapon, six shall hurt four in the body: In Law four shall obtain his suit, which is for money or merchandise.

35. Four to seven; now seven will beat four with short weapon, and hurt him on the knee and face: In Law seven shall obtain his suit.

36. Four to eight: with short weapon four shall hurt eight in the breast and arms: In Law four shall obtain his suit.

37. Four to nine; here nine shall beat four, and hurt him in the side: And in Law nine shall obtain his suit.

38. Five to six; by long weapon six shall hurt five on the head and face: In Law six shall obtain his suit, which is for money lent.

39. Five to seven; here five shall kill seven: In Law five shall win his Fathers inheritance.

40. Five to eight; with long weapon five shall be hurt in the side, and in the hands: In Law eight shall obtain his suit.

41. Five to nine; now five shall hurt nine on the shoulder: In Law five shall obtain his suit.

42. Six against seven; here seven shall have the choice of weapon, and hurt six on the head: In Law seven shall win, and the suit is for Merchandise.

43. Six to eight; here six shall be hurt on the head by eight: In Law six shall obtain his suit; they be both good men, and will at last be friends.

44. Six to nine; now nine shall beat six and hurt him on the arm, and on the leg: In Law nine shall obtain his suit.

45. Seven to eight; here eight shall beat and hurt seven on the breast and heart: In Law eight shall obtain his suit, which is for garments, or moveable goods left by kindred.

46. Seven against nine; by short weapon nine shall be hurt in the face: In Law nine shall obtain his suit.

47. Eight against nine; with nine eight shall be overcome, and hurt in the body: In Law nine shall obtain his suit.

> A Rule abridged, to know which of the two that fight, or go to Law, shall have the Victory.

♂ in ARIES 1, against ☉ in ARIES 1, the lesser conquers.
♀ in TAURUS 2, against ☽ in TAURUS 2, the bigger conquers.
♂ in GEMINI 3, against ☿ in GEMINI 3, the lesser conq.
♃ in CANCER 4, against ☽ in CANCER 4, the greater conq.
♀ in LEO 5, against ☉ in LEO 5, the lesser conq.
☿ in VIRGO 6, against ♀ in VIRGO 6, the higher conq.
♀ in LIBRA 7, against ♄ in LIBRA 7, the lesser conq.
♂ in SCORPIO 8, against ☿ in SCORPIO 8, the greater conq.
♃ in SAGITTARIUS 9, against ☽ in SAGITTARY 9, the lesser conq.
♀ in TAURUS 1, against ☽ in TAURUS 2, two shall conq.
♂ in GEMINE 1, against ☿ in GEMINI 3, 1 shall conq.
♀ in LEO 1, against ☉ in LEO 4, 4 shall conquer.
☉ in ARIES 1, against ♂ in ARIES 5, 1 shall conq.
☿ in VIRGO 1, against ♀ in VIRGO 6, 6 shall conq.
♀ in LIBRA 1, against ♄ in LIBRA 7, 1 shall conq.
♂ in VIRGO 1, against ☿ in VIRGO 8, 8 shall conq.
♃ in SAGITTARIUS 1, against ☉ in SAGITTARIUS 9 conq.

♂ in GEMINI 2, against ☿ in GEMINI 3, 3 shall conq.
♃ in CANCER 2, against ☽ in CANCER 4, 2 shall conq.
♄ in LIBRA 2, against ♀ in LIBRA 5, 5 shall conq.
♀ in SCORPIO 2, against ☿ in SCORPIO 6, 2 shall conq.
☿ in GEMINI 2, against ♂ in GEMINI 7, 7 shall conq.
♄ in CAPRICORN 2, against ♂ in CAPRICORN 8, 2 shall conq.
♀ in LEO 2, against ☉ in LEO 9, 9 shall conquer.
☉ in ARIES 3, against ♂ in ARIES 4, 4 shall conq.
☿ in GEMINI 3, against ♂ in GEMINI 5, 3 shall conq.
♀ in VIRGO 3, against in ☿ in VIRGO 6, 6 shall conq.
♄ in AQUARIUS 3, against ☿ in AQUARIUS 7, 3 shall conq.
♀ in VIRGO 3, against ☿ in VIRGO 8, 8 shall conq.
♂ in ARIES 3, against ☉ in ARIES 9, 3 shall conq.
♀ in LEO 4, against ☉ in LEO 5, 5 shall conquer.
☿ in VIRGO 4, against ♀ in VIRGO 6, 4 conquers.
♄ in CAPRICORN 4, against ♂ in CAPRICORN 7, 7 conq.
♃ in SAGITARIUS 4, against ☉ in SAGITTARIUS 8, 4 conq.
☿ in GEMINI 4, against ♂ in GEMINI 9, 9 conq.
☉ in LEO 5, against ♂ in LEO 6, 6 conquers.
♀ in ARIES 5, against ☉ in ARIES 7, 5 conquers.
♃ in SAGITTARIUS 5, against ☉ in SAGITT. 8, 8 conq.
☿ in GEMINI 5, against ♂ in GEMINI 9, 9 conquers.
♂ in ARIES 6, against ☉ in ARIES 7, 7 conquers.
☿ in VIRGO 6, against ♀ in VIRGO 8, 6 shall conq.
☿ in GEMINI 6, against ♂ in GEMINI 9, 9 conq.

☉ in LEO 7, against ♀ in LEO 8, 8 conquers.

☉ in ARIES 7, against ♂ in ARIES 9, 7 conquers.

☿ in VIRGO 8, against ♀ in VIRGO 9, 9 conquers.

And these be the reasons of the Rules going before, which you must observe in every Medicine you make.

Another Rule more brief, according to the Numbers and Names going before.

These Numbers				The Conquerors of		The Conqueror is of			
2	4	6	8		1		3	5	7	8
3	5	7	9		2		1	4	6	8
1	4	6	8		3		2	5	7	9
2	5	7	9		4		1	3	6	8
1	3	6	8		5		2	4	7	9
2	4	7	9		6		1	3	5	8
1	3	5	8		7		2	4	6	9
2	4	6	9		8		1	3	5	7
1	3	5	7		9		2	4	6	8

Unity is ascribed to the Sun, 2 is ascribed to the Moon, 3 ascribed to Jupiter, Sol and Venus, 4 is of the Sun, 5 is ascribed to Mercury, 6 is attributed to Venus and Juno, 7 belongs to Saturn, 8 is attributed to Jupiter and Vulcan, Cybele and Bacchus; some attribute it to the three Ladies of Destiny; 9 belongs to the Moon, and the nine Muses; 10 belongs to the Sun and Janus; 11 is attributed to the Moon, 12 is attributed to the World.

Chapter XVII.

The Resolutions of all manner of questions, and how by these Numbers you may be happy, etc.

1. Whether a person shall live long, or not.
2. If a person shall be healthful or sickly.
3. If one shall find the party at home one would speak with.
4. Whether one absent be dead or alive.
5. Whether a Ship shall come home safe.
6. If a man shall be rich.
7. If reports be true or false.
8. If find again the thing lost.
9. If a man shall enjoy the Estate of his Father.
10. If it be good to hire or take the Farm or House desired.
11. If good to remove from one house to another.
12. If one shall have Children.
13. Whether the Father be dead or not.
14. If the Child be right Fathered, or a Bastard.
15. Whether a Town Besieged shall be taken.
16. If there be any ill company in the way a man would go.
17. If it be good to put on new Clothes.
18. If a promise made shall be performed.
19. If the Earth shall bring forth plenty of fruits, or not.
20. If a sick party shall live or die.
21. If a servant shall get free from his Master.
22. If it be good to take Physick.
23. If it be good to visit the sick person, or not.
24. If a man shall marry.
25. If he shall marry well or ill.

26. If a man be wise, or a fool.
27. If a woman be rich or not you would marry.
28. If agree after Marriage, or not.
29. Whether a Damsel be a Maid, or no.
30. Whether a Woman be honest to her Husband, or not.
31. If beasts lost, be dead or alive.
32. Whether a Thief shall be taken, or not.
33. Whether the thing lost be stolen, or not.
34. If a City, Town, or Castle shall be taken, or not.
35. If a party absent be dead or alive.
36. Whether the man shall die a good death.
37. If the Wife's Portion shall be obtained.
38. If it be good to call Angels in matters of Love, or not.
39. If the Spirit be good or evil, that appears, and whether it be an Angel of Heaven; or a Devil of Hell.
40. If the wind shall blow fair.
41. If get the Philosophers Stone, etc.
42. If Dreams be for good or evil.
43. If the Parson shall obtain the Benefice or not.
44. If one shall obtain the preferment desired.
45. If it be good to go to Battle, or no.
46. If the King, Pope, Prince, or Lord sick, shall amend, or no.
47. If love betwixt two shall continue: If it be good to go to the Court or not.
48. If thy Friend be faithful, or a Traitor.
49. If one shall be imprisoned or not.
50. If a party be bewitched, or not.
51. Whether one shall enter into the favor of the King.
52. If the Prince shall have the Victory in War.
53. If there shall be peace betwixt England and France.
54. If the Captain be valiant, or not.

55. If the Horse shall win the Race.
56. If a Prisoner shall come out of prison.
57. If a sickness shall be a long or short one.
58. If you shall enjoy the woman desired.
59. If it be good to take a journey.
60. If the child shall be fortunate or not.
61. If it shall be a plentiful year.
62. If it be good to trade in Merchandise.
63. If it be good to take a Wife.
64. If friendship shall take good effect, or not.
65. If a man shall be fortunate in house.
66. If a man has secret Enemies, or not.
67. The way to Happiness, and how to obtain it, etc.
68. The Prolongation of Life.
69. The Restitution of Youth in some Degree.
70. The Retardation of Age.
71. The Curing of Diseases counted Incurable.
72. The Mitigation of Pain.
73. Easier and less Loathsome Purgings.
74. The increasing of Strength and Activity.
75. The increasing of Ability to suffer Torture or Pain.
76. The Altering of Complexions: And Fatness, and Leanness,
77. The Altering of Statures.
78. The Altering of Features,
79. The Increasing and Exalting of the intellectual Parts.
80. Versions of Bodies into other Bodies,
81. Making of New Species.
82. Transplanting of Species into another.
83. Instruments of Destruction, as of War, and Poison.
84. Exhilaration of the Spirits, and Putting them in good Disposition.

85. Force of the Imagination, either upon another Body, or upon the Body itself.
86. Acceleration of Time in Maturations.
87. Acceleration of Time in Clarifications.
88. Acceleration of Putrefaction.
89. Acceleration of Decoction.
90. Acceleration of Germination.
91. Making Rich Composts for the Earth.
92. Impressions of the Air, and raising of Tempest.
93. Great Alteration; as in Induration, Emollition, & etc.
94. Turning crude and watery Substances into Oily and Unctuous Substances.
95. Drawing of New Foods out of Substances not now in use.
96. Making new Threads for Apparel; and new Stuffs; such as are Paper, Glass, & etc.
97. Natural Divination.
98. Deceptions of the Senses.
99. Greater Pleasures of the Senses.
100. Artificial Minerals and Cements.

All which you shall find in the Books, in order; first choose a Number, and Telesmatically engrave it at a convenient time for your work; elect a proper hour, and you cannot after err, but perform incredible, extraordinary things; understand well this book, for the easier opening the rest, and God prosper the work.

And this you may do of all other Questions, whereof you would be resolved.

And now that you may better understand this Figure, and all things, and the Resolution of the demands you would propound, you must first of all choose a Number, what you will at your discretion, as five, seven, or nine, or any other more or less; this done, take the Number of the day, as you shall find in order, and then take the Number which you find in the second Chapter, or that you find in the Globe upon the first Letter of your Name, as you were Christened.

For example, if your Name be Francis, you must take F. and the number which is over it, and you shall find all in order in the Schema; and gather all those Numbers into one sum, and divide them by thirty, reserving the rest as remains; and search in the Figure; and if you find it above in the upper half, your matter shall speed well; and if it be in the neither half, it shall be evil: And thus you may know all that you desire, and be it love which alters the Humor, as Ulysses was altered by the Music of his Mistress.

> When to her Lute Penelope sings,
> Her voice enlivens the leaden string,
> But when of sorrows she doth speak;
> Even with her sighs the strings do break;
> And as her Lute doth live or die,
> Led by her passions, so do I.

For to know whether you shall enjoy your Love, or not; take the number of the first letter of your Name, the number of the Planet, and of the day of the week, and all these Numbers ye shall put together, and then divide them by thirty, as you did before, and take your remainder, and see in the upper part, if it be there, you shall have your request; if it be in the nether half, it is contrary: And thus may you be resolved of all things you would know; you must observe the Numbers in the Figure exceed not thirty, as you shall find them beginning with one, two, three, and four, and so consequently to thirty.

The Numbers of the Planets, and their Characters.

7.	3.	9.	10.	6.	5.	2.
Saturn	Jupiter	Mars	Sol	Venus	Mercur.	Luna
♄	♃	♂	☉	♀	☿	☽

Numbers of the days of the Week.

Sunday	106
Monday	52
Tuesday	52
Wednesday	102
Thursday	31
Friday	98
Saturday	45

Thus, have we showed you the Numbers of the Planets, and the days of the Week, and their Numbers. Now that nothing may be wanting to this Art, here follows the names of the Idea's, Rulers, and Angels thereof, according to the Method of God.

EHEIA, JOD, JEHOVA, JEHOVA ELOHIM, EL, ELOHIM GIBOR אביבלוד יהוקש, ELOHA, JEHOVA, SABAOTH, ELOIM SABAOTH, SAIDAI, ADONOY MELEY.

KETHER, HOCHMAH, BENAH, HESED, GEBURAH, ZEPHERETH, NEZAH, HOD, JESOD, MALCURH.

AMBRIEL, ASMODEL, MALTHIDIEL, METT, BARCHIEL, CAMBIEL, HANAEL.

*** ♊ *** ♉ *** ♈ ***. JOPHIEL, *** ♒ **** ♑ ***.

6. 5. 1. 3. 2. 1. ♄ ZAPHKIEL, 900. 800. 700. 600. 500. 400.

ר ת ר כ כ א י ♃ ZADKIEL ד ו ה ו ה ד
400. 500. 600. 700. 800. 900. ♂. CAMAEL, 1. 2. 3. 4. 5. 6.

Υ Φ Χ Ψ ⊙ ANIMA MUNDI, A B Γ Δ E
1. 2. 3. 4. 5. 6. ♀ HAVIEL, 7. 8. 9. 10. 11. 12.
α β γ δ ε ζ ☿ MICHAEL, η θ ε κ λ μ .
400. 500. 600. 700. 800. 900. GABRIEL, 1. 2. 3. 4. 5. 6.
Y Z I V Hi Hu   Issim, A B C D E F.

ALECTO, MAGERA, ARACUS, ACTEUS, MAGALEZIUS, LUCIFER, READAMANTUS, NICON, LICOS, MIMON, CTESIPHONE.

DAPSA, BEEMOTH, OGIA, LEVIATHAN, CORRITIA, OPHALIA, EGIN, THESMEPHORIA, AMAIMON, HORMA.

ADVACHIEL, ENEDIEL, ZURIEL, HANALIEL, VERCHIEL, MARIEL.

* ♐ **** ♏ *** ♎ ****** ♍ **** ♌ **** ♋ ***.

300. 200. 100. 90. 80. 70. 60. 50. 40. 30. 20. 10. 9. 8. 7.

ש.ר.ק.צ.פ.ע.ס.נ.מ.ל.כ.י.ט.
ת.ח.

7. 8. 9. 10. 20. 30. 40. 50. 60. 70. 80. 90. 100. 200. 300.

Z.H.Θ.I.K.Λ.M.N.Ξ.O.Π.P.Σ.T.

13. 14. 15. 16. 17. 18. 19. 20. 21. 22. 23. 24.

ζ.ε.ο.π.ρ.σ.τ.υ.φ.χ.τ.ω.

7. 8. 9. 10. 20. 30. 40. 50. 60. 70. 80. 90. 100. 200. 300.

G. H. I. K. L. N. N. O. P. Q. R. S. T. U. X.

And this will wonderfully advance your preparations and knowledge of diseases.

Here lies a wonderful virtue, worth, and efficacy in Numbers, as well to good, as to bad; and they say, Angels say he as frequently conversed with as Devils, by the direction and help of the Figure before; and the eminent Philosophers do unanimously teach, and learned Doctors, both in Divinity, in the Law, and Doctors of Physick, and in occult mysteries in Chemistry, and in Rosie Crucian secrets practice.

As St. Hierom, Austin, Origen, Ambrose, Gregory Nazianzen, Athanasius, Baesilius, Hillarius, Rubanus, Bede, and many more, as R. Lully, Diodorus Siculus, etc. confirm. Hence Hillarius in his Commentaries upon the Psalms testifies, that the seventy Elders, according to the efficacy of Numbers, brought the Psalms into order: Rabanus also a famous Doctor, composed an excellent book of the virtues of Numbers. But now how great virtues Numbers have in nature, is manifest in the herb which is called cinquefoil, i.e. five leaved-grass, for this resists poison by virtue of the Number five; it drives away Devils, conduces to expiation, and one leaf of it taken two times in a day in Wine, cures the Fever of one day; three the Tertain Fever, four the Quartan; in like manner four grains of the seed of turnsole being drunk, cures the Quartan, but three the Tertian: In like manner Vervain cures Fevers, being drunk in Wine with aurum potable; and the third joint cures the Tertian, the fourth the Quartan; a Serpent if he be once struck with a spear teeth; if twice, recovers strength.

These and many others we read of in several Authors; we must know now whence these are done, which certainly have a cause, which is a various proportion of various Numbers amongst themselves: there is also a wonderful experiment of the Number of seven, that every seventh Male born without a Female coming betwixt, has power to cure the Kings Evil by his touch alone, or word: Also every seventh Daughter that is born, Rosie Crucians say, wonderfully helps forward the birth of Children; and so doth the Sun give the like virtue to aurum potable, as Dr. Culpepper often experienced; neither is the natural

Number here considered, but the formal consideration that is in the Number: And these Numbers are not in vocal, or Numbers of Merchants, buying and selling, but in rational, formal and natural: These are distinct Mysteries of God and Nature; but he that knows how to join together the vocal Numbers and natural with divine, and order them Telesmatically into the same harmony, shall be able to work, and know wonderful things, as the Rosie Crucians have said this Book teaches. The Rosie Crucians prognosticate many things by the numbers of names, and you must know, that simple Numbers signify Divine things: Number of ten Celestial, number of an hundred Terrestrial, number of a thousand, those things that shall be in future age; besides seeing the parts of the mind are according to an Arithmetical mediocrity, by reason of the identity, or equality of excess, coupled together; but the body, whose parts differ in their greatness, is according to Geometrical mediocrity compounded: but an Animal consists of both, viz. soul and body, according to the mediocrity which is suitable to Harmony: Hence it is that Numbers do work very much upon the Soul, Figures upon the Body, and Harmony upon the whole Animal: And one says Numbers

> Have in their natures a most fiery force,
> And also spring from a celestial source.

God gave to man mind and speech, which are thought to be a gift of the same virtue and immorality: The Omnipotent God has by his Providence divided the speech of men into divers languages, which languages have, according to their diversity, received divers and proper characters of writing, consisting in their certain order, number, and figure, not so disposed by chance, nor by the weak judgement of man, but from above, whereby they agree with the celestial and divine bodies, and virtues; but before all motes of languages the HEBREW is most sacred in the figures of characters, points, of vowels, and tops of accents, as consisting in matter, form, and spirit.

The position of the stars being first made in the seat of God, which is Heaven, after the figures of them are most fully formed the letters of the celestial mysteries, as by their figure, form and signification, so by the numbers signified by them, as also by the various harmony of their conjunction; he therefore that will find them out, must by each joining together of the Letters so long examine them until the voice of God is manifest, and the framing of the most sacred Letters and their Numbers be opened and discovered; for hence voices and words have efficacy in Magical works, because that in which Nature first exercised efficacy, is the Voice of God: But of these you may read largely in my *Temple of Wisdom*, a Book of Telesmet and Geomancy.

The Letters in the Figure of the World going before, have double Numbers of their Order, viz. extended, which simply express of what number the Letters are, according to their Order, and collected, which recollect with themselves, the Numbers of all the preceding Letters; also they have integral numbers, which result from the name of Letters, according to their various manner of numbering, the virtues of which numbers he that shall know, after our *Axiomata*, shall be able in every tongue to draw forth wonderful mystery, by their Letters engraved, call down Angels, Spirits, and Souls of men. And Eugenius brings in a Rosie Crucian, that brought him acquainted with Ethereal men, and him doth Theodidactus thus bring in speaking of himself.

> Force me befits, with this thick cloud I drive,
> Toss the blue Billows, knotty Oaks up rive;
> Congeal soft snow, and beat the earth with hail,
> When I my brethren in the air assail,

> For that's our field; we meet with such a shock
> That thundering skies with our encounters rock,
> And cloud-struck-lightning flashes from on high,
> When through the top of all the world I fly,
> I force death in her hollow caves, I make
> The Ghosts to tremble, & the ground to quake.

Solomon knew by the *Axiomata* how the world was made, and the operation of the Elements, the beginning, ending, and the midst of times, the alterations of the turning of the Sun, and the change of seasons, the circuit of years, and the position of Stars, the natures of living Creatures, and the furies of wild Beasts, the violence of winds, and the reasoning of men, the diversities of plants, and the virtues of roots; what things have been past, and what things are to come. There are also other mysterious Truths; Happiness, Knowledge, long Life, Health, Youth, Riches, Wisdom and Virtue; how to alter, change, cure and amend all Diseases in young or old, and the Art of preparing Rosie Crucian Medicine, and their Rules to raise the dead; all which they have experienced and fitted to the several Complexions of men. But I shall teach you these in the following Book. Wherefore according to the Doctrine of our LORD AND SAVIOR JESUS CHRIST, FIRST SEEK YOU THE KINGDOM OF God, AND ALL THESE THINGS SHALL BE GIVEN YOU.

Si Tu JEHOVAH, DEUS MEUS, ILLUMINAVERIS ME, LUX FIENT TENEBRA MEAE.

(End of The Holy Guide, Part 1)

The Holy Guide, Part 2

> To The
> Truly Noble
> By All Titles,
> Sir John Hammer,
> Baronet.
> Sir,

Your Worthiness and grateful acceptance of this kind of Learning, which I promised your honorable self, I would put forth, is now flown to your Temple of Safety, Knowledge, Perfection, or acquired parts for refuge and protection, from the wickedness of itinerant scandalous Pulpit Sycophants, school-sophisters, and some of my own profession,

Lawyers: I mean the Phanatick Rabble of Gownmen, that rage against the King and Bishops, whom God preserve out of their power: these contend against me continually, and contemn that which they do not know. But take heed ye unwise among the people; O ye fools, when will ye understand? they Judge they know not what, and condemn without evidence. This *Holy Guide,* which about ten years past, with some others of affinity thereto, for my private exercise and satisfaction I had at leisure, composed; which being communicated unto one, it became common unto many; and was by transcription successively corrupted, until it arrived in a most depraved copy to Doctor Nicholas Culpepper, and from him many had Copies, which Bone highly esteemed, and others abused: it came to pass about seven years past, I showed my true Virgin invention in manuscript to the learned Mathematician Mr. John Gadbury, who was then in company with

Captain George Whorton and other Gentlemen, of which one had a Copy, but imperfect; and therefore knowing me to be the Author, intreated me to publish mine: I suspected my ability, because it was set down many years past, when I was very young, and was the sense of my Conceptions at that time, not an immutable law unto my advancing Judgement at all times; and therefore there might be many things therein plausible unto my passed apprehension, which are not agreeable to my present self; therefore unwilling any work of mine should be printed. But at last I was persuaded; Now the enemies of King Charles & the Bishops, very proudly, with full mouth, bitter hatred, envy, malice and calumnies, hindered me from putting of it forth. Hence I began to be at a stand, whether I should put forth the rest of the book, or no; whilst I did doubt that I should by this means expose myself to public censure, and as it were cast myself out of the smoke into the fire, a certain rude fear seized upon me, lest by putting then forth I should seem more offensive then officious to you, and expose your Worship to the envy of malicious carpers and tongues of detractors, whilst these things trouble me with a various desperation, the quickness of your understanding, exact discretion, uprightness of Judgement, Religion without Superstition, and other most known Virtues in you, your authority and integrity beyond exception, which can easily check and bridle the tongues of slanderers, removed my doubting, and informed me to set upon that again more boldly, which I had almost left off, by reason of despair: Therefore (most honored Sir) take in good part, this book, in which we show the mysteries of Astromancy and Geomancy, Art and Nature, Celestial and Terrestrial, all things being opened and manifested; which experienced Antiquity makes relation of, and which came to my knowledge, that these Secrets of R. Crucis (hitherto neglected, and not fully apprehended by men of later tines), Nay with your protection be by me, after the showing of Natural Virtues proposed to them that are studious and curious of these secrets: by which let him that shall be profited, and receive benefit, give you the thanks, who have been the occasion of this publication, and setting of it at liberty to be seen abroad, wearing the honorable Title of

May 1.

    1662. SIR,

Your humble Servant,

John Heydon.

Book III. Chapter I.

The Way to Long Life.

1. How to make one live two hundred years; 2. John Macklain's our Country-man and others; 3. Policy to prevent occasions; 4. Helps from Egypt and Arabia; 5. Nothing can beget and work upon itself; 6. The heads of doing causes; 7. The wisdom of God; 8. A team of Fire; 9. Moistness; 10. Of male and Female stuff; 11. Mixtures; 12. Of the stuff clothed with wind; Clean air and heat of Heaven; 14. The Secret heat; 15. The starry fire and fat Æther; 16. Earth and Water; 17. lire and Fire; 18. Differences of heads; 19. Of Hair and Hoof; 20.Examples; 21. Of making and perishing; 22. The means to Long Life; 23. The food of Life; 24. The cause of Long Life; 25. The truths of Nature; 26. The Justice of God and End of Man; 27. Natural Mysteries; 28. Of the clearness of man's body; 29. The Justice of Nature; 30.The ways of Nature; 31. Methusalem; 32. 1 long Race; 33. Helps to Long Life; 34. The life of Giants; 35. King Argathon's life; 36. Plato's Commonwealth; 37. enacted by the Law of Nature, what, etc.; 38. Th. food of Stars; 39. Hungry spirits; 40. Mixt bodies, and their four enemies; 41. The changeable world and course of Creatures; 42. Natural means to Long Life; 43. Soul, Life and Heat of natural things; 44. Of the Element of Fire; 45. Of the nature of Æther; 46. Of the food of Æther; 47. Of the unseen first Moisture and Being of Life; 48. Of the first stuffs of the fine Oil of the food of life; 49. Of a plains pattern of adjournment of life; 50. Natures pattern not counterfeit, or the blood and flesh of seed; 51. Cause of Life; 52. Instruction and nourishing; 53. An example of Cardanus; 54. Our single Oils ; 55. Natures Works equal in weight and truth.

1. Here we have met with the common argument, wherewith the unlearned use to deface this goodly sequence; we must go forward and encounter with the learned, who because their great deeds & effects promised, that is, to make all happy, knowing, long lived, healthful, young, wise, blessed and virtuous, are above their skill, or of their Ancestors; The Grecians rate both the works impossible, and the workmen's way false and guileful; I mean, I say, prove, according to my task appointed, that those great acts and deeds may be done & performed by other and weaker means then Hermes Medicines; And this I must do with more pains and diligence, because this way and entry once made in their hearts, the great marvelous truth of this secret, may the more easily come in and take possession.

2. But of such variety of hard and slippery matter, where were it best to set out? Which way first to take? Were it not meet the means and helps unto pleasure should be first cleared and read before we come to pleasure itself? And among then to give long life the foremost place, if not for his worthiness, yet for his behoof and necessary, being needful in all Commonwealths and private persons; first to seek to live, before to live well, though that unto this end: then let us see what is long Life, and how all men may reach unto John Macklain.

3. But why do we make such great haste? We had need be slow and advised in so great a matter, and to look before we venter on so long away, and of so many days journey, that we be well provided and furnished of all things: wherein I hope, if I have not of mine own; or if after the thrifty manner, when I am well stored myself; yet I borrow to prevent lending, although I took upon trust so much as would serve this turn, it shall be no stain unto my credit; but be rather deemed a safe and wary way, to cut off occasion of robbery, both at home and

abroad, especially if I take it up of such men as are most famous and well beloved.

4.   These should be my friends of Egypt and Arabia (though we have their secret help now and then) the best able indeed, and the nearest unto me, if they were so well known and beloved in the world; but because they be not, I will fly to that other side of Greece, and to the most renowned there, and best liked: Hypocrites, Plato, and Aristotle, whom I doubt not to find very free and willing in this matter: Let us see then awake our old studies out of sleep, and lye to them, what need many words? After greeting, and the matter broken, they make me this answer jointly together: God, because he was Good, did not grieve to have others enjoy his Goodness, that is, to be, and to be well, meaning to make a world (though Aristotle withdrew his hand herein) full of all kind and everlasting changeable things, first made all, and blended them in one whole confused lump together, born up by his own weight bending round upon itself.

5.   Then seeing it lay still, and that naught could beget and work upon itself, he forced out and sundered away round about, a fine lively piece (which they call Heaven) for the Male-Mover and Workmen, leaving still the rest (as gross and deadly) fit for the Female, to contain the working and fashioning, which we term the four beginnings (or Elements) earth, water, air, and fire, and thereof sprung the love which we see yet between them, and the great desire to be joined again and coupled together.

6.   Then that there might be no number and confusion of workman and doing causes, but all to flow from one head, drew all force of working, and virtue of begetting, into a narrow round compass, which we call the Sun, from thence to be sent out, spread and bestowed all about the world, both above and below, which again meeting together,

made one general heat, light, nature, life and soul of the world the cause of all things.

7. And because it becomes the might, wisdom and pleasure of such a Builder, to make and rule the infinite variety of things here below, and not evermore one self-same thing; he commanded that one light in many to run his eternal and stintless race, to and fro, this way and that way, that by their variable presence, absence and meeting, they night fitly work the continual change of flitting Creatures.

8. This Soul, which Plato calls the ever moving mover, quite contrary to Aristotle, which he himself construed, a moveable mover, (that we may marvel how Tully could translate it, as to make it all one with Plato, unless Lucians Gallores misled him, which is found in some copies that he might be an eternal mover, is, in Nature) and being a moat subtile and small bean, a spark of heavenly fire, in property and quality, by his cleanness, light, and fineness, hot; and for his moistness, withal temperate, as appears to him that bends his mind upon it.

9. If you doubt of this moistness, think nothing is made without mingling, which is by drawing in, and breaking small together the whole stuff, when a dry heat draws out and scatters the fine from the great, and thereby wastes and narrows all things, making nothing: As for example, dung hatches an egg, and quickens anything apt to receive life, when warm ashes will never do it; what need we more? Imagine a heavenly flame by a good burning water, which flaming upon the hand on a dry cloth, heats them both gently, without heat or punishment; and yet this Sunny beam is not moist of itself, before it is tempered with the moistness of his wife, the Moon, to make it apt for generation. Then Hermes calls the Sun and the Moon the Father and the Mother of all things.

10. Now the stuff and female, to be fit to suffer working, must be first open, that is, soft and moist, and then not one, nor yet many like things, least in both these cases they should stand still the same, and not when they be stirred by the workman, rise and strive, bruise and break one another, fitly by continual changes, until at last they come unto a constant rest and stay; and that upon small occasion the same consent might jar again, and come and change the wished end and purpose of the work; And therefore God cast in at first, the known four fighting enemies: yet in the soft and open stuffe, there are but two of them, Earth and Water in one mixture, seen and extant at the beginning, before the painful, soul draws out and works the rest, Fire out of Earth, and out of Water that breath-like and windie thing called Air.

11. So that if there be much Earth, little Water, and great heat to mingle them, fire Will show itself and bear the sway; if but small heat upon the sane measure of Earth and Water, Earth will rule the roost; if on the other side, upon small store of Earth, and much Water, but a small heat of working; the thing will fall out to be raw and waterish; if upon the same quantity, and stronger heat, there arises an Airie, which is termed a fat and oily body.

12. Wherefore when the Soul comes down by the Aspects of Stars (Read the *Harmony of the World*) upon the stuff, clothed with a fine windy coat of the cleanest Air next unto Heaven, called Æther (without the breaking of which means, the two extremes and unacquainted strangers would never bargain and agree together) by his moat mild heat it moves it, and alters it very diversely, making many sorts and kinds of things, differing according to the strength of the one, and the obedience of the other.

13. And so by reason in that separation of that fine and male part, at first, the stuff was thoroughly tossed and mingled, and the heat of Heaven thereby (like a hot Summer, after a wet Spring) very fitly; all which, man and all were made alike, without any seed sown, otherwise then by the great Seedsman of Heaven, upon the common stuff Earth and Water, and is still seen in the Common tillage, yet used in those lame and untilled Wights, which some call Startups, and sprung out from themselves, As we may be easily led to think, if we Consider how, not only all kinds of plants, without all setting or sowing, grow up by themselves in some places; and some kind of Fish in the Sea are only Females; but also what plenty of fish there abounds in that frozen Country, for the great heat and fatness of the waters; and Chiefly that upon the slimy and the hot lands of Egypt, there are yet some bloody and perfect Land-sights (as Hares and Goats, etc.) all made and fashioned.

14. But because afterward the well mingled and fat fine stuff, and strong working heat failed (as it must needs in time) and yet the great Lord would have the continual flitting, change, and succession hold the same, and fit causes were daily kept by continual succession within the body of the perfect Wight, the stuff in the she, and the heat in both, yea, and as far as need required in seeded Plants also.

Now we must understand as well, that this heavenly Soul, when it is so clothed with that windy body, is called spirit (not only moves and worketh with his heat) but also for food wastes the stuff; for nothing that is made, is able to bear up his state and being without his proper and like food and sustenance. See my *Harmony of the World*.

15. Then as our gross fire here below feeds on weather and wind, called Air, as upon his lightest meat; and as it in his due place, is too thin and scattered, spreading the figure so far as it follows his food,

until at last it vanishes to nothing, unless it be plentifully heaped and crowded up together, and so kept in a narrow shell of water, which is called oil or fatness; even so it is between the fine starry fire and his like food, the fine fat of Æther, for that cause besides the divine purpose abovesaid, it cometh down in post into these quarters, to find and dress himself store of meat, as appears by his tarrying; for as soon as his food is spent he flies away as fast, and leaves his Host at six and seven uncared for. I was about to tell you the course of the divers sorts and suits of these lower Creatures, but that there was a great puss of matter came between and swept me away. This now being passed over, I will, go forwards.

16. Then if the suffering stuff be gross, foul and tough, and the making heat very small and easy, as it is within and under the ground, things are made, which they call Metals, or rather by the Arabic word, Minerals, little, broken, altered, or changed; but the gross beginnings, Earth and Water (Earth especially) rule still; and the life and soul, as it were, in a dark dungeon, fast shut up, and chained, as not able to stir and show itself at all. When the stuff is finer and softer, with greater heat upon it, then will arise a rooted and growing thing, called a Plant, better mingled, and smaller, and further broken from the low and foul beginnings; and the life of Heaven shall have more scope, because Wind, or Air, or Water (and yet Water chiefly) sways the matter.

17. But if the Soul be yet more mighty, and the stuff yet finer, he is able (Air and Fire) but that above this exalted, to show himself a quicker workman, and to make yet a finer piece of work, moving forward, and by mighty force perceiving; but by reason these two causes, passing by those degrees, to mount and rise at last, there is an excellent and fiery kind contrived, over our kind, I mean, most thoroughly, and fair, and finely wrought, even so fat indeed, that he

may not easily seem made at all of these all-making seeds, the four beginnings: Whence it is, that when a Corpse is consumed with fire, there are found scarce six ounces of clear earth remaining; which fineness of body gives occasion to the greatest quickness and freedom. of the Soul, and ability to perform (as his duties of life) moving and perceiving; yea, and shall I put in understanding also? For albeit God has inbreathed us with another finer and cleaner mover, called Mind, for a special and divine purpose; yet that mind, as well as the soul above, is all one of itself in all places, and working diversely, according to those divers places, as we shall see more at large hereafter.

18. Then you see all the differences of the four great heads and kinds, which contain all things; yea and of many lesser degrees and steps lying within every one of these, which I named not before; as also of sundry sorts (not worth the naming) of doubtful and needless things, touching and partaking on each side of the four great ones (or between the first two, stones budding like herbs in the Scottish Sea; between Plants and Beasts, that sprung Apes, or rather hairy wild men, between beasts and us) to proceed from the divers mixtures of their bodies. If you cannot quickly perceive the matter, behold at once the outward shapes and fashions, as they here go down a short pair of stairs before you.

19. Do you not see man alone, through his exceeding fine and light body, carried up and mounted with a mighty heat of Heaven, of an upright stature and carriage of himself, that his divine wit might be freed from the clog of the flesh? when other Wights, from the contrary cause, (which the gross or earthly leavings, or excrements, of hair, hoof, and such like declare) are quite otherwise disposed, as we see, towards the ground, their like companion; and so the less hot and fine they be, that is, the like the earth, the nearer they bend unto her, being

less of stature still; and after that many-footed to support them; but at length footless and groveling, until they come to their heads downward, and there they stay not, but pass quite over, and degenerate from Wights to Plants, and from thence, if I might tarry about them, I would send them down still through all the steps of them and Minerals, until they come to their main rest and stay, from whence they all sprung clean Earth and Water.

But I think it be now high time to take my leave of these philosophers and physicians, and to set forwards as soon as I have packed up my stuff round together, especially the best and most precious things, my Medicines.

20.   Then we gather by that enlarged speech, one chief and notable rule of learning, that the shape, nature, being, perfection, and all the difference in all things here below, springs from the mixture and temper of the stuff. and beginnings; the doing, making and working cause, that makes, mingles, broaches and sets all a running, to be a piece of the finer part of the whole, parted and packed up together in the Sun; of which finer part, some remains still in the raw and rude stuff secretly hid and placed: othersome more freely, in the half-made stuff, called seed; and in finer seed yet more lively, and in man most at liberty, excepting where I said it was free indeed from all kind of body; and yet all, these but one and the self-same thing, called soul, life, heavenly and natural heat.

21.   Thus, means divine Hippocrates when he says, naught is made, and nothing perishes, but all are altered, and changed up and down by mingling: And again, that no Wight can die, unless all tall; where he is most agreeable, and jumps with these grounds and rules, and with the whole web of our Rosie Crucian Physick. If any man doubt of the other two, Plato and Aristotle, let him read their books with heed, and he

shall find them, where they speak naturally, and by the light of human reason, to draw still towards this head and point of truth, though they come to stay sometime, misled, I think, by the over weening wisdom of Astronomy, to the Infinite variety of divers natured and conditioned Stars above, and such like Influences causing the like endless odds, and differences of all things.

22.   Let us now, I say, set forward in our first day's journey to long life, unfolding first what it is, and the cause thereof, and lastly, the common and high way to it.

It seems hard for a man to appoint what bounds of life are large and long enough for Man, unless God (who knows best both the measure of pleasure and happiness fit for him, and the race of time meet for him) first set and marked them; so that the greatest age and farthest time that the lustiest men and best disposed bodies, both by kind and diet, have at any time reached and lived, may well, by the great and good will of our great Landlord, be set the bounds, stint and end of life, large enough to hold all the pleasures meet for mankind, and the mark which we may all aim and level our endeavors at, yea and with sun, hope to hit and reach it, and no further, is about a hundred and fifty years, as you shall hear anon.

Now if there do three causes meet to the making up of things, and thereon leans all their being; the stuff, the mover, and the meat of the mover, which is the fatness of the stuff, then sure the Cause of their long being and continuance in their estate can be nothing else but the favor and goodness of those three causes.

23.   The soul and heat of heaven is good and favorable to Wights (to let the rest go far more dark and further off my purpose) when she pours herself plentifully upon them; for there can be no other odds in

one and the selfsame thing in all places, but the fat food of life which they call the first moisture, and is the finest piece of all that is lying hid and unseen in the sound second part of Wights, and yet by skill to be fetched out and set before us, must not only be plentiful and great in store, to match the feeding soul, but also fast and fine, that by his fineness he may be both friendly and like to live, and Airy, or rather Ætherial (we must leave these words without handling) to keep himself both in cold and heat flowing, and that through his fatness and closeness, (which they call in Latin, *densum* or *solidum*) that through his much stuff. in a narrow room he may be more lasting and fit to continue, tow the stuff. and body is beet when it is fast, and fin, also, to hold and hang all together, and that other to give free scope without stopping or let, unto the continual and wise race of life.

24.   Then to make a sun of all, the cause of long life is a fast-fine body, sprinkled and seasoned with much like fine moisture, and store of heavenly heat. If this matter needed any further proof, I could easily by cutting up the nature of things, so lay it open before you, as your own eyes should witness and see the same; but if it need to some, they shall see something, and that sufficient to content them.

For the first, Aristotle said, and we find it true by experience, that they live longest in hot Countries for their dry, sound, soft, and fine bodies; but chiefly for their fineness, yielding free recourse and passage unto life; for age and kindly death come of rottenness, which flows from the stillness of heat, and slackness to salve and refresh the parts.

Touching the rest, to wit, that much heat and much good fatness are a cause of long life: mark the short life of all those Wights, that either want them by kind, as the maimed and imperfect ones, or waste them by motion, as the male Greyhound of Lacedemon was, against the course of kind, shorter-lived then the Bitch, for his pains in running;

and the gelt male Round, and spayed female, hunt better, and live longer than others. And the Cock-Sparrow lives but half so long as the Hen, and yet this but three years for their venery; the world is full of such examples: and behold again, the Elephant on the other side, for the great help and favor of all the causes above the rest, as may appear by their great fruit and effects in him, that is strength, bigness, and stomach, being able to bear the ground work of a Castle of fifteen armed men, to eat 9. bushels at a time, and to drink 14. sirkins (to endure and hold out much longer than any of the rest, and to live (Aristotle is my Author in the story)) three hundred years in all.

Now we know what long life is, and the cause thereof, let us see whether all men may reach it or no, and then which way they may reach it.

25. At the first all mankind by the will and appointment of kind, was found, and lusty, and lived long, and all the failing and corruption nowadays (which falsely seems a weak condition of our nature) crept in through disorder in ourselves, by little and little, & so by sowing still the like children, it spread itself at last deeply rooted over all, and made it, as it were, a certain state, nature and kind of men; wherefore by good order in ourselves, it may be reformed and brought back again unto the ancient Estate; but how may we prove this? If God and Nature have ordained man unto a divine end above the rest; and yet some beasts (as Theophrastus for a wonder complains) live longer than our common rate, yea and longer than any bounds above set; certainly we ought to do as much and more, by the rate of nature, and of all right and reason, and some did at first, before we fell by our default, which may be mended.

26. But least I may happen to deal with some, who will neither grant the Justice of God, nor yet yield to the end of man; with some, I

say, that have so far put off all humanity, I will bring them to natural causes; I will open and lay before them, both the sorts and fruits of Wights, I mean of men and beasts; that they being a monstrous doubtful kind between both, that is, Beasts within, clothed only with outward shape of Men, nay the better Judge of both (as in like case they formed of the like misshapen Monsters the Poets know my meaning, it is not worth the flourish of a chaste and modern Pen) which has in kind the more cause to live long; that seeing at last the worse lights to overgo us in life, and to run to the very goals itself, and yet to have received less cause from nature, they may be driven by force of reason to yield, that we have a better kind and worse custom, and that we did and might live long, but for our own fault, which may be reformed.

27.   To begin with the soul and natural heat for his worthiness, let us see which of them is endued with more store of him, that is, of the chief cause of long life; man walketh upright, when the rest are thrown to the ground, because they lack the force of this light and ascending heat, to bear up the weight of their bodies, which we have abundantly; but if we leave the outward shape and look into them, we shall by the great foresight of natural lights, which are hot and full of blood, have against the root and spring thereof, to root and temper the same, a contrary in place and property set, the brain, I mean some more and some less, still according to the behoof and request of the heart; in so much that they that have less blood and small heat within them, as not needing any cooler, have no brain at all.

28.   Then by certain race and coarse of kind, if that be true which all Physicians & Philosophers hold, that a man has the greatest brain of all lights; it must needs follow, that he has the greatest store of heat also: but inter further into them, and you shall see man by bow much more

he goes beyond a beast in wit, so much to burn in heat above him: for wit springs out of the clearness of the body. And this out of heat, as I will prove in his place hereafter.

29.   Now if this first point be done and granted, the next is quickly made, even as one match is made by another. It stands with the justice of nature that makes not in vain, to match this greedy heat with store of good meat, that is, of fast and fine Etherial first moisture suitably, or else sure, says Heraclitus, the officers of Justice, the Fairies would soon apprehend her. To be short, both this and that, and the third likewise, a close fine body and all is cleared, if it be so that man in making is moat clear and finely mixt, and broken of all the lower creatures, as we heard even now decreed in the Council of the best Philosophers and Rosie Crucian Doctors; for if naught makes but heat, then naught makes well but much heat; if there were no other odds in souls, as was abovesaid: and if the beginnings be well and finely mingled, and the concoction hold, they must needs gather themselves in close together also to make another cause, yea, and the last; for what is fine oil and fatness, but water wherewith we flow, as our brain declares, thoroughly mingled and raised into an airy, or rather into an Ethereal close substance; but if you will not stand to this degree, then once for all consider and weigh but this one example, that albeit man be more given to lust, then any other light, and thereby drying up the body, it plainly appears more than in any other, and weakens all the help of long Life together, both the moisture, that knits and holds the frame, and that which feeds our heat, and this all; and so the sum of life, which is yet due by nature, he pays before his day to his own wantonness, yet he lives and holds out longer than almost any other; that we may plainly see, that if he lived as chastely, and in other points as orderly as the rest, he might far pass and overrun them all, in this race of life and continuance.

But methinks I hear then whisper, that I forgot myself, and the bounds of my long Life, when I make men able to live as long, and longer than any beast; for to let pass that Hart, Badger and Raven, which overtake the longest life of our old men; since the Elephant, as we have heard, goes far beyond the very bounds of age, especially the Raven, whom Euripides will have to live nine of our ages.

30.   There may seem some matters, but chiefly the last uncurable, and yet they are indeed light and easy, and the last most of all; I mean the Raven; for if there was never yet man of Sound judgement and knowledge in the wales of nature, that allowed the story (and Aristotle by name condemns it, when he gives the Elephant the longest life of all Wights, and man next to him) what? should we search after Poets Records? Besides, doth not one among them confess himself, they are not to be believed, and held as witnesses? doth not Plato, once a Poet, and then a wise Philosopher, chase them up and down in all places? And in one place sayeth, they are besides themselves, when they sit on their Muses stools, and run like a spring pouring out all that comes? Are they not all, in wise men's account, the greatest enemies to God, good manners, and all right and true knowledge, that ever the world or the Devil bred?

31.   But I slide too far unawares; and if we must of force receive this aged Raven, yet perhaps there shall be no great hurt received: and I cannot see why we may not match him with Methusalem, and some other aged Fathers in holy writ, reported to have lived as many years as nine of our ages cones to, with advantage, it is not enough to say that which some say, those years are to be meant for months, and not as we account them; for albeit I know the Egyptians reckon (so we may see in Pliny, where some of them are said to live a thousand years apiece, that is, so many months) yet is agreed among the Divines, men

best skilled in these matters, that the Jews account was otherwise, even as we and all other Nations make it. But if this ancient story of our holy men be a thing in doubt, or certainly untrue, or to be meant of months, yet your aged Raven may go with it, and the Father of the tale together; and we may, when we will, pass to the Elephant. Aristotle indeed is the Author of this stone, that the Elephant lives three hundred years; how then shall we mistake in like manner of this man, and refuse his witness? I cannot tell what to say; it is a very hard matter that he says: and again I know, that when the power and purse of his King and Scholer, Alexander, who gave him eight talents of Silver, a huge sum, to that rise, he heaped up a rabble of all kinds of reports and hearsay into some of those books (by some called πολυτάλαντα.) and some false and untrue tales might creep in among them; yet I owe much to the man's worthiness; and again the books have ever held the place of a true Record; and besides this matter of the Elephant, both for the forecouched causes, and for his wit and manners, somewhat near our nature, say reasonably well agree with the Sound of reason. Row then? I say again, methinks I feel my mind to ebb and flow within me: And yet suppose it true, that this Beast should live so many years; the islanders of Tell near Colecut, and the inhabitants of the kill Atho, both of them commonly and usually reach our appointed time of am hundred and fifty years, by the favor of the mire only and soil where they dwell, taking (besides, for ought I can know) the common rate and course of the world; that we may lawfully deem, if they lived as chastely as the Elephants, who comes but once in two years to Venery, and swallowed his other good Orders of life as well, that they might easily draw forth their age longer, and come to the days of the Elephant.

32. For as we in our less happy soils, by our own ill diet and crooked customs, have cut off and lost the better half of our time, so it

may seem to them; for we must not think in this disorder of the world, that any man fulfills the time of nature, but all are swept away with the blast of untimely death.

33. But it say chance that long race of life, which the Author sakes the beast to run, was no common and ordinary course in that kind, but of some odd and rare example; and then, no doubt, as there be some amongst us which by their diligence, and I know not by what good hap, double the common tern, so there be not wanting in those places, which sometimes prove aged men, and which live twice as long as the common sort, that is, as long as the Elephant.

34. Wherefore, for all this, or ought else that can be cast against us, let us conclude, that man, if he kept the good and kindly diet and order of life, which other nights, void of reason by the true and certain guide of Nature keep, having more helps and means unto it, might live longer than any of them; yea, and with ease reach the bounds of long life appointed, and perhaps further also; but we have stayed in the midst and mean, as it were, because it seems to obey the secret Will of God the better, and yet withal to fill the whole desire of Nature.

Then say you, it were good to learn the order of life which Beasts do use to keep and follow, if it were meet and seemly for men to lead a beastly life; do not so take the meaning of a good thing, with the snare of a foul and filthy word; a man is not one and single as they be, but double and two things, and partly a light, nay a Beast (be spoken with reverence) and partly a more divine thing; and therefore albeit, according to his divine part and reason, he ought to follow the divine pattern and form of life above set; yet as he is a Wight, and an earthly Creature also, it is not uncomely, nay it is necessary to do as they do, after a sort; and if it were altogether so, it were better, and more agreeable with the will of Nature, who knows best what belongs unto life, that is, unto herself; for kind leadeth them still after one due and orderly manner, when great variety of wit and device guides us against Minerva's will, as they say, and quite besides the way of Nature, unto a thousand by and foreign Customs, which is the only cause of our degeneration from our ancient and first whole and second estate. Wherefore if a company of picked and lusty men and women, would agree to live together in some wild, open, clear and sweet air, scattered like a country village, and not like a close and smothered city (which one thing prevents a thousand diseases and deaths alone) and to live together to the right end of nature, that is, for children, and not for pleasures sake (for this was made to the right purpose) and in as seldom and due course, as the better sort of Beasts, the ready way to preserve life and forestall diseases, but especially to get good children, and to bring up their children in labor and hardship, mingled with much mirth and sleep together, no small help to long life and health, as the directors themselves confess and know.

But for their meat and diet (wherein those Beasts offend and fail greatly) if they would consent to take no physick, but in great danger cast in by misfortune (in which case the Beasts do not want their remedies) never to drink wine, the shortener of life; and to be short, not to take any meat and drink that the fire has touched (for it sunders the fine from the gross, that is, the best from the worst, which we now choose) but as Nature has left them, and other lights use them; if these things, I say, were duly kept and performed, I am fully persuaded that within three or four generations and offspring, it would come to pass, that we should see this people prove a Nation of Giants, not only passing the age of Beasts, and the bounds of long life afore set, but wholly recovering and restoring all the blessings of the first estate of the body.

35. And this I gather, not by our own contrary customs only, taking effects as cross and contrary, but chiefly by the life and use of giants and lusty people in times past, and some other yet at this day, which was and is the very self-same race and course which I described: And sure for the Inhabitants of Teill and Atho, which I brought in even now, filling the term of our long life, although I am not certain of their use and custom, and where I find the story, I know the cause is laid open, the goodness of the soil in the first place (for it is thought to be the blessed paradise) and upon the goodness of the Air in the next, for the height of the bill, without all wind and rain, two great troubles of men's bodies; yet I am led to think that they do keep the sane orderly and kindly form and rule of life, or at least to draw near unto it, because albeit clean Air, by cleaning and quickening the spirits, and searching the body, be not little helps and comforts in this Journey (as we shall easily see, if we mark how among all Creatures, that lead their lives in the cleaner Element, do live the longer; Fish then Worms, and land Wights then these; and winged ones yet longer, because the higher, the better air still; insomuch as Cardan dares think, that if any dwell in Æther, as Plato's Heir affirm, they live forever); yet if ill diet went withal, it would mar as much as the other made, and greatly cloy and hinder, yea and cut short the race of their long life.

36. I am of the same mind for all other odd and private persons of great age and long life recorded, (as for some Italians in Plinies time, registered of one hundred twenty-four years) and such other aged men in Authors; a man might let in here a sea of examples; but I must be short; neither would I name King Arganthon, that lived an hundred and twenty years, and reigned eighty thereof; nor yet the old Knight of our Country, Sir Alington, and Parre, etc. Yet twenty years older; but that is so strange in Nobility, that they come, as it were, unto that kindly course of life, as unto the gale and end of long life.

Then we see at length that it is not impossible, as they say, but an ordinary and easy matter to strengthen the weak nature of mankind, to enlarge the straights of his life, and so lead him on still to the ancient age and long life appointed.

37. But I see them start up and say the like as Cato in affairs of state, used to give counsel (unwisely, though never so well) as if he had been in Plato's Commonwealth, and not in the dregs of Romulus: So in matter of diet and order of body, speak as if we lived in the former golden Age, which, as poets fain, was under saturn, and not in the corruption of jupiter's kingdom; and that with the world, as it now goes, cannot be brought (without a kind of divine power, to raze out the old, and make a new world, and that in long time) unto the first and kindly custom of life; I must, if I mean to do wisely, take the men as I find them, and prove that all such weakness as now is among them, may by man's endeavor and skill of healing be upholding and led forth unto those bounds, and the end of long life aforeset.

Albeit I have done as much as reasonably may be required at my bands in this place, which was allotted out to show the possibility of the matter, yet because I count it better by plainness of speech to do good, which is the end of my writing, then by subtleness of Argument to obtain my purpose, I will come unto you, and venture upon that point also, be it never so hard and desperate, hoping not that fortune will favor bold men, but God good men.

Then as there are three causes of life and being, the life and soul itself, and his food the first moisture, and the frame and temper of the body that holds them both; so, let us take them all in order, and see how they say be preserved, and kept together, beginning first with the last, because it is least and lightest.

38. It is enacted by the law of Nature, that no body, mixt or simple, shall or say live and preserve his estate, and being without two helps or stays, that is, meat and exercise, each like his kind, and of his nature; as in lone and simple and subtle bodies (for it is plain in the first row, especially if they be living, as they term them, though all things indeed have life and souls, as we heard above) the hot ones crave fiery seat and moving exercise; moist ones, wind and water, flowing food and exercise; cold and dry things like an earthly, sustenance and rest for exercise, which is also like, and preserves their state and being.

39. But if all alone and simple things be within the compass of this Law, then Heaven may not be free, nor exempted; and they speak not altogether fondly, that say, the stars feed upon the sea; and for that cause, by good advice of Nature, the Ocean so rightly placed under the course and walk of the Sun; for although the water be yet so far off, and unlike them, yet their power and strength is such, as they are able by their Labor easily to refine it, and turn it first into Air, and then into Æther, a weaker like thing, and their proper food.

40. That this is so, the hungry Souls (which are but Imps slipped off the heavenly body) makes it plain here below unto us, when we see them still unwilling to tarry, and unable to live among us without meat, as they bewray themselves by the plain expense and waste of the first moisture: Nay take this one way, if you would mark well, and all lie on the ground: then there is old coil and fighting here below for meat and exercise, that is, for life and being (which sakes the cause of all action and doing, rest and change, and every one runs easily and gladly to his like; and if his strength be never so little greater, he subdues, digests, and turns him into his own nature, and is strengthened by him; but if he misses of his like food at hand, and be much stronger, he dares encounter, and is able to equal unlike things

also; as I find of the Stars, mightiest things, giving might to all things in the world.

But in case the unlike and contraries be of equal power and matches, then neither devours nor consumes each other, but both are named, dulled, and weakened, which they call consent, and temper, and mixture; for example, fire extra hot and somewhat dry withal, and water very cold and somewhat wet, meeting together in even powers and proportions of strength, are both impaired, but neither lost and destroyed; but if this mature chance by the heat of heaven to be taken in hand, and turned into an airy and fat substance, though there be now two monsters set against the drought of fire, yet because of the heat of weather and Heaven abounding, it is now become partly like fire, his weaker foe and enemy yielding himself for food unto it, and increasing his strength and nature. But if on the other hide air add unto his exceeding moisture, matching the draught of fire, yet some strength and watery coldness (as appears in thick and foggy weather) it is able easily to overcome the fire, and eat him up.

41.   Now for a mixt body (which is a consent and dulling of the four first famous enemies, made and kept in tune and awe, by the force and skill of an heavenly and natural heat upon them) it has the same reason; for when either for lack of meat, or driven by violence, this heat departs, the friends begin to stir and fight for food and freedom, until someone stands out above the rest, and recovers some part of his former power, which puts those that can feel to the worst, and breeds diseases, and at last gets the whole Lordship, and rules over all, and turns them all into his own nature; them the old consent, knot, and body is broken, lost and spoiled, and a new made and gotten, still going downward, until they return to earth, from whence they all came: for example, and that near hose: In the fiery frame of man's

body, when the soul for want of food fails and flits away, they straight retire and run back in order: First, fire waxes moist and lukewarm, supt up with air, and this soon after thick and cold, that is, waterish, and water muddy, still more and more thick and dry, till at length it be most dry and heavy, and all be devoured and brought to earth, from whence they all, set forth before. And this is the natural dissolution and death of our body; forcible death and destruction is by disease (to bear out other force, which no man can warrant) when either breath or seat, distempered in some quality, do feed and nourish someone their like beginnings above the rent, and make him strong and able to vanquish them, and bring in the Jar of the musical consent aforesaid; as when by waterish heat and air all the beginnings are changed into water, through hot and dry into a fiery temper, and so forth; or else when the body wants the exercise which is owing and due unto him, which is quick motion, to preserve the air and tire in the fine frame of man, from the sloth and idleness of the slow and rusty beginnings.

For in a Disease called the yellow jaundice, when all the blood is converted into choler, if there be not a way to convert that choler back into blood, how can the man live? For if all the blood converted into choler be let out, he must needs die; so he must also if there be not a way left in nature to transmute this choler back again into blood: I might instance the like of the Dropsy; but I should make too long a marginal note; study Nature, and she will make thee a better Physician then Galen himself was, so shall you learn to fortify that quality of the body that is weak, and almost eaten up by its adverse quality, as a Musician winds up that string that is slacked, till it makes a harmony in the rest, but he winds it not too high, least it sound overtop them. By which grounds laid, we see the way to uphold the temper of our body made plain and easy; no more but to feed and cherish it with clean and temperate Air and meat continually; that all the beginnings served and fed alike, one say not be more proud, strong, and able then another, to subdue the rest, and overthrow the State; and therefore poison kills us, because it is extreme cold and dry, (for we may shut out all rotten, as also fiery and watery tempers from the name of poison) feeding and strengthening the dregs, but devouring the fine liquor of the body, as venomous Juices the like Plants, and these noisome Beasts, and one of these another; nay which is very strange, I have read of such natured men of India, that used to eat Toad. and Vipers: And Albertus says, he saw a girl of three years old, that fed greedily upon spiders, and was never hurt, but liked greatly with it.

42. Do not think it any discord, when I said above, fatness and raw temperateness upholds the body; all i. one; it cannot be fat, unless the earth and water be well and evenly mixt; nor fine, except fire and air bear as good a stroke, as rule among then; but you will say, that Nature has given her creatures a walk of course, not to stand still in one stay and place for ever, but to nave and walk up and down, to and fro, from one side to another; that is, as was said before, God has made a changeable world, and therefore that frame and building of man's body, cannot ever hold and hang together, but must needs one day be loosened and fall asunder. I grant, it must needs be so by the course of nature, because to fulfil the will of her Lord, she has appointed stronger means and causes to work, either the want and absence of the inward friendship and keeping of the soul, in those which the common sort call living things, or in the rest, the presence of some ravenous and spoiling enemy: but if cunning Art and Skill (which by the help of mature is above the course of nature) by knowing of the due food for life, and defense against the enemy, may be able to defend the one, and keep off the other, then, no doubt, the frame and temper of both dead and quick may last forever.

43.   The way I found already, and known by certain people for the one; I mean, that Art has often, by keeping off the failing enemy with a strong contrary, preserved and upheld a dead thing, of slippery state and soon decay forever; as a Corps by Balm or water of Salt, Timber by the oil of Brimstone, and such like: Why then should the next prove impossible? To wit, by giving store of fit food still to life and natural, heat (for the other helps of meat and exercise are easy) to undershore or keep upright our weak and falling frame forever.

The Greeks hold, that our natural heat and life (because it feeds upon, and wastes the most fine and unseen oil (called first moisture) daily, which no food of Air and Meat is fit and fine enough to repair, must need faint and fail withal, and cannot be restored: Let us see what may be said to this, yea and bend all our force unto it; for this is all.

44.   The soul, life and natural heat of things is often and fitly compared and likened unto the other gross and fierce, hot and dry body, called fire; to feed and maintain this, his weak-like, that is, air cannot be wanting; and because it in his due place is too thin and Scattered, dividing the fire to naught in pursuit of his food and sustenance, it must needs by heaps be crowded up in a shell of water, called oil; if much heat and oil meet together, the work is great and busy, and thereout rises a smoke as a leaving of the meat, and the fire follows as far as the smoke has any fatness, which makes a flame.

45.   Albeit the nature of fire continues as long as it has food enough, & craves no great exercise, and will in a close place as under ashes, yet a flame being more than fire, a hot breath or smoke besides, desires open and clear air, both to receive the thick, the refuse, which else would choke him, as also for his like weaker food, that he be not starved, which two are enough, besides a little motion for his exercise; that we may not marvel at those men, which be in cooling for another

needful thing in this business, whereas the kind of fire and air abhors cooling as his contrary, as it is engraved in the nature of things, still to fly from that which hurts it.

Now in like manner to come to the purpose, if the fire of life and natural heat be not great, a little fine oil, and first moisture, will serve to feed it, and out of the slack working small store of refuse breath and smoke arises to make need of fresh and open air to clean and feed it, as appears by those lights, which are able to live in their places without help of wind, breath, and air: the little parted Vermin (called in Latin *Isecta*) lives anywhere; and Fish in the water, nay in the Bound earth sometimes; Toads in close Rocks, as Agricola says; and Flies in the most secret Miners fire, as Aristotle reports: but when the heat, on the other side, is great and lively, like a flame, as in the hotter fish, and other, no light can want fresh air and fine breath, both by his clearness to purge, and his weaker likeness to nourish the Ætherial smoke, and spirit that carries it.

Now this, no more than flame, needs cooling to preserve his being; but to temperate the kind of his proportion, fit for wit and weighty perceiving, which, I say, before I brought, and not the air performed.

46. That Æther is stronger than air, and able to consume it, it is plain in reason by his warmth and moistness, passing air in his own nature; and yet gross and thick air, as bent towards enmity and contrariety with it, will stand in combat against it, and overcome it, and thence it is, that in deep Mine-pits, and Caves underground, where the air is thick, corrupt, and unkind for want of flowing, no light nor light can draw breath and live, unless by sly desire the way be found to move and nourish the same air, and make it kindly.

47.   Then to draw near the matter, if the Stars do feed on Æther, and this upon clean and spotless air, as on their weaker lights, and our soul and life is of a starry kind, even a slip and spark thereof (as you may read at large in my *Harmony of the World*) as is aforesaid, then it so floweth forth to feed our Æther, the Carrier of our soul, with good air, which is round about us; that will serve the turn, but to nourish life and heat it self. Either it self must be the food, or this body which is so high and past our reach, except this spark of heavenly fire were able, like the whole body and spring above, by his power over our meats, to turn the water first into breath, and this into Æther, which it is not, and can go no further then to air, and to make a common oyl and fatness fit to nourish an elemental, as they term it, but not a heavenly fire.

48.   Where then shall our life find food and sustenance, say you, fit to bear it up, and maintain his being? In that fine oyl, and unseen first moisture and fat, and call you that Ætherial? How can that which was once seed, and before that blood, and first of all a plant, become a body so fine, clear and Æthereal? Especially when one weak Star, and soft fire of heaven, is not able to make so fine a work, so fair and highly sundered; I say, this is the secret and depth of all, which because the Greeks never founded, I do not marvel if the means to preserve life did escape then; but us shut up every word, and help them in this helpless matter, yea although we be driven to open the things that have long lain hid and covered over with great darkness.

When our life in the lusting parts is by the bellows of thought stirred up, and moved unto work, it sends forth out of every part, the hot natural spirits and breath of begetting clothed with the shell of seed, cut out from the dewy part of our meat, ready to be turned into our body (or at least already, and now turned into earth) and not from the refuse and leavings of it, as some say, when I could show it, if time would suffer, the best juice in all our body.

49. This is the furthest and finest workmanship of our meat and food of body the very beginning and first stuff of the fine oil, the food of life, after the remaining forty days in heat, before it come to perfection, being wrought, as we know, with the double natural heat of the begetting breath, and the womb, forty days before it be fully framed and fashioned into the form and shape of a man, ready to draw food or nourishment (be it milk or menstrue, received by mouth or navel, I cannot stand to reason) from the mother, to the increase of the tough and sounder parts: but the first moisture is now at his full growth and perfection, and from thence feeds life, being unfed itself, and wastes daily against the grounds and rules of Physick, for the child has now received all that the workman can give, & is put over for the rest, which is his nourishment, unto his mother's payment; but what has she to give for food unto the food of life? Naught, as I showed before, else we might live forever.

Then we see what the first moisture is, and how it excels the food of the body, & why it cannot be maintained by it, because it is the most airy piece (for the rest go everyone his way, and make his own part from whence he came) of all the seed mingled, wrought, purged, raised, and refined, and then closely thickened and driven up close together, forty times more & above our meat, which in one day is ended and ready to be turned into earth, and therefore unfit in any wise to increase and cleave to our first moisture, the food of life, even as unmeet for all the world, as water is to Æther, oil or fatness.

50. And by this to come to the point, we have a plain pattern, (if we be wise and careful) and way to work the great mystery of *Adjournment of life*; for if it be so, as I proved above, that all the

moisture of the matter lies in the maintenance of our natural heat, and it, as our men, and all reason teaches, follows the steps of common fire, waxes and wanes, is quick and faint, according to the store of his food, and first moisture; then sure we can make an oil as fine and close as this, nay in all points all one with this; it will easily mingle and join with our first moisture, and so feed, nourish, and increase, and like withal; even in as good and plain reason, as the same oil dropped still into the fire augments both food and flame; yea put case the same natural fire of ours, should not only impair his strength, for lack of meat, and slack his force, but abate his bigness also, as some Physicians hold: yet there were no great hurt done; for this second spark and slip of the great and common fire of Nature, being a piece of the finer part of the whole (which is all one in all things) and fellow to his like in us, when it is made free and close in these fine Ætherial Medicines, would restore the heap, and mend the matter.

But how shall we get the like fine oil and fat first moisture?

51. The matter is drawn so far, that there is all the hardness; I showed the pattern, even as Nature got the sane before you, by the like stuff and food, and by the like heat and moving workman: this by certain proof of all our men is easily to be found, even a gentle, continual, equal, and moist, that is, rotting heat. But the seed seems hard and unable to be matched, because a kind of strange and hid proportion and temper of our body (which no man by counsel and knowledge, much less by hand and workmanship, can reach and counterfeit, no not if he boiled all the mixtures in all the heats that all the wits in the world could devise, made It thus after his own fashion.

52. Then how if we take the same frame aid temper not by us, but by kind proportioned? I mean the same blood, or flesh, or seed, if we will (which the men of Germany choose, and command it, above all,

and call it Mumia) would it not be very natural? For if the Doctors hold it good, if any part about us fail in his duty, to correct and help him with the like part of some beast, passing in the property; as to mend fainting lust with the yard of a lusty Beast; the womb that cannot hold, with the womb of a quick Conceiver; narrow breathing with the lungs of a long-winded light. See the *Harmony of the World*, & etc.

Than consider with how much more kindly consent we might with our own parts finely dressed help ourselves in our diseases. But for my part I cannot unwind the bottom of this great Secret of Germany; for we mean not to make a Man, which is to be feared in the course, if his rules be true, but a fast moisture only; and then with all things are made of the same stuff, by the same workman, and differ but by mingling only, it boots not where we begin the same mingling, and form it the last, which Art is able in time to do at once, she may do often, and so reach the end of Nature.

53. What need I say more? Is not the matter clear enough, that another fast fine oil and first moisture nay be made, in all like to our own, and able to maintain, or repair it and the natural together? And then that by the same (though other easy means would serve) because it is so temperate, the body may be brought and held in square and temper, and so by reason all the causes meet and flock together, the life nay be preserved, I dare not say forever, for fear of the stroke of destiny which God has made, and will have kept, but unto the term, and those bounds above set, and beyond them also, if ever any man have gone beyond them. See the *Temple of Wisdom*.

54. But if it should chance any of our chosen children (to use the phrase of our Family) be unable yet, for all this teaching, to take and digest this food of learning, what is to be done? Shall we cast them off for untoward Changelings, as the foolish women think? Or else for

Bears and Apes, as Galen did the Germans? No, that were inhumanity; let us rather nourish them still easily and gently, hoping that they will one day prove men; and give it unto them, that all the most wise and cunning men in the world, I mean all the hosts of Hermeticists, have from age to age ever held (but under veils and shadows) somewhat covertly, and taught for certain, that such a first fine oil, whereof I spoke, and which they call a fifth nature, Heaven, or by a more fit name, Æther, is able alone to hold together the brittle state of man, very long above the wonted race both in life, health, and lustiness: nay, for fear there be yet some suspicion left in their Authority, I will go further; As many in the other side of Greece, as had travelled in these matters, and seen something (though not with eyes, but with minds I think) confess the same; as (besides them which perhaps I know not) Fernel in part, and altogether Fecinus and Cardanus (who were as wise and learned men as any time has brought forth) do openly declare in their writings: But if this soft and easy kind of delivery will not yet serve the turn, and they must feed their eye as well as their belly, as the proverb goes, then let them tell me by what diligence did Plato so order himself and school his own body (to use his words) as he could be able to cause nature to end his days at his pleasure? And by departing or dying on the same day eighty-one years after his birth, to fulfil of purpose (but I know not of what purpose) nine tines nine, the most perfect number; Might he not have had some such Medicines? Nay, is it not like he had them when he was in Egypt among the Priests and Wise men, and brought home some great learning from among them? And when he speaks so much and often in disgrace of his own Countries Physick, though Hippocrates himself then reigned? But it is for certain written in divers of our Records, that many of Egypt, the spring of this water of life, have before and since Plato, by the self-same water, kept themselves alive twice as long as Plato; if I might bring in their witness, or if this whole kind of proof (which I like full ill) were not counted by the Art of People unskillful.

55.   Then let this one example told by Cardan, a man allowed among them, serve for all; That Galenus of late Charles the fifths Physician, by this Heaven of ours, beset with Stars (as some do term it) increased the spirits of herbs, by an easy seat put into them, and so preserved himself in lusty sort until one hundred and twelve years.

56.   Neither think that mixture better than our single oil, (though Lully, Rupersis, Paracelsus, and some others allow it so) but rather worse in reason for too much heat in a weak and loose body; I mean for long life; by his over greediness in eating up too fast his own and our first moisture; it may be better because it is stronger against diseases; even as the Leaches judge between the dunghill and a garden herb for the same cause.

57.   But I think the device not good in either, nor agreeable to the Justice of Nature, which more evenly weighs her works; nor yet to the kindly skill of Hermes, who, to the great advantage of his Medicines, has a most fast, tough, and lasting stuffe, according as we shall show in that which followeth. Now is it time to rest, we have made the Third a long day's journey.

Chapter II.

1.2. Of the accurate structure of man's body: 3. Of joy and grief, and difference of wits.

1.   I admire the goodness of God towards us in the frame and structure of our bodies; the admirable Artifice whereof, Galen, though a Naturalist, was so taken with, that he could not but adjudge the honor of a hymn to the wise Creator of it. The continuance of the whole, and every particular is so evident an argument of exquisite skill in the Maker, that if I should pursue all that suits to my purpose, it would amount to too large (yet an entire) Volume. I shall therefore write all that is needful to be known by all men, leaving the rest to be supplied by Anatomists: And I think there is no man that has any skill in that Art, but will confess, the more diligently and accurately the frame of our body is examined, it is found the more exquisitely conformable to our Reason, Judgement and Desire; so that supposing the same matter that our bodies are made of, If it had been in our own power to have made ourselves, we should have framed ourselves no otherwise then we are: To instance in some particulars, As in our Eyes, the Number, the Situation, the Fabric of them is such, that we can excogitate nothing to be added thereto, or to be altered, either for their Beauty, Safety, or Usefulness; But as for their Beauty, I have treated largely of it in my youthful merry Poems, and now am not minded to transcribe my tender nice subject, and couple it with my severer style; I will only note how safely they are guarded; and fitly framed out for the use they are intended: the Brow and the Nose saves them from harder strokes; but such a curious part as the Eye, being necessarily liable to mischief from smallest matters, the sweat of the forehead is fenced off by those two wreaths of hair, which we call the Eyebrows; and the Eyelids are fortified with little stiff bristles, as with Pallisadoes, against the assault of Flies and Gnats, and such like bold Animalcula; besides, the upper lid presently claps down, and is as good a Fence as a Portcullis against the importunity of the Enemy; which is done also every might, whether there be any present assault or no, as if mature

kept Garrison in this Acropolis of man's body, the Head, and looked that such Laws should be duly observed, as were most for his safety.

2.   And now for the use of the Eye, which is sight, it is evident, that this Organ is so exquisitely framed for that purpose, that not the least curiosity can be added: For first, the Humour and Tunicles are purely transparent to let in light, and colors unfolded, and unsophisticated by any inward tincture. And then again, the parts of the Eye are made convex, that there might be a direction of many rays coming from one point of the object, unto one point answerable in the bottom of the eye, to which purpose the crystalline humour is of great moment, and without which, the sight would be very obscure and weak. Thirdly, the tunica uvea has a musculous power, and can dilate and contract that round hole in it, which is called the Pupil of the Eye, for the better moderating the transmission of light. Fourthly, the inside of the uvea is black like the wall of a Tennis Court, the rays falling upon the Retina again; for such a repercussion would make the sight more confused. Fifthly, the tunica arachnoides, which envelops the crystalline humor, by virtue of its processus ciliares, can thrust forward, or draw back that precious useful part of the Eye, as the nearness or distance of the objects shall require. Sixthly and lastly, the tunica retina is white, for the better and more true reception of the species of things (as they ordinarily call them) as white paper is fittest to receive those images of ink; and the eye is already so perfect, that I believe it is not needful to speak any more thereof; we being able to move our head upwards and downwards, and on every side, might have unawares thought our selves sufficiently well provided for; but Nature has added Muscles also to the Eyes, that no perfection might be wanting; for we have oft occasion to move our Eyes, our heads being unmoved, as in reading, and viewing more particularly any object set before us; and that this may be done with more ease and accuracy, she has furnished that

Organ with no less than six several Muscles; and indeed this framing of Muscles, not only in the Eye, but in the whole body, is admirable; for is it not a wonder, that even all our flesh should be so handsomely formed and contrived into distinct pieces, whose rise and insertions should be with such advantage, that they do serve to move some part of the body or other? and that the parts of our body are not moved only so conveniently, as will serve us to walk and subsist by, but that they are able to move every way imaginable that will advantage us; for we can fling our Legs and Arms upwards and downwards, backwards, forwards, and round, as they that spin, or would spread a Molehill with their feet.

To say nothing of respiration, the constriction of the diaphragm for the keeping down the Outs, and so enlarging the thorax, that the Lungs may have play, and the assistance of the inward intercostal muscles in deep suspirations, when we take more large gulps of Air to cool our heart, overcharged with love or sorrow; nor of the curious Fabric of the Lainix, so well fitted with Muscles for the modulation of the voice, tunable speech, and delicious singing: You may add to these the notable contrivance of the heart, its two ventricles, and its many valvulae, so framed and situated, as is most fit for the reception and transmission of the blood, and its sent thence away warm to comfort and cherish the rest of the body; for which purpose also the valvulae in the veins are made.

3. But we see by experience, that joy and grief proceed not in all men from the same causes, and that men differ very much in the constitution of the body, whereby that which helps and furthers vital constitution in one, and is therefore delightful, hinders and crosses it in another, and therefore causes grief. The difference therefore of Wits has its original from the different passions, and from the ends to which

the appetite leadeth them. As for that difference which arises from sickness, and such accidental distempers, I have appointed them for the second Part of this Book, and therefore I omit the same as impertinent to this place, and consider it only in such as have their health, perfection of body, and Organs well disposed.

Chapter II.[18]

1. Of the perfection of the Body, 2. And then of the Nature of the Senses. 3. Of Delights. 4. Pain. 5. Love. 6. Hatred. 7. Sensual Delight. 8. And Pains of the Body. 9. Joy. 10. And Grief.

1. Other things I have to say, but I will rather insist upon such things as are easy and intelligible even to idiots, or such physicians that are no wiser, who if they can but tell the joints of their hands, or know the use of their teeth, they may easily discover it was Counsel, not Chance, that created them; and if they but understand these natural Medicines I have prepared in this Book for their example, they will know that they shall be cured of all Diseases, without pain or any great cost; and Love, not Money, was it that made me undertake this Task. Now of the well-framed parts of our body, I would know why we have three joints in our Legs and Arms, as also in our fingers, but that it was much better than having but two or four? And why are our fore teeth sharp like Chisels, to cut, but our inward teeth broad, to grind? But this is more exquisite then having then all sharp, or all broad, or the fore-teeth broad, and the other sharp; but we might have

---

[18] This is called CHAP. II. in the original printed book. -pnw

made a hard shift to have lived, though in that worse condition. Again, why are the teeth so luckily placed? Or rather, why are there not Teeth in other bones as well as in the Jaw-bones, for they might have been as capable as these. But the reason is, nothing is done foolishly, nor in vain. I have showed you how to prolong life, and to return from Age to Youth; and how to change, alter and amend the state of the body: To keep the body in perfect health is my present design, and to cure all Diseases without reward, for there is a divine Providence that orders all things. Again (to say nothing of the inward curiosity of the Ear) why is that outward frame of it, but that is certainly known that it is for the bettering of our hearing?

2. I might add, that Nature has made the hindmost parts of our Body (which we sit upon) most fleshy, as providing for our ease, making us a natural Cushion, as well as for Instruments of Motion for our Thighs and Legs; she has made the hinder part of the Head stronger, as being otherwise unfenced against falls and other casualties. She has made the Backbone of several vertebrae, as being more fit to bend, more tough, and less in danger of breaking then if they were all one entire bone, without those gristly Junctures. She has strengthened our Fingers and Toes with Nails, whereas she might have sent out that substance at the end of the first and second Joints, which had not been so handsome and useful, nay, rather somewhat troublesome and hurtful. And lastly, she has made all bones devoid of sense, because they were to bear the weight of themselves, and of the whole body; and therefore, if they had had sense, our life had been painful continually and dolorous.

3. And now I have considered the fitness of the parts of man's body for the good of the whole, let me but consider briefly his senses and his nature, and then I intend more solidly to demonstrate the

cause of all Diseases, and with that the Cure, because I intended a *Holy Guide* in my *Harmony of the World*, and other Books. By our several Organs we have several Conceptions of several qualities in the objects; for by sight we have a conception or image composed of Color and figure, which is all the notice and knowledge the object imparts to us of its nature, by the excellency of the eye. By Hearing we have a conception called Sound, which is all the knowledge we have of the quality of the object from the Ear: And so, the rest of the Senses are also conceptions of several qualities or natures of their objects.

4.   Because the Image in vision consisting of Color and shape, is the knowledge we have of the qualities of the object of that Sense, it is no hard matter for a man to fall into this opinion, That the same Color and shape are the very qualities themselves; and for the same cause that sound and noise are the qualities of a piece of Canon or Culvering charged with sulphureous Powder, fired, or of the Air: And this opinion has been so long received, that the contrary must needs appear a great Paradox. The same qualities are easier in a bell; and yet the introduction of species visible and intelligible; (which is necessary for the maintenance of that opinion) passing to and fro from the object, is worse than any Paradox, as being a plain impossibility. I shall therefore endeavor to make plain these points.

5.   That the subject wherein Color and image are inherent, is not the object or thing seen.

6.   That there is nothing (really) which we call an Image or Color.

7.   That the said Image, or Color, is but an apparition unto us of the motion, agitation, or alteration, which the object works in the brain, or spirits, or some internal substance of the Head.

8. That as in vision, so also in conceptions that arise from the other Senses, the subject of their inherence is not the object, but the continent.

9. That Conceptions and Apparitions are nothing really, but motion in some internal substance of the Head, which motion not stopping there, of necessity must there either help or hinder the motion, which is called Vital; when it helps it is called Delight, Contentment, or Pleasure, which is nothing really but notion about the Heart, as Conception is nothing but motion ii the Head, and the objects that cause it are called, Pleasant, or Delightful, and the sane Delight, with reference to the object, is called Love; but when such motion weakens or hinders the vital motion, then it is called Pain, and in relation to that which causes it, Hatred.

10. There are two sorts of pleasures, whereof one seems to affect the corporeal Organ of the sense, and that I call sensual, the greatest part whereof is that by which we are invited to give continuance to our Species; and the next by which a man is invited to meat, for preservation of his individual person. The other sort of Delight is not particularly any part of the body, and is called, The Delight of the Mind, and is that which we call Joy. Likewise of Pains, some affect the Body, and are therefore called, The Pains of the Body; and some not; and those are called Grief.

Chapter III.

1. Of the nature of the soul of Man: 2. Whether she be a mere Modification of the body: 3. Or a substance really distinct: 4. And then whether corporeal, or incorporeal: 5. And of the temper of the body.

1.   Here I am forced to speak what I have in my book called *Familiar Spirit*, and it is not impertinent to my purpose; therefore if we say that the soul is a mere modification of the body, the soul then is but one universal faculty of the body, or a many faculties put together; and those operations which are usually attributed unto the soul, must of necessity be attributed unto the body; I demand therefore, To what in the body will you attribute, spontaneous motion? I understand thereby a power in our selves of wagging, or holding still most of the parts of our body, as our hand, suppose, our little finger: If you will say that it is nothing but the in mission of the spirts into such and such Muscles, I would gladly know what does emit these spirits, and direct then so curiously; Is it themselves? or the brains? or that particular piece of the brain they call the pine-kernel Whatever it be, that which doth thus emit them and direct them, must have Animadversion; and the same that has Animadversion has Memory and Reason also: Now I would know whether the spirits themselves be capable of Animadversion, Memory and Reason; for it indeed seems altogether impossible; for these animal spirits are nothing else but matter very thin and liquid, whose nature consists in this, that all the particles of it be in motion, and being loose from one another, trig and play up and down according to the measure and manner of agitation in them.

2.   I therefore, demand, which of these particles in these so many loosely moving one from another, has Animadversion in it? If you say

that they all put together have; I appeal to him that thus answers, how unlikely it is that that should have Animadversion that is so utterly uncapable of Memory, and consequently, of Reason; for it is impossible to conceive memory compatible to such a subject, as it is bow to write Characters in the Water, or in the Wind.

3. If you say the brain emits and directs these spirits; how can that so freely and spontaneously move itself, or another, that has no Muscles? Besides, Doctor Culpepper tells you, that though the Brain is the instrument of Sense, yet it has no sense at all of itself; how then can that that has no sense direct us spontaneously and arbitrarily, the animal spirits into any part of the body? An Act that plainly requires determinate sense and perception: But let the Physicians and Anatomists conclude what they will, I shall, I think, little less than demonstrate that the brains have no sense; for the same in us that has sense, has likewise animadversion; and that which has animadversion in us, has also a faculty of tree and arbitrary Fancy and Reason.

4. Let us now consider the nature of the brain, and see how compatible those alterations are to such a subject; verily if we take a right view of this Laxe, pith, or marrow in man's head, neither our sense nor understanding can discover anything more in this substance that can pretend to such noble operations, as free imagination and sagacious collections of Reason, then we can discern in a lump of fat, or a pot of honey; for this loose pulp that is thus wrapped up within our Cranium, is but a spongy and porous body, and previous, not only to the animal spirits, but also to more juice and liquor; else it could not well be nourished, at least it could not be so soft and moistened by drunkenness and excess, as to make the understanding inept and sottish in its operations. Wherefore I now demand, in this soft substance which we call the Brain, whose softness implies that it is in

some measure liquid, and liquidity implies a several motion of loosened parts; in what part or parcel thereof does Fancy, Reason and Animadversion lie? In this **laxe** consistence that lies like a Net, all on heaps in the water; I demand, in what Knot, Loop, or Interval thereof, does this faculty of free Fancy and active Reason reside? I believe not a Doctor in England, nay, not Doctor Culpepper himself, were he alive, nor his men, Doctor Freeman, and the rest, can assign me any; and if any will say, in all together; they must say that the whole Brain is figured into this or that representation, which would cancel Memory, and take away all capacity of there being any distinct notes and places for the several species of the things there presented, but if they will say there is in every part of the brain this power of Animadversion and Fancy, they are to remember, that the brain is in some measure a liquid body, and we must enquire how these loose parts understand one another's several Animadversions and notions; and if they could (which is yet very unconceivable) yet if they could from hence do anything toward the emission and direction of the animal spirits into this or that part of the body, they must do it by knowing one another's minds, and by a joint contention of strength, as when many men at once, the word being given when they weigh Anchor, put their strength together for the moving of that nasal body, that the single strength of one could not deal with; but this is to sake the several particles of the brain so many individual persons, a fitter object for laughter, then the least measure of belief.

5.   Besides, how come these many Animadversions to seem but one to us, our mind being these, as is supposed? Or why if the figuration of one part of the brain be communicated to all the rest: does not the same object seen situated both behind us, and before us, above and beneath, on the right hand and on the left; and every way, as the impress of the object is reflected against all the parts of the brains? But

there appears to us but one Animadversion, and one sight of things, it i.e. a sufficient Argument that there is but one; or if there be many, that they are not mutually communicated from the parts one to another, and therefore there can be no such joint endeavor towards one design; whence it is manifest, that the brains cannot unit or direct these animal spirits into what part of the body they please.

Chapter IV.

1. How a Captain was killed: 2. Of spontaneous motion: 3. Of the external Phenomena: 4. Of the nature of the Essence: 5. Of the Soul herself: 6. What it is: 7. And whether it be corporeal: 8. Or incorporeal.

1.   Now I must tell you, that the brain has no sense, and therefore cannot impress spontaneously any motion on the animal spirits; it is no slight argument, that some being dissected, have been found without brains: and this I saw, a Captain in Chrisley, in Arabia, that was accidentally killed by an Alcade and an Arabian; the stone is pleasant, but not pertinent to our purpose; but this man had nothing but a limpid water in his head, instead of brains; and the brains generally are easily dissolvable into a watery consistence, which agrees with what I intimated before. Now I appeal to any free Judge, how likely these liquid particles are to approve themselves of that nature and power, as to be able by erecting and knitting themselves together for a moment of time, to bear themselves so, as with one joint contention of strength, to cause an arbitrarious obligation of the spirits into this or that determinate part of the body; but the absurdity of this I have sufficiently insinuated already.

2.	The Nerves, I mean the Marrow of them, which is the same substance with the brain, have no sense, as is demonstrated from a catalepsy, or catochus; but I will not accumulate Arguments in a matter so palpable. As for that little sprunt piece of the brain, which they call the conacion, that this should be the very substance, whose natural faculty it is to move itself, and by its motion and nods to determine the course of the spirits into this or that part of the body, seems to me no less foolish and fabulous then the Stone of Thomas Harrington, Culpeper's man, who tells a Tale of his Masters Ghost, & etc. If you heard but the magnificent story that is told of the little lurking Mushroom, how it does not only hear and see, but imagines, reasons, commands the whole fabric of the body more dexterously then an Indian Boy does an Elephant: what an acute Logician, subtle Geometrician, prudent Statesman, skillful Physician, and profound Philosopher he is! And then afterwards by dissection you discover this worker of miracles to be nothing but a poor silly contemptible Knob, or Protuberance, consisting of a thin Membrane, containing a little pulpous matter, much of the same nature of the rest of the brain.

Spectatum admissi risum teneatis amici!

3.	Would you not sooner laugh at it, then go about to confute it? And truly I say the better laugh at it now, having already confuted it in what I have afore merrily argued concerning the rest of the brain.

4.	I shall therefore make bold to conclude, that the impress of spontaneous motion is neither from the animal spirits, nor from the brain, & etc. Therefore, that those operations that are usually attributed unto the soul, are really incompatible to any part of the body; and therefore, as in the last chapter I hinted, I say, that the soul is not a seer modification of the body, but a substance distinct therefrom.

5. Now we are to enquire, whether this substance distinct from what we ordinarily call the body, be also itself a corporeal substance, or whether it be incorporeal? If you say it is a corporeal substance, you can understand no other than matter more subtle and tenuous then the animal spirits themselves, mingled with them, and dispersed through the vessels and porosities of the body; for there can be no penetration of dimensions: But I need no new arguments to confute this fond conceit; for what I said of the animal spirits before, is applicable with all ease and fitness to this present case; and let it be sufficient that I advertise you so much, and so be excused from the repeating of the sane things over again.

6. It remains therefore that we conclude, that that which impresses spontaneous motion upon the body, or more immediately upon the animal spirits: That which imagines, remembers, and reasons, is an immaterial substance, distinct from the body, which uses the animal spirits and the brain for instruments in such and such operations. And thus, we have found a spirit in a proper notion and signification, that has apparently these faculties in it, it can both understand and move corporeal matter.

7. And now this prize that we have won will prove for our design in this new method of Physick and Philosophy of very great consequence; for it is obvious here to observe that the soul of man is as it were ἄγαλμα Θεῦ, a compendious statue of the Deity; her substance is a solid Effigies of God; and therefore, as with ease, we consider the substance and motion of the vast Heavens on a little sphere, or

Globe, so we may with like, facility contemplate the Nature of the Almighty in this little Model of God, the soul of man, enlarging to infinity what we observe in our selves when we transfer it unto God, as we do imagine these Circles which we view on the Globe, to be vastly bigger while we fancy them as described in the Heaven.

8.   Wherefore we being assured of this, that there is a spiritual substance in ourselves, in which both these properties do reside, viz. of the understanding, and of moving the corporeal matter; let us but enlarge our minds so as to conceive as well as we can of a spiritual substance that is able to move and actuate all matter whatsoever, never so far extended, and after what way and manner soever It please, and that it has not only the knowledge of this or that particular thing, but a distinct and plenary cognizance of all things; and we have indeed a very complete apprehension of the nature of the eternal and invisible God, who, like the soul of man, does not indeed fall under sense, but does everywhere operate so, that his person is easily to be gathered from what is discovered by our outward senses.

Chapter VI.

Of Plants, that the mere motion of the matter may do something, yet it will not amount to the production of Plants. That it is no botch in Nature, that some phenomena be the results of Motion, others of substantial forms. That beauty is not a near fancy, and that the beauty and virtue of Plants is an Argument that they are made for the use of our bodies from an intellectual principle.

1.   How weak is Man, if you consider his nature, what faculties he has, and in what order he is in respect of the rest of the Creatures? And indeed, though his body be but weak and disarmed, yet his inward

abilities of Reason, and artificial contrivance, is admirable, both for finding out those secret Medicines, which God prepared for the use of Man, in the Bowels of the Earth, of Plants and Minerals.

2. And first of Vegetables, where I shall touch only these four Reads, their Form and Beauty, their Seed, their Signatures, and their great use, as well for Medicines as sustenance; and that we may the better understand the advantage we have in this closer contemplation of the works of Nature, we are in the first place to take notice of the condition of the substance, which we call matter, how fluid and slippery, and indeterminate it is of itself; or if it be hard, how unfit it is to be changed into anything else; and therefore all things rot into a moisture before anything can be generated of them, as we soften the wax before we set on the seal.

3. Now therefore, unless we will be foolish, as because the uniform motion of the Air, or some more subtle corporeal Element, may so equally compress or bear against the parts of a little vaporous moisture, as to form it into round drops (as we see in the dew, and other experiments) end therefore because this more rude and general motion can do something, to conclude that it does all things: We must in all reason confess, that there is an eternal Mind and Virtue, whereof the matter is thus usefully formed and changed.

4. But mere rude and undirected motion, because naturally it will have some kind of results, that therefore it will reach to such as plainly imply a wise contrivance of counsel, is so ridiculous a sophism, as I have already intimated, that it is more fit to impose upon the inconsiderate souls of fools and children, then upon men of mature reason, and well exercised in Philosophy, or the grave and well-practiced, seraphically illuminated Rosie Crucians. Admit that Rain, and Snow, and Wind, and Rail, and Ice, and Thunder, and Lightning,

and a Star I mention for example, that may be let in amongst Meteors, by some called Hellens Star, and is well known at Sea, I have seen it melt Copper Vessels aboard a ship; it cometh of an heap of such vapors as are carried by violent cross Winds up from the Earth; and such like Meteors may be the products of heat and cold, or of the motion and rest of certain small particles of the matter; yet that the useful and beautiful contrivance of the Branches, flowers, and Fruits of Plants should be so too, (to say nothing yet of Minerals, and the bodies of men) is as ridiculous and supine a collection, as to infer, That because mere heat and cold does soften and harden Wax, and puts it into some shape or another, that therefore this mere heat and cold, or motion and rest, without any art and direction, made the silver seal too, and graved upon it so curiously some Coat of Arms, or the shape of some Bird or Beast, as an Eagle, a Lyon, etc. nay indeed this inference is more tolerable far then the other; these effects of Art being more easy, and less noble then those other of Nature.

5. Nor is it any deficiency at all in the works of Nature, that some particular phenomena be but the easy results of that general motion communicated unto the matter from God; others the effects of more curious contrivance, or of the Divine Art, or Reason (for such are the *rationes seminales*) incorporated in the Matter, especially the Matter itself being in some sort vital, else it would not continue the motion that it is put upon, when it is put upon, when it is occasionally this or the other way moved; and besides the Nature of God being the most perfect fulness of life that is possibly conceivable, it is very congruous, that this outmost and remotest shadow of himself, be some way, though but obscurely vital: Wherefore things falling off by degrees from the highest perfection, it will be no uneven or unproportionable step, if descending from the top of this utmost Creation, man, in whom there is a more fine conception, or reflexive Reason, which hangs on, as

every man has so much experience as to have seen the Sun, and other visible Objects, by reflection in the Water and Glasses, and this as yet shall be all I will say for this reason; I will give you more than I promised in the Contents, by four Propositions concerning the nature of Conceptions, and they shall be proved; and also of the main deception of Sense, that Color and Image may be there where the thing seen is not: But because it may be said, That notwithstanding the Image in the Water be not in the object, but a thing merely phantastical, yet there may be colors really in the thing itself; I will urge further this experience, That divers times men see directly the same object double, as two Candles for one, which may happen from distemper, or otherwise without distemper if a man will; the Organs being either in their right temper, or equally distempered, the Colors and Images in two such Characters of the same thing, cannot be inherent therein, because the thing seen cannot be in two places.

6. One of these Images therefore is not inherent in the Object; but the seeing, the Organs of the sight are then is equal temper or distemper; the one of them is no more inherent then the other, and consequently, neither of them both are in the Objects, which is the first proposition mentioned in the precedent number.

7. Secondly, that the Image of anything by reflection in a Glass, or Water, or the like, is not anything in, or behind the Glass, or in, or under the Water, every man may grant to himself; which is the second Proposition of Des Cartes.

For thirdly, we are to consider, first, that every great agitation or concussion of the brain (as it happens from a stroke, especially if the stroke be upon the eye) whereby the Optic Nerve suffers any great violence, there appears before the Eyes a certain light, which light is nothing without, but an apparition only; all that is real being the concussion or motion of the parts of the Nerve; from which experience we may conclude, that apparition of light is really nothing but motion within. If therefore from Lucid bodies there can be derived motion, so as to affect the Optic Nerve in such manner as is proper thereunto, there will follow an Image of light somewhere in that line, which the motion was at last derived to the eye, that is to say, in the Object, if we look directly on It, and in the Glass or Water, when look upon it in the line of reflection, which in effect is the third proposition, namely, That image and Color is but an apparition to us that motion, agitation, or alteration, which the object works in the brain, or spirits, or some internal substance in the head.

4. But that from all lucid, shining, and illuminate bodies, there is a motion produced to the eye, and through the eye, to the Optic nerve, and so into the Brain, by which the apparition of light or Color effected, is not hard to prove. And first, it is evident that the Fire, the only lucid body here upon Earth, works by motion equally every way, insomuch as the motion thereof stopped or inclosed, it is presently extinguished, and no more fire. And further, that motion whereby the fire works is dilation and contraction of itself alternately, commonly called Scintillation, or glowing, is manifest also by experience; from such motion in the fire must needs arise a rejection, or casting from it self off that part of the medium which is contiguous to it, whereby that part also rejects the next, and so succesively one part beats back another to the very eye, and in the same manner the exterior part of the eye presss the Interior (the Laws of refraction still observed.) Now the interior coat of the eye is nothing else but a piece of the Optic Nerve, and therefore the motion is still continued thereby into the Brain, and by resistance or reaction of the Brain, is also a rebound into the Optic Nerve again, which we not conceiving as motion or rebound from within, do think it ii without, and call it Light, as has been already

shown by the experience of a Stroke: We have no reason to doubt that the Fountain of Light, the Sun, worketh by any other ways then the Fire, at least in this matter: And thus all vision has its original from such motion as is here described; for where there is no light, there is no sight; and therefore Color must be the same thing with light, as being the effect of the lucid bodies, their difference being only this, that when the light cometh directly from the Fountain to the eye, or indirectly by reflection from clean and polite bodies, and such as have not any polite bodies, and such as have not any particular motion internal to alter it, we call it light; but when it cometh to the eye by reflection, from uneven, rough and course bodies, or such as are affected with internal motion of their own that may alter it, then we call it Color; Color and light differing only in this, that the one is pure, and the other perturbed light; by that which has been said, not only the truth of the third Proposition, but also the whole manner of producing light and Color is apparent.

5. As Color is not inherent in the object, but an effect thereof upon us, caused by such motion in the object, as has been described; so neither is found in the thing we hear, but in ourselves; one manifest sign thereof, is, that as man may see, so also he may hear double and treble by multiplication of Echoes, which Echoes are sounds as well as the Original; and not being in one and the same place, cannot be inherent in the body that makes them; nothing can make anything which is not in itself; the Clapper of a Bell has no Sound in it, but motion, and makes motion in the internal parts of the Bell; so the Bell has motion and not sound, that imparts notion to the air; and the air has motion, but not sound; the Air imparts motion by the Ear and Nerve into the Brain; and the Brain has motion, but not sound; from the Brain it rebounds back into the Nerves outward, and thence it becometh an Apparition without, which we call sound. And to

proceed to the rest of the Senses, it is apparent enough, that the smell and taste of the same thing are not the same to every man, and therefore are not in the thing smelt or tasted, but in the men; so likewise the heat we feel from the fire is manifestly in us, and is quite different from the heat which is in the fire; for our heat is pleasure or pain, according as it is great or moderate; but in the cool there is no such thing: By this the last is proved, vim, that as in vision, so also in conceptions that arise from other Senses, the subject of their inherence is not in the Object, but in the Sentinent[19]: And from hence also it follows, that whatsoever accidents or qualities our Senses make us think there be in the world, they be not there, but are seeming and Apparitions only; the things that really are in the world without us, are those motions by which these seeming's are caused; and this is the great deception of sense, which also is to be by sense corrected: for as sense tells me, when I see directly, that the Color seems to be in the object; so also sense tells me, when I see by reflection that Color is in the object. But now I am out of the way from the outward Creation of Man, in whom there is a principle of more fine and reflexive reason, which hangs on, though not in that manner, in the more perfect kind of Brutes, as sense also (10th to be curbed with too narrow compass) lays hold upon some kind of Plants, as in those sundry sorts of Zoophyta, but in the rest there are no further footsteps discovered of an animadversive form abiding in them; yet there be the effects of an inadvertent form

($λόγ\̣Θ$ $ἔνυλ\̣Θ$) of materiated or incorporated Art or seminal Reason; I say, it is no uneven jot to pass from the more faint and obscure example of Spermatical life, to the more considerable effects of general Motion in Minerals, Metals, nor yet to say anything of the Medicines

---

[19] Exact spelling as in the printed edition. -pnw

extracted, mortified, fixt, dissolved, and incorporated with their proper Veagles[20], because we have intended it our last business, to return to Minerals, Metals, and sundry Meteors, whose easy and rude shapes have no need of any particular principle of life, or Spermatical form distinct from the rest, or motion of the particles of the matter.

10.   But there is that curiosity of form and beauty in the more noble kind of Plants, bearing such a suitableness and harmony with the more refined sense and sagacity of the soul of Man, that he cannot choose (his intellectual touch being so sweetly gratified by what it deprehends[21] in such like objects) but acknowledge that some hidden cause, much a kin to his own nature that is intellectual, is the contriver and perfecter of these so pleasant spectacles in the world.

Nor is it at all to the purpose to object, that this business of beauty and comeliness of proportion is but a conceit, because some men acknowledge no such thing, and all things are alike handsome to them, who yet notwithstanding have the use of their eyes as well as other folks; for I say, this rather makes for what we aim at, that pulchritude is conveyed indeed by the outward senses unto the soul, but a more intellectual facility is that which relishes it; as an astrological, or better, a geometrical scheme is let in by the eyes, but the demonstration is discerned by reason: And therefore it is more rational to affirm, that some intellectual principle was the Author of this pulchritude of things, then that they should be thus fashioned without the help of that principle: And to say there is no such thing as pulchritude, and some say, there is no way to felicity: The first, I answer, is, because some men's souls are so dull and stupid. The first cannot relish all objects alike in that respect: The second knows not happiness, nor the way to

---

[20] Spelled exactly as in the printed edition. -pnw
[21] Spelled exactly as in the printed edition. -pnw

long life, nor the means to health, nor how to return from age to youth, & etc. Which is as absurd and groundless, as to conclude there is no such thing as reason and demonstration, because a natural fool cannot reach unto it. But that there is such a thing as *The Holy Guide*, long life, and a certain way to health, not as yet known in England, I will demonstrate: the way to health I shall show you anon in this Book, the rest in another Part, as I promised you.

12.   Now that there is such a thing as Beauty, and that it is acknowledged by the whole generations of men, to be in Trees, Flowers, and Fruits, and the adorning of buildings in all Ages, is an example, and undeniable testimony; for what is more ordinary with them, then taking in flowers and fruitage for the garnishing of their work? Besides, I appeal to any man that is not sunk into so forlorn a pitch of Degeneracy, that he is as stupid to these things as the basest of Beasts, whether for example, a rightly cut tetradrum, cube or icosaedrum, have no more pulchritude in them, then any rude broken bone lying in the field or highways: Or to name other solid Figures, which though they be not regular properly so called, yet have a settled Idea, and Nature, as a cone, sphere, or cylinder, whether the sight of these do not gratify the minds of men more, and pretend to more elegancy of shape, then those rude cuttings or clippings of Freestone that fall from the Masons hands, and serve for nothing but to fill up the middle of the wall, and so to be hid from the eyes of Man for their ugliness: And it is observable, that if Nature shape anything near this geometrical accuracy, that we take notice of it with much content and pleasure, as if it be but exactly round, as there be abundance of such stones upon Mesque, a hill in Arabia; I have seen them there, ordinarily Quinquangular, and have the sides parallels, through the Angels be unequal, as is seen in some little stones, and in a kind of Alabaster found here in England, and other pretty stones found upon

Bulvertonhill near Sidmouth in Devonshire, and near Stratford upon Avon; and in Tins Grove at Colton, and at Tardebick, Stony-Hill, the Shawes and Quarry Pit, Razlehill, and Ash-hill in Warwickshire, are found such stones that grow naturally carved with various works, some with Roses, others with Lions, Eagles, and all manner of delightful works; these stones, I say, gratify our sight, as having a nearer cognation with the soul of man that is rational and intellectual, and therefore is well pleased when it meets with any outward object that fits and agrees with those congenite ideas her own nature is furnished with: For symmetry, equality, and correspondency of parts, is the discernment of Reason, not the object of Sense, as I in our *Harmony of the World* have in another place proved.

13. Now therefore it being evident, that there is such a thing as beauty, symmetry, and comeliness of proportion (to say nothing of the delightful mixture of colors, and that this is the proper object of the Understanding and Reason; for these things be not taken notice of by the Beasts) I think I may safely infer, that whatsoever is the first and principal cause of changing the fluid and undetermined Matter into shapes so comely and symmetrical, as we see in flowers and trees, is an understanding Principle, and knows both the nature of man, and of those objects he offers to his sight in this outward and visible world, and would have men search and find out those secrets by the which he might keep his body in health many hundreds of years, and at last find the way our *Holy Guide* leadeth; for these things cannot come by chance, or by a Multifarious attempt of the parts of the matter upon themselves; for then it were likely that the species of things, though some might hit right, yet most would be maimed and ridiculous; but now there is not any ineptitude in anything, which is a sign that the fluidness of the matter is guided and determined by the overpowering counsel of an eternal mind.

14. If it were not needless, I might instance in sundry kinds of flowers, herbs, and trees; but these objects being so obvious, and every man's fancy being branched with the remembrance of roses, marigolds, gilliflowers, peonies, tulips, pansies, primroses, ferneflowers and seed, orange flowers, the leaves and clusters of the vine, & etc. Of all which you must confess, that there is in them beauty, and symmetry, and use in Physick, and grateful proportion; I hold it superfluity to weary you with any longer induction, but shall pass on to those considerations behind, of their seed, signature and usefulness, and shall pass through them very briefly, and then I shall come to mineral Medicines; those observables being very necessary first to be known by way of an introduction, and as ordinary and easily Intelligible; but for your better instruction in the understanding of this Book, read the *Harmony of the World*, and the *Temple of Wisdom*. You must remember our design is to prove both the Theory and Practical Parts of these Mysterious Truths.

Chapter VII.

1. 2. 3. 4. 5. Of the Seeds: 6. 7. 8. and Signatures of Plants: 9. 10. 11. And wherefore God made them.

1. Every plant has its seed; Rosie Crucians therefore say there are secret Mysteries lie hidden in them, which should be our delight to find out; for Divine Providence made all good for the use of man: And

this being no necessary result of the motion of the matter, as the whole contrivance of the plant indeed is not; and it being of great consequence that they have seed for the continuance of propagation of their whole species, and for the gratifying of man's Art also, industry and necessity (for much of Husbandry and Gardening lies in this) it cannot but be the act of Counsel to furnish the several kinds of Plants with their seed, especially the earth being of such a nature that though at first for a while it might bring forth all manner of Plants, (as some will have it also to have brought forth all kinds of Animals) yet at last it would grow so sluggish, that without the advantage of those small compendious principles of generation, the Grain of seed would yield no such births, no more than a Pump grown dry will yield any water, unless you pour a little water into it first, and then for so many Seasons full, you may fetch up so many Tankards full.

2.   Nor is it material to object, that stinking weeds and poisonous plants bear seed too, as well as the most pleasant and useful; for even those stinking Weeds and poisonous Plants have their use in Rosie Crucian Medicines, as you shall know hereafter; besides our common Physick-mongers often use them as their fancy guides them, grounded upon no other reason than woeful and deadly experience; sometimes the industry of man is exercised by them, to weed them out where they are hurtful; which reasons, if they seem slight, let us but consider, that if human industry had nothing to conflict and struggle with, the fire of sans spirit would be half extinguished in the flesh, and then we shall acknowledge that that which I have alleged, is not so contemptible nor invalid.

3.   But secondly, who knows but it is so with poisonous Plants, as vulgarly is fancied concerning Toads, and other poisonous Serpents that lick the Venom from off the earth? So poisonous Plants may well

draw to them all the malign Juice and nourishment, that the other may be purer and more delicate, as there are Receptacles in the body of man; and Emunctories to drain off superfluous Choler and Melancholy, etc.

4. Lastly, it is very well known by them that know anything in Nature and Physick, That those Herbs that the rude and ignorant world call Weeds, are the materials of very Sovereign Medicines; that Aconitum Hiemale, or Winter Wolfs Balm, that otherwise is rank poison, is reported to prevail mightily against the biting of Vipers, Scorpions, and sad Dogs, which Sir Christopher Heydon assents unto; and that that Plant that bears death in the very name of it, Solanum Lethiferum, prevents death by procuring sleep, if it be applied in a Fever; nor are those things to be deemed unprofitable, say the Rosie Crucians, whose use our heavy ignorance will not let us understand; but they will teach us as follows.

5. We come now to the Signatures of Plants, which indeed respects us more properly and adequately then the other, and is a key (as the Rosie Crucians say) to enter man into the knowledge and use of the Treasures of Nature; I demand, therefore, Whether it be not a very easy and genuine inference from the observing that several herbs are marked with some mark or sign that intimates their virtue, what they are good for; and there being such a creature as Man in the World, that can read and understand these signs and characters; hence to collect that the Author both of man and them, knew the nature of them both; and besides Divine providence would only initiate and enter mankind in the useful knowledge of her Treasures, by the Seraphical illuminated Rosie Crucians, leaving the rest to employ the vulgar that they might not be idle; for the Theatre of the world is an exercise of man's wit, and therefore all things are in some measure obscure and

intricate; that the sudulity of that divine spark, the soul of man, may have matter of conquest and triumph, when he has done bravely by a superadvenient assistance of God.

6. But that there be some plants that bear a very eminent signature of their nature and use, for example, Capillus Veneris, Polytrichon, or Maidenhair; the Lye in which it is sodden or infused, is good to wash the head, and make the hair grow in those places that are bare; the decoction of Quinces, which are a downy and hairy fruit, is accounted good for the fetching again hair that has been fallen by the French pox; the leaf of balm, or allelujah, or wood-sorrel, as also the roots of anthora, represent the heart in figure, and are cardiacal.

7. Walnuts bear the whole signature of the head; the outward green cortex answers to the pericranium, and a Salt made of it is singular good for Wounds in that part, as the Kernel is good for the Brains, which it resembles.

Umbelicus Veneris is powerful to provoke Lust, as Doctor Culpepper affirms; as also your several sorts of satyrions, which have the evident resemblance of the general parts upon them; aron especially, and all your orchisses, that they have given names unto, from some beast or other, as cynosorchis, orchis, miodes, tragorchis, & etc. the last whereof notorious for its Goatish smell, and Tufts not unlike the beard of that lecherous Animal, is of all the rest the most powerful incentive to lust.

8. The leaves of hypericon are very thick prickled, or pointed with little holes, and it is a singular good Wound-herb, as useful also for deobstructing the pores of the body.

9. Scorpioidhes, echium, or scorpions grass, is like the crooked tail of a scorpion; and ophioglossum, or adders tongue, has a very plain and perfect resemblance of the Tongue of a Serpent; as also

ophioscoroden of the entire head and upper parts of the body; and these are all held very good against poison, and the biting of Serpents; & generally all such plants as are speckled with spots like the skins of Vipers, or other venomous creatures, are known to be good against the stings or biting of them, and are powerful objects against poison.

10.   Thus, did Divine Providence by natural Hieroglyphics, read short Lectures to the rude wit of vulgar man; others of the seraphically illuminated fraternity being entered, and sufficiently experienced of these, found out the rest, it being very reasonable that other herbs that lad not such signatures, might be very good for Medicinal uses, as well as they that had.

11.   Rosie Crucians have quickened and actuated their Phlegmatic natures to more frequent and effectual venery; for their long lives, health, and youthfulness, shows they were not very fiery, to say nothing of their happiness, knowledge, riches, wisdom and virtue, because I have in this Treatise spoken of it largely.

Chapter VIII.

1, 2. Of the usefulness of Plants: 3, 4. And of the Works of God.

1. You shall now briefly take notice of the usefulness and profitableness of Plants, both for Physick and Food, and then pass on to the consideration of the inspired Rosie Crucians, what their Medicines are: As for the common uses of Plants Herbals teach you something; but I refer you to the singular medicines of Rosie Crucians in my Book of *The Harmony of the World*; for the salvation of your health; Animals know as much by instinct and nature; and that which is most observable here is this, That brute Beasts know as much as many Physicians do that are taught by Herbals only; and these deny the Power of God in the works of Nature, and the power of Nature in the skill of Man, that it should be impossible to make trees bear fruit in December, and apple trees to grow, to blossom, and bear apples, contrary to kind, in March.

2. Beasts have knowledge in the virtue of Plants as well as Men; for the Toad being overcharged with the poison of the Spider, (as is well known) has recourse to the Plantain-leaf. The Weasel, when she is to encounter the Serpent, arms herself with eating of Rue. The Dog, when he is sick at the stomach, knows his cure, falls to his grass, vomits, and is well. The Swallows sake use of celandine, the lennet of euphragia, for the repairing of their sight. And the Ass, when he is oppressed with melancholy, eats of the herb asplenium, or miltwast, and so eases himself of the swelling of the Spleen. The Raven makes use of cinquefoil for the prolongation of his life, to sometimes six or seven hundred years; and therefore I think it is, that Rosie Crucians prescribe the oil of Ravens, Swallows, and Harts, for the use of man to

anoint himself, to continue his fresh and well—complexioned body from wrinkles and lameness: and Dictamnum Cretense is much used, as I told you in my *Wise Man's Crown*, and *Temple of Wisdom*: Cretian Dittany cures Wounds of what nature soever.

Which thing I conceive no obscure indignation of Providence; for they doing that by instinct and Nature, which Men, who have free Reason, cannot but acknowledge to be very pertinent and fitting, nay such, that the most skillful Physician will approve and allow; and these Creatures having no such reason and skill themselves as to turn Physicians, it must needs be concluded by virtue of that principle that contrived them, and made them of that nature they are, enabled them also to do these things.

3. Let us now consider the fruits of the Trees, where I think it will appear very manifestly, that there was one worker of miracles, and inspirer of Rosie Crucians; I might now reach out to Exotic Plants, such as the Cinnamon-tree, the Balsam-tree, and the Tree that bears the Nutmeg, enveloped the Mace, as also the famous Indian Nut-tree, which at once (as the Rosie Crucians say) affords almost all the necessaries of life; for if they cut but the Twigs at Evening, there is a plentiful and pleasant juice comes out, which they receive into Bottles, and drink instead of Wine, and out of which they extract such an aqua vitae, as is very sovereign against all manner of sicknesses; the branches and boughs they make their Houses of, and the body of the tree being very spongy within, though hard without, they easily contrive into the frame and use of their canoes, or boats; the kernel of the Nut serves them for bread and meat, and the shells for cups to drink in; and indeed they are not mere empty cups, for there is found a delicious cooling milk in them; besides, there is a kind of hemp that encloses the Nut, of which they make Ropes and Cables, and of the

finest of it Sails for their ships; and the Leaves are so hard and sharp pointed, that they easily make Needles or Bodkins of them for stitching their Sails, and for other necessary purposes; and that Providence may show herself benign as well as wise, this so notable a Plant is not restrained to one Coast of the World, as the East Indies, but is found in Africa, Arabia, and in all the Islands of the West Indies, as Hispaniola, Cuba, where our men are victors, and several other places of the newfound World.

4. But I though fit to insist upon these things by way of Proof and Instruction, but to contain myself within the compass of such subjects as are necessary for our knowledge, and familiarity and ordinarily before our eyes, that we may the better (these things understood) take occasion from thence to demonstrate the Rosie Crucian way to health, and their ordinary Medicines which to us are not yet known, & etc.

Chapter IX.

1. The Rosie Crucian way how to get health. 2. The causes why we eat food. 3. Of the first nature of the World. 4. A measure of raw and temperate meat. 5. And the cause of the fiery, and scummy Gall. 6. And needless muddy bowels the Milt. 7. Nature careless of making the reins of Urine drawers. 8. Drinkless Animals have none at all. 9. How to cleanse yourself from these idle Bowels. 10. And avoid all Diseases.

1. Do you not consider the weakness of man, what faculties he has, and in what order he is in respect of the rest of the Creatures;

Rosie Crucians observe, though his body be weak and disarmed, yet his inward abilities of reason, and artificial contrivance is admirable; he is much given to search out the medicinal virtues of plants, wights, and minerals, and have found out those that were of so present and great consequence, as to be Antidotes against poison, that would so quickly have dispatched mankind; it were good for us to demonstrate the Rosie Crucian medicines, now our land is afflicted with a sickness called the new disease, of which all sorts die, without remedy, for none as yet have prescribed a medicine; for young men that desire to live, and for old men that wish for health, without which no life is sweet and savory; then let us bend ourselves to cure our brethren first, and endeavor to show the means (besides the common Collegian Doctor's drenches, or Culpepper's way, how every man may get and keep his health, that is something strange, but a vowed truth; the consent and equal (I mean agreeable to kind) temper and dulling our four first beginnings, the staff of our bodies; for if this knot be broken, and they loose towards their former liberty, they wax proud and strong, end fight; for their nature is together by the ears, and put us to palm, and lets the rule of nature, and this they call the disease.

2. Then to handle one at once, as our manner is, and will keep our custom still, to keep our health and body in temper, seems no such matter to me as the world would make it, even plainly impossible, when I know all the ways and entries to let in diseases and distempers of the body, may by small heed be stopped and sensed.

3. We need to draw breath and eat meat; for the cause I shall speak of it in its place; and as this is not all clear and agreeable, so nature has her leavings; and again, labor and rest are needful, and perhaps we cannot choose but be moved in mind with joy, grief, fear, hope, and such like passions, though the stoics deny necessity, says Des Cartes.

4. By so many ways and gates diseases may enter, if they be not well watched and looked unto, which may be done in reason, and has been done often, as they assure us that have lived long without all diseases and sicknesses, as John Harding relates of a Minister, called John Macklaine, to have continued for these five-score years last past together in health, and after his hair, teeth, eyes, and flesh renewed, and became young again; and such like stories are to be found enough, if we might stay to seek them; some are contented for all but air and meat, but these say they have often seeds of diseases lye hid in them, unable to be foreseen or prevented, as we find those meats that make the finest show (as Wine, and Sugar, and such enticing baits) to have hid in them most hateful diseases and dregs in the bottom; so the sir, when it seems the best and highest, yet is sometimes infected and poisoned with venomous breath, sent out and thrust into it, either from below, or from the Stars of Heaven, and as the cause is hidden and unknown to us, so the hurt impossible to be avoided and prevented.

5. If I list to let my speech run at large, especially in other men's grounds, I could find that Division is false; first, (to come to meat anon) and then if it were true, yet the cause of that infection not unable to be foreseen and warded; but I am so sorry for the fault above, that I can the better take heed hereafter; yet methinks it is a grief to hear the harmless and glorious divine things above, so defaced with slander, and no man makes answer for them.

6. Gentle Reader, be pleased to stay a little, if the stars have no light, and so no power but from the sun, that most wholesome and prosperous creature, then they hurt him most wrongfully, and reprove themselves very rightly. And again, if they be but a piece of the finer part, and first nature, as it were, of the World, as I have showed in my

Book of *The Temple of Wisdom*, then they be the most wholesome things in the World, so far be they from poisoned slander: And so let their Lights be never so grossly mingled in their meetings, and thereby that State of the Weather changed suddenly, and from thence our bodies troubled and turned into Disease, because they were not prepared and made ready for it, yet the things are good and prosperous; and by knowledge of Astrology, or influences of the Planets, and races of the Stars, we may prepare ourselves, and prevent all, if we cannot have that happiness to converse with our Guardian Genius. Now for lower reflection, it is not worth the answering, when there is so much waste ground in the World; then let us pass over to that other breach; may we not shun the leaving baits in our Diet, and take such meat as is most temperate and near our Nature, and then dress the same after the most kindly and wholesome manner, seasoning it well with labor, mirth and sleep?

7. And to be plain, I have showed in another part of this Treatise of mine, so much noted by our Writers, what a jewel of health it were to use all raw and temperate meat, or because we be wise and virtuous, and this Diet would perhaps change our Nature of fire, but like philosophers a quite contrary way, taking the best, when as none is lost, and leaving worst, which is that we now take, a Way I say, to strip of all grossness and foulness of bodies, the only hurt of themselves, and is the Food of Diseases.

8. I will tell you another way which you will think strange, but you shall find it true, if the meat be temperate, as I bid you choose it, there is no hurt can come thereby, (if you keep measure in your selves) save from your leavings; these in so clear a Diet first will be very few; but if you would be ruled by my counsel which nature taught me, those few should never hurt you.

9. Of all the Leavings in the body, there are three which the liver makes most troublesome unto us (for the rest are easily dispatched) a light and easy, or rather a fiery (as some call it) choler; a cold and heavy mud, called melancholy; and the third is urine, which I will treat of in the next Chapter, but those two the worse; and this fault is not in themselves, but all by reason of the needless and hurtful bowels in our bodies, (as the Seeds-man uses to sow good and bad together) which being of the same kind and quality with those humors, do draw and pull them still unto them (as all other parts and things do) for their Food and nourishment; and so by the narrow passages to and fro, their greediness in pulling and holding, and a hundred such means, subject to great mischances, have brought in as many mischiefs, whereas Nature the great expeller of her unlike, and Enemies, if she had free choice and liberty, would otherwise with ease, and without hurt, expel those Leavings, especially so small a number of the better sort in so clean a Diet, nay, set the malice of those parts, (those parts are milt, gall, and reins) if there be not sufficient store of other foul meat at hand, like a poisoned or a purging Medicine, they use to draw good Juices, and to make Food of them; what is not manifested in this chapter, shall methodically be demonstrated in the ensuing, for I intend to be serious in this part of my Book, and will show you what Nature taught me.

Chapter X

1. Rosie Crucian Medicines made plain by examples, and those are above controlement. 2. That the wet Sun-beams declare some fine and forrain fatness to nourish mankind. 3. How to live twenty years without Food, as many creatures do. 4. Use and Custom a second Nature. 5. The Bird called Manuda Diaca, and the singing Dog. 6. That the Camelion never eats food. 7. An experienced Medicine, and how to apply it with Paracelsus, and the Rosie Crucian new Art of Healing.

1.   Of Aristotle it is reported, that he is the witty Spy of Nature, and as if he had been made in this matter, he shows the need and use of the greater Entrails and Bowels of Wights, and says very truly and wisely. The Heart and Liver as the spring of Life and Food, be needful for all Wights, adding to the hotter one the brain to cool, and the lights to cleanse the heat, staying there as if he thought the other three unprofitable; nay for one of them in the same Book, (I say) telling the stories of the Hart and Camel, and giving the reason why they be both so swift, healthful, long-lived and other good proportions above the rest enfeoffed, vouched in plain terms, the want of the fiery and scummy Gall, as a great Enemy to them, for the Milt that muddy Bowel, that it may be left out as needless in the bodies of the better creatures. The Meadows near Cortina and Maggadere declare when by a strange and hidden virtue they bereave the Beasts thereof that graze upon them, of it; the Herb is called Asplenium, as I told you in the preceding Chapters, nay, that the Milt is not only idle, but hurtful, which all experience, even in ourselves, has taught it.

2.   The Turks light Footmen, (I say, which I know not by what example unless it were the want of the same in the Camel, making the

Beast able to travel a hundred miles a day, and so without drink fifteen days together) being in their childhood purged of their Milt, prove thereby the most light, swift, sound and lasting Footmen in the World.

3. As for the veins of Urine-drawers, as drinkless creatures have none at all, so some men have but one of them, as if nature passed not to make any at all; if we could forbear our drink (as these Beasts do by kind, and some men by custom) we might the better spare them, and avoid many miechiefs in our bodies.

4. Therefore the odd man, Paracelsus, I know not by what light, if not of the Rosie Cross, (cast in I think from Seraphical illumination) not only sees these faults, but also finds ways to amend them, and to cut the mischiefs off all these three noisome parts, not with any yielding Craft, but with Rosie Crucian divine kind of Healing, with aurum potable, etc. so that to avoid all diseases that spring of the leavings, take of aurum potable one ounce; one pound of the Oil of Ravens; two round of miltwast, or asplenium, a handful of cinquefoil, of dictamnum cretense, ophioglossum and scorpiodes, echium, of each a like quantity, and observe the Ascendant and his Lord; and the Moon, and Lord of the fixed, at your discretion, and take the quantity of a Walnut every night and morning, and anoint the face and hands, and (if you will) the rest of the body: Rosie Crucians have other healing and yielding Medicines; you shall know them in their places; this is such an experienced Medicine, that you know where to find it; I need not chew you to put out the sway and power of these idle bowels, or perhaps it should not need, and in a stock that eases our clean Diet Nature herself as she doth in those Meadows, by other creatures, would also quite raze and dispatch them within a few generations.

5. But I will go further, hear a Rosie Crucian new and unheard of opinion, and yet let not your judgement run before you see good

reason; what if we could fast for ever, and live without all food? Might not all hurt and danger of meat be then forestalled? If other Creatures; whose life hangs upon the same hold, by the sufferance, nay by the commandment of God and nature, do last forever, there is no reason but the same common nature will at last suffer it in us; Let us see. And to step over the chameleon, because it is a cold and bloodless creature; what say we to a bird which is an hot and perfect one? A bird in the Molucco Islands, Manuda Diaca by name, that has no feet at all, no more than an ordinary fish, as Mr. Moore says, and I have seen her; the bigness of her body and bill, as likewise the form of them, is much what of a swallows, but the spreading out of her wings and tail has no less compass then an eagles; she lives and breeds in the aire, born up by the force of wind with more ease then archytas his dove, and comes not near the earth but for her burial; for the largeness and lightness of her wings and tail sustain her without Lassitude, and the laying of her Eggs and brooding of her young is upon the back of the male, which is made hollow, as also the breast of the female, for the more easy incubation, taking no other food (as alas how should she?) then there is found: but whether she lives merely of the dew of heaven, or of flies and such like insects, I leave to others to dispute.

6. Nay, have you not heard of the little dog in the West Indies, which sings so sweetly all the night long, neither night nor day eating anything? But there be examples in our kind as well; then it is certain above controlment: Sir Christopher Heydon says there is a mouthless and so a meatless people or kind of men about the head of Ganges, which lives by the breath of their nostrils, except when they take a far Journey, they mend their diet with the smell of flowers: and lest you may think I lean upon bare Authorities without the stay of reason, all the matter rests upon this reason I told you before, that our life lay in the hand (beside a little exercise) of two like meats, one for soul and

natural heat which is within us, and the finest and first moisture in our body; the other is without any meat, of the same Temper with our body as near as nay be, to uphold the frame and building of the same which I said to be a fine airy and fiery flame.

7.   And we are now grown so out of order, and so much estranged from our ethereal first moisture and the life of God, that we creep downward towards the earth through diseases, before we can reach the life of the vehicle; within six-score years we die, and are hidden from the sensible approach of renewing life.

Chapter XII.

1. Of Nature and her medicines experienced by Rosie Crucians. 2. Of the occult virtue of Mysteries. 3. Of the healing and consuming medicines. 4. Of their use. 5. Of the Gout, Leprosy, Dropsy and Falling Sickness, etc.

1.   Now the air itself, especially when it is evermore as the wet sunbeams declare, so sprinkled with some fine foreign fatness, nay seem sufficient food to nourish the finer part of our frame, wherein the temper of mankind, and his life (touching that point) standish, which is as much as any meat can do to life, (for it is not fed by common food, as I said above) though not enough for strength, because the grosser, sounder, and tougher parts wherein the strength lies, shall want food

in this Diet, and fail no doubt greatly; yet life shall last still, as long as air and first moisture holds, in my opinion: or if we think that too spare a Diet, we may mend it (as the mouthless people do) with smell of flowers: or rather, as we know nature is able to draw airs and other food which she desires through the skin into all places of the body; so if she had meat applied to the stomach, she would no doubt satisfy herself that way most finely, without the heap of hurts let in at the broad and common gates, as we see by example for drink, that all the while we sit in water, we shall never thirst: And for meat, I have heard Rosie Crucians say, by applying of wine in this sort they fasted without all hunger for two years together.

2. And in like manner I have experienced this, and fasted two days when I first studied the nature of the guardian genii: But if that would not serve the turn, and we must needs receive in meat at the common gate, yet we may let it pass no further than the gate, and make the stomach in the mouth, which is the use of some Rosie Crucians when they are Seraphicalli Illuminated; and to provide enough for life and strength, and a great deal better for our health, then we do, because the clearer part alone should be received. And moreover I say, for the clear dispatch of that our ordinary trouble and annoyance which your reverence will not suffer me to name, although I might among Physicians, but they know my meaning: But it shall not need to steal shifts and holds if you will believe the Rosie Crucians, that we nay easily fast all our life (though it be three hundred years together) without all kind of meat, and so cut off all doubts and dangers of diseases thereof springing; and for my own part, I know some that have fasted and lived in the holy orders of the fraternity without all food ten years space together. What need we say more? if you be both so hard of belief, and dull of sight; and reports of good Authors, nor my own experience will sink into you, nor yet can you see the light of

reason shining before you; take here a few of ordinary matters in the life and use of men, and weigh one with another; is it not as common in use, and indeed needful, to spit, and avoid another nameless leaving? And to drink, but to sleep especially? If some of these, nay all may be spared, why not our meat also? let us see a little, and by example, because Reason is both too long and too open to cavil.

3.   To leave think, which many have all their lives left; Elizabeth Drewe a Devonshire gentlewoman, is reported never to have spat, nor the Indian nation. Sir John Heydon says, he knew one that kept the nameless matter forty days together. And although this answers not the question, yet it shows the truth of the former holy story; for if in so foul and gross a thing as diet is, he could so long want it; why not these men forever, so clear and fine a diet, almost empty and void of all leavings? For the grosser sort, which make up this foul and shameful, one we left before, as you heard, and the finer in the passage from the stomach through the former Gates were drawn all away to the liver, as the like is ever in us and voided otherwaies. To close up all, I was at Sea with one that slept not one wink for these three years last past, and Mr. John Knotsford is a Witness to this truth, and Captain Windsor.

4. And thus we see these strange things fall out in proof; but how, I cannot stand to show. First, nature suffers them, then use and custom, another Nature, brings them in; yet we may well believe the like in this matter of meat we have in hand; For as the Bear (according to the guise of many Beasts that Lurk in Winter) fasts forty days, so Eugenius Theodidactus, the reported Rosie Crucian tells of a Scottish young man, David Zeumons, that waited on him, that by use brought himself to fast three dales together, which by use might have been three hundred as well, if he had ordered himself thereafter by slow and

creeping custom, as Captain Copeland calls it, and by such means as I Bet down before.

5. So we see, I say, great wonders prove plain and easy truths in the sight of Wisdom: you have read of the wonderful works of God in the accurate structure of mans body, of his soul, of his senses, of plants, of minerals, and Rosie Crucian medicines shall be that which I will insist upon, and that by the means aforesaid (where are more than one, if this like them not, they may take another) it is possible for all men by kind and custom to keep their health forever: Let us come to the next point, that is, as well to be recovered if it were lost, and that all diseases may be cured. This is a point much harder than the first, even so beset and stopped with all kind of lets and encumbrances, that a man can scarce tell which way to set his foot forwards. First appears Asculapius, Hippocrates, and Plato, the chief among the Grecians, bearing in hand sundry diseases of both kinds (both come by descent, and gotten by purchase) hopeless and past recovery, and giving over the men that owe them, for troublesome to themselves and to the Commonwealth: Then you may see Galen, his soft and fine Company with him, that follow these as Gerard and Riverius, and Culpepper, and these with a long train of hedge-doctors; and among these stand the Bill-men, that daube their Medicines upon every post, with caterers and cooks, laden after them with all kind of dainty Drugs, stand forth and cry, they have these many Ages devoured heaps of Books, and took endless pains in searching out the Nature of single Medicines, and making mixtures of the same, and yet could hardly cure some Agues, and other lease diseases: But for the four great diseases, viz, the Gout, Leprosy, Dropsy, Falling Sickness, they could never heal them, and have therefore for Oracles set them down incurable.

Chapter XIII.

1. That the knowledge and virtue of Medicines are secretly hid from vulgar understanding: 2. How they may be gotten: 3. And of what lies couched in the Oil of Bodies: k. Of the use, and how to fetch it out by skill, the Haven of Medicine.

1. What is left to be done in this Matter? what shall we set against the weight of so many great men's authorities? Equally put them in the balance, as we have done hitherto, and weigh them with truth and reason: But where shall we find it? say they; As it is everywhere, as Mr. Hobbs said, drowned in the deep, so in this matter it is scattered all about, and largely spread withal; for there be three things, and every one full of under-branches, belonging to the Rosie Crucian art and way of healing; the first is knowledge of the diseases, the second the remedies against them, and the third of the appliance of the remedies, all which shall be traversed in this Methodical mysterious Treatise: but it shall not need I hope; nay we must take heed how we enter into so long and large a race, in so short and narrow a compass of time appointed, especially being never run before by any man but ourself, not one of the wise Egyptians, nor our ancestors, the holy company of Moses and Elias, whose steps we strive to follow, and their successors; for when they have once hit the mark they have shot at, and gotten the great and general Medicine Caput Mortuum A.P. curing with ease all diseases, they think it strait enough, and an empty and needless labor, as it is indeed, to trouble themselves and their children with large rules about innumerable signs and causes of infinite diseases, and about other small particulars in appliance. Neither would I have you set Sendivogius, Paracelsus, and their heirs upon me, and say they have

taken great and goodly pains in this field; you will then force me to speak my thoughts.

2. Though these men (to let the Bill-men go, as too young and childish yet) by great light of Wit wherewith they flowed, and by long proling[22] both with eyes, ears, and hands, in the mysteries of Egypt, saw and performed many of the Rosie Crucian deep secrets, yea and there got most of their worldly praise, although I think a number feigned, yet Paracelsus his new Art and Rules of Healing are not good in my opinion; For first, against the example of the Rosie Crucians, from whom he had received all things, and then in despite and disgrace of Galen, for miscalling his Countrymen, as you have heard, but chiefly carried away with a mad and raging desire of fame and honor, which Culpepper always despised, yet the stars favored him, when I assisted to set up that new, famous and strange work of physick, now well-known and practiced, which Paracelsus took in hand, a man unfit to do it, to pull down and raze the old Work, and to set up our new experienced secret, which he could never do all his life.

3. Then we see how it is performed; he sets down some false rules, some waste, idle, and some wanting, and all inconstant, disordered, and unlearned; when he doth well (as be doth sometime) he doth no more than was done before him, and brings in the same thing disguised with new, odd, cross, and unheard of names, such as may move wonder at the first, but when they be scanned, Laughter, as Mr. Moor says of Philalethes his like devices of his Welch Philosophy. And that I do not slander them where there is no cause, I could prove, if this place would admit a Volume. Wherefore let us follow the true and right Rosie Crucians, as easily you may know them by their Actions, if ever you fortune to see them and be acquainted with them; and leave

---

[22] Spelled exactly as in the 1662 printed edition. -pnw

Paracelsus, and the rest in this ill matter, and Light and Apish, as he makes it; and why should we spend all our care and thought about a small matter? you have a good medicine and remedy against diseases, when old wives in the Country, and some good women, amongst other Dr. Culpepper's late wife, and simple men, on our side (I mean Simple in respect of the Grecian subtilities about nothing) when these people have healed most, nay, even all diseases, and with womanish Medicines indeed; the German Doctor (let us give him his due praise) has quite slain the Grecian Physick, and here done much for mankind, by describing and describing and dispatching our close and secret enemy, which under Color of friendship and fighting against our enemies has this long time betrayed us and done us much mischief; which thing one of their best Captains of their State, Fernelius by name, after he had been a while in Egypt, begun to smell at last, and began to repent himself of all his former pains (which we know were great) bestowed in that kind of healing, saying it to be but words, and the whole force and weight of this art to lean upon the knowledge and virtues of Medicines, secretly kid and couched in the midst and oil of bodies, to be fetched out and gotten by the skillful means of alchemists; even so of that Art, which is so much condemned of his fellows before and since him, have fled and do daily fly from the daily toil and trouble of their fruitless and barren Dead Sea: No. let us shift our sails, and fly further too, I hope of wind and tide and all, which we have.

4.   But let us mount up to the main-mast top of our Knowledge, and see if we can describe the raven of Rosie Crucian medicines, and see what marks it has, and how it differs from other Creeks adjoining, lest at our journeys end we miss with more shame and grief, and suffer shipwreck. A medicine is that which kills the face of that which hurts

us; and this it doth many ways, and yet also to one end (which is the end of doing and working as I said before) for his food and sustenance.

Chapter XIV.

1. Of Medicines. 2. Of Witchcraft. 3. How to cure those that are afflicted thereby. 4. Although their bodies be possessed with evil spirits. 5. That cause them to vomit up Needles, Thimbles, 6. Pots. 7. Glasses. 8. Hair. 9. And shreds of cloth. 10. Which by the Devil were conveyed into the Body. 11. That Winds and Tempests are raised by Witches upon mere ceremonies of Medicines. 12. Of Poisons; with the examples also of other supernatural effects of unclean Spirits. 13. Of imagination. 14. How to cure a Witch. 15. And to take away her power.

1. A servant of God and secretary of nature, must be well advised of what he writes, especially in this age, and of this matter (viz.) of the Rosie Crucian physick, lest he should, as I said before, fail in this design, and so it may be a shame that he should be reproved, by the pretenders to those wise truths he alone has opened to public view; then let us come again and sort our speeches.

A medicine heals us and kills our enemy, either by dulling or consuming it; for when it meets with a contrary of even strength (as when oil and poison, etc. join) then in strength they neither eat up nor destroy one another, but both are dulled and weakened, and make one heavy thing, which Nature casts out for an unlike and unkindly dead thing, which they call an excrement, or leaving; but in case it be of more strength and power then our enemy, then it quite destroys, devours, and turns him into his own nature. And this consumer is either like the thing that hurts us, in which sort even as every herb of sundry qualities draws and feeds upon his own juice in a Garden, so one poison cures another, and all purging and drawing things do heal us, and all Rosie Crucians hid and divine properties do work by plain reason; or else it is unlike and contrary to their custom; after which manner, as dry sticks, and tow, and vinegar, quench wild fire, or other fat fires, before water, whose fatness feeds it, for the strong contrary quality quelling and eating up the weaker; so doth any cold and dry thing, as bolearminick, terra lemnia, etc. cure a rotten Poison; and so are a great number of cures done; which only course, in a word, the Rosie Crucians use for Physick, and not indeed without good success; we heard even now of two hindrances of healing, which our common Physicians did take unawares, and Paracelsus pretends to have found out before me, gave any hint to the World of our experienced inventions, of Gold dissolved and made potable, being incorporated with its proper veil, which we now use by the name of Aurum Potable; but Paracelsus strays much in the making of it, and knows it not no more than Thomas Harrington Dr. Culpeper's Man, whether in their poisons, on the other side, when they think all Cures thereby performed.

2. Now when the consuming Medicines have done their duties, Nature expels them for poison and unlike strange things, according to the *Holy Guide*, as well as the Grecian Rules, because all their Medicines were not approved by the Fraternity, and were by their confession such: But if they had either thought of the dulling Nourisher, which as I told you, takes the nature of the leavings or excrement, or had known the Rosie Crucian wholesome Medicine, they would have made another reckoning: But let them go, and let us see out in time towards the raven of health. If the *Art of Healing* be nothing but destroying hurtful things, and their stronger enemies (but equality will sometimes serve the turn) or likes together; and the world

be full of both these kinds of Creatures, following the nature of their Parents of four beginnings, which are, as we see, some like, and some contrary one to another.

3.   Then sure the Rosie Crucian *Art of Healing* is not (as some may say) impossible; truly it wants nothing but a man well skilled in the Nature of things, a servant of God, and secretary of nature by name; for (I think) I need not put in a physician, to know what other part the Causes of the diseases, which must be known and matched, because as Sir Christopher Heydon the seraphically illuminated Rosie Crucians, and learned astrologer well says, he that knows the changes and chances of things in the great World, may soon find them in the Little.

But our naught-healing Bill-men, that daube Medicines upon every wall and post, and some Leaches will step in and say, Diseases are in some so great, and in all so many, and man's wit is so weak and shallow, and the Medicines so hid and drowned in the deep of Nature, that it is not possible to find them all; or if they were found, to apply them with such discretion, as Nature might abide those poisoned Fraies and Battels within her. And again, admit all this untrue, yet there be some diseases sent from witchcraft and sorcery, and other means which have their cause, and so their cure. I have read of some that have vomited up pieces of cloth with Pins stuck in them, Nails, Needles, and such like stuff; and this is ingested into the Stomach by the prestigious sleights of Witches: Others I have seen vomit up Hair, Glass, Iron, and pieces of Wood with Pins stuck in it; another's Corps was dissected, and ripping up the ventricle, there they found the cause of the diseases, which was a round piece of wood, four knives, some even and sharp, others were indented like a saw. Others do miracles by casting flintstones behind their backs towards the west, or striking a river with broom, or flinging of sand in the air, the stirring of urine in a role in the ground, or boiling of hog's bristles in a pot; some by whispering some words in the ear of an horse, or wild stag, could direct rim a journey, according to their own desire. But what are these things available? To gather Clouds, and to cover the Air with darkness, and then to make the ground smoke with peals of Hail and Rain, and make the Air terrible with frequent Lightning and rattling claps of Thunder: But this is from the power of the Devil (as some fancy) which he has in his Kingdom of the Air.

4. For the remedy of these mischiefs, I have seen a man was present, when some have vomited up needles, thimbles, shreds of cloth, pieces of pots, glass, hair; another would suffer himself for money to be run through with a Sword, when I was not there, but it appeared to me a Fable. I have seen a Rosie Crucian physician cure those afflicted People. But if you will say, there is a touchstone whereby we may discern the truth of Metals, but that there is nothing whereby we may discover the truth of Miracles recorded everywhere in History. But I answer there is, and that is this:

5. First, if what is recorded, was avouched by such persons who had no end nor interest in avouching such things.

6. Secondly, if there were many eyewitnesses of the same matter.

7. Thirdly and lastly, if these things were so strange and miraculous, leave any sensible effects behind them; though I will not acknowledge that all those Stories are false that want these conditions, yet I dare affirm, that it is mere humor and sullenness in a man to reject the Truth of those that hear them; for it is to believe nothing but what he sees himself, from whence it will follow, that he is to read nothing of history; for there is neither pleasure nor any usefulness, if it deserve no belief.

8. Another Remedy for these supernatural diseases is, let one watch the party suspected, when they go home to their house and presently after, before anybody go into the house after him or her, let one pull a handful of the Thatch, or a Tile that is over the Door, and if It be a Tile, make a good Fire, and heat it red hot therein, setting a Trivet over it; then take the parties water, if it be a Man, Woman, or child, and pour it upon the red hot Tile, upon one side first, and then on the other, and again put the Tile into the Fire, and make it extremely hot, turning it ever and anon, and let no body come into the house in the meantime.

9. If they be Cattle that are bewitched, take some of the hair of every one of them, and mix the hair in fair water, or 'wet it well, and then lay it under the Tile, the Trivet standing over the Tile, make a lusty fire, turn your Tile oft upon the hair, and stir up the hair ever and anon; after you have done this by the space of a quarter of an hour, let the fire alone, and when the ashes are cold, bury them in the ground towards that quarter of Heaven where the suspected Witch lives.

10. If the Witch live where there is no Tile, but Thatch, then take a great handful thereof, and wet it in the parties' water, or else in common water mixed with some salt, then lay it in the fire, so that it may moulter and smother by degrees, and in a long time, setting a Trivet over it. Or else take two new Horseshoes, heat them red hot, and nail one of them on the Threshold of the Door, but quench the other in the Urine of the party so bewitched, then set the Urine over the fire, and put the Horseshoe in it, setting a Trivet over the Pipkin or Pan wherein the Urine is; make the Urine boil with a little Salt upon it, and the Horse nails, until it is almost consumed, viz, the Urine; what is not boiled fully away pour into the fire: Keep your Horseshoe and Nails in a clean cloth or paper, and do likewise three several times; the

operation will be far more effectual if you do these things at the very change or full Moon, or at the very hour of the first or second Quarter.

If they be Cattle, you must mix the hair of their Tails with the Thatch, and moisten them being well bound together, and so let them be a long time in the fire consuming.

11. You have heard the Cause of some of these diseases, and have heard the Cure; but these are, without the compass of Nature, and so let them pass with our sickle standing, which is daily and hourly so beset with destinies, that no man can warrant nothing.

12. Truly destinies are so deep and bottomless (to return straight Homer-like upon them, and therefore it were best indeed to let them go, and the applying of the Medicines with them) the rather because the other (I mean the former) is so slight a matter to a discreet Physician, such a one as is pointed out by their old and famous Leader Hippocrates, who both in this, and all other' duties of his Art, made such speed, and so far passed all his fellows (as none since, which is a good time, could ever overtake him) no nor yet come so near as to keep the sight of him whom they had in chase and followed.

13. Then for those supernatural causes, which I shall not stand here to search (for so they are called) if they flow from unclean and wicked Spirits (as some think) they are not the stuff of the things that hurt us, though sometimes they dwell in and possess the body, but windy matters, much like unto those fierce and sudden changes of the Weather, proceeding from the Influences of the Planets and fixed Stars, and working the like effects in men's bodies, so that the nearest cause is natural, let the rest be what they will, and the Cure be done by natural means, as we see by experience amongst us: And therefore E.A. that pretends this, and puts the fault in the faith of the wicked,

which is a thing as far above Nature, yet holds its Cure with a natural Medicine, which we call a quintessence.

14. Although I am not willing, that sometimes this sickness is such, as he bids us sometimes withstand it with another as strong a belief set against it, but for my part, I cannot reach it with my conceit (let deeper heads then mine, or the Vice-Chancellor of Oxford, Doctor Owen, think upon it) how these beliefs and imaginations, and other parts and powers of the soul or mind of man, can so fly out of their own kingdom, and reign over a foreign body, when we know the soul and mind is so fast bound ii the body in durance, and so like to be, until it be the great pleasure of the omnipotent and the omniscient God, the chief good, who has committed them, to let them loose at once, and set them still at liberty; and this may be disputed with grace and knowledge on my part; let this man therefore buzz against my knowledge, which he would have to be more than grace, I appeal to the natural faculties of any free judge, whether there be not as much grace in me as there is honesty in him, that was Oliver Cromwell's Creature, and appointed to examine and judge me he did not understand? All men censure as they like of Stories; so, let them pass amongst old wives' tales for me; we will severely follow our task. That if the effect do not cease which the object has wrought upon the Brain, so soon as ever by turning aside of the Organs the object ceases to work, viz, through the since be past; as the stroke of a stone, a blast of wind, puts standing water into motion, and it doth not presently give over moving as soon as the wind ceases, or the Stone settles; so the Image or Conception remains, but more obscure, while we are awake, because some object or other continually plies and soliciting our eyes and ears, keeps the mind in a stronger motion, whereby the weaker doth not easily appear. And this obscure conception is that we call phantasy, or imagination being (to define it) conception remaining,

and by little and little decaying from and after the act of sense, etc. If some of these diseases spring, as Doctor Culpepper and some others hold, and with good reason, from neither of both these two roots named, but from a foul and venomous breath, sent forth from a poisoned temper of the Witches body, through the windiness of hateful eyes: For Thought fashions the Blood and Spirits almost at his pleasure; then all the causes being ordinary, and agreeing to the course of Nature, they may be cured and put to flight by the same course and means: which opinion, if you please to bear with my tarrying, it is worth the handling, taketh hold upon this reason, because (as Rosie Crucians do witness) some beasts of ranker venom, do witch and hurt after the same manner; as an old Toad by steadfast view, not only prevails, but benumbs a Weasel, but kills a young Child. And by the same means the Beaver hunts the little Fish, and takes his prey: But most fiercely and mischievously of all Creatures in the world, the two Monsters in kind, the cockatrice and apoblepas: again, for that the eye of a menstruous woman (as all report) spots the glass which it beholds: And moreover Eugenius Theodidactus, in the *Wise Man's Crown*, tells of many folk that through a poisoned prerogative, which a monstrous Mark of a double-sighted eye gave unto them, were able to bewitch to death all those upon whom that Eye was angerly and surely set and fastened; but chiefly because we see them that use this wicked Trade, to be by kind of a muddy and earth-like complexion and nature, brought by age, as they be most commonly, long life, and gross diet, to the pitch of Melancholy, that is, to a cold and most dry nature in the world.

15.   For certain proof whereof, bring one of them out of that beastlike life, brought unto merry company, and fed full with dainty Diet, and within twenty days, as has by a Rosie Crucian been tried a truth, the whole state and nature of her body will be so changed, as it

shall not suffer her to bewitch and hurt again, as you may read in my *Familiar Spirit* or *Guardian Genius*, and in my Book called *The Temple of Wisdom*.

Chapter XV.

1. 2. The Natural effects of Medicine. 3. The force and power of minerals in diseases. With examples also that every disease-breeder has the cure or remedy in it. 5. Examples that poison prepared cures poisoned people. 6. Rosie Crucian Medicines. 7. The virtue and power of the Planets and heavenly Stars poured through the influence of the moon upon the Lower Creatures. 8. Of Hot Stomachs. 9. Of the Ethereal first moisture of man. 10. Examples also of Rosie Crucian Natural and Supernatural cures. 11. Of the understanding of these experienced truths by the wit of man.

1.   Let us come to the next and chiefest point; And there we must not say for shame, that these helps and remedies lye bid in nature, too far for the wit of man to find, unless we will accuse our own sloth and dullness: For nature has brought them forth and laid them open as well as the Poisons and hurtful things, or else she were very cross and ill willing to him for whose sake it seems she does all things.

2.   Any further her good will is such, as she has not only laid them open, but given us ways to come by them, and means of speech, hands and wit also, far above all other living creatures. And yet she has not left us so, but lest by chance we might go wide and miss them, to show her motherly love and affection towards us, she has guided many

witless Beasts, even by common sense, unto their speedy helps and remedies in their diseases: That we by the plainness and shame of the example might be taught and moved to seek out the mysterious truths of nature in celestial bodies, as well as beasts that seek and find us Medicines helpful in the like diseases, for our terrestrial tabernacle. As to name a few not unworthy meaning; she makes the beast hippocampus in time of his fulness and fatness to go to a reed, and by rubbing a vein to let himself blood, and to stop it again by laying mud upon it; A sick dog to seek an Herb and purge himself; and the bear to do the same after his long fast in winter; she leads the panther, when he is poisoned, to her foul and nameless leaving; and the Tortoise, after he has eaten a Viper, to Summer Savory: And the Hedgehog is so good a natural Astronomer, that he fortifies his hole against foul weather; the Hog will gather Moss and straw to cover himself a little before it rains; The Dog knows the influence of Mars when he doth sleep by the fire, and will not go out adores when he is in any evil position: and many such like examples has nature laid before us for our instruction; by which at last wise Plato, Philo, Apollonius, Pythagoras, and painful men of Greece, as they themselves report, be they Elias or Elisha from whom the order of the Rosie Cross came, (as some say) or else as others will have it, from Moses, or Ezekiel, or whosoever, and by laying reason and further proof together, first made the Art and rules of healing, to know whence diseases came, and bow to recover them. And then seeking all about for remedies to serve each turn, by little and little they matched the most part of the lesser rank with single Medicines, and the greater ones they doubled and coupled many together, insomuch as at last, which was in Hippocrates time, they were able to heal all (saving four,) of the greatest & deepest diseases, the gout, the dropsy, the leprosy, the falling sickness, which are now healed by the Rosie Crucians only. But this race is below the seraphically illuminated fraternity: now not a Physician that is lined with Plush in England, Spain, Germany, or France, but holds that long-life, Health, Youth, not attainable, they therefore with one consent, amongst the other four, call them impossible.

3.   But to come to the point; what wrong this was both to skill and nature, they do easily see and laugh at, which know that in this Labour, they did not only oversee and skip the Minerals, the stoutest helps in the whole storehouse of Nature (although they could dig them out well enough to other and worse uses) but also, which is in all, did let the Rosie Crucian skill of preparing Medicines, whereby weak things are made almighty, quite escape them.

4.   Wherefore to make up the Rosie Crucian *Art of Healing*, and to make it able (as they say) to help and cure all diseases came in, or rather went before, into man's body; The Egyptians in great favor too with nature both for their soil and bringing up, so notably commended above all nations, (having for example, to move and teach them even the great Wight of the world as Sir John Heydon says) for wits to devise, and bodies to put in practice.

5.   Whereby in short time they unfolded the knot why the Minerals were of greatest force and power against diseases; and soon after, which was a divine light, aid insight, they perceived the huge labor of seeking such a huge sort of singles end mixtures to be vain and empty, and pitiful among wine men.

6.   Because first, there is nothing hurtful and a breeder of disease, but it has the heal and remedy for the same about him: For the wings and feet of cantharides, the fruit of the root bezar, the Ashes of Scorpions, Toads, and Vipers, and divers other stronger poisons, both by nature and skill dressed and prepared, do cure and heal their own and all other Poisons; nay as all stronger likes do cure their likes

throughout the whole world of diseases, even so when a man has found out a thing that hurts him, he may by easy skill mingle and break the temper of the same further; that is, make it able to eat up and consume itself as easily, without any further doubt, toil and labor; But especially because there is no one thing in the world, take what you will, that has lot the virtues of the planets arrested and fastened upon it, and also of the qualities thereof within itself, that is not as good as all, and may serve instead of all, and that is not able to cure all diseases; which thing weighed, and with discourse of wit and reason fully reached, they went to practice, and by the like sharpness of wit, they found out the kindly and ready way to dress and make fit these three kinds of Medicines aforesaid, which contain all the Art of healing; all the rest are but it was words and grievous toil, to tire a world of wits about a bootless matter, as says Des Cartes. But especially they rested in the last, which is enough alone, and yet not without great forecast, to chase one of the best, and that the very best of all, for their ease in dressing. Though Dr. Culpepper of late was not content with this, but ran through the rest, as well to spite his enemies, the Collage of Physicians, as to make himself famous in taverns and ale-houses, as Paracelsus in his time did: whose steps he strove to follow against the rule of Rosie Crucian wisdom and virtue, and the example of his ancestor's.

7. But has everything all the virtues and influences taken from the planets and stars, by the moon, to the earth? That is, all the curing and healing power of all the things in the world? Very well you must remember that I proved above all the virtues and powers of heaven, poured down through the Influence of the Moon upon these lower creatures, to be nothing else (as Cap. George Whorton truly says) but one selfsame life and Soul, and heavenly heat in all things, and again, that all diseases flow from distemper, and as it were discord of the

Natural consent of the body; then that thing which is endued with store of life, and with exact and temperateness, seated upon both a subtle and strong body, (which the thing in the bottom is) able alone by subduing his weaker enemies, those distempered diseases, by strengthening his fellow life, aurum potable, in our bodies. And lastly, by orderly binding together the frame that was slipped out of order, to do as much as all the powers and forces of all the Plants, Wights, and minerals in the world, that is, to put to flight all troubles of diseases, and restore the body to perfect health and quietness.

8. But how is all this done? We talk of high things, and huddle up too many great matters together. It were good for us to work them out distinctly; when this aurum potable we speak of, and strong tempered medicines, Blip into the stomach, it stays no long digestion, being already digested, nor looks for any ordinary passages to be opened into it, but as soon as it is raised out of sleep by his fellow, the natural heat, by and by he flies out, and scour's about, as fast as the dolphin after his prey, or as nature herself, whom Mr. Tho. Keydon as I take it, says to pierce bounds, and all to the purpose, that is to seek his like food, and sustenance, whereby to preserve his state and being, which is the purpose of all things in the world, as was said above.

9. Now there is nothing so like and near a perfect temperature in the world, as the ethereal first moisture in man; but what is this, you may read in my book entitled *Ventus Magnus*.

10. This is best and most in the heart, the root of life, then thither it hyeth[23] and preys upon that part first, and that is the cause why it presently restores a man half dead, and as it were, pulls him out of the throat of death; then it runs to the rest all about, increasing by that

---

[23] Spelled exactly as in the printed edition. -pnw

means the natural heat, and first moisture of every part of the body; when this is done, he turns upon the parts themselves, & by encountering with them in the same sort, according to his might, upon them, and brings them a certain way towards his own nature, even so far as we will by our usage suffer; for if we take it with measure and discretion, it will bring our body to a middle man and state, between his own exact temperature, and the distemper of diseases, even a better state then ever it had before; if we use it out of measure, it takes us up too high, and too near his own nature, and makes us unmeet for the deeds of the duties of an earthly life. But in the mean while in the midst of this work, we must know that by his exceeding heat and subtleness which is gotten by Rosie Crucian skill, and which makes up the strength above all things, it divides and scatters, like smoke before the wind, all distempered and hurtful things, and if they cannot be reconciled and turned to goodness, nature throws them out as dead and unfruitful leavings.

11. But how do we talk so much of exact and perfect temper, when by the verdict of all the quest in these cases there is no such thing found in nature, but in heaven only? Neither heard you me say that it floated aloft, but was sunk to the bottom of all nature; notwithstanding by a true and Holy Rosie Crucian to be founded and weighed up. For as heaven was once a gross and distempered lump (as I told you in my book of the nature and dignity of angels) by the divine art of God that ordered all things (as you have read in the Introductory part of this book,) refined and sundered away round to the place and nature where it now stands; even so one of our gross bodies here below, being a piece of the same lump also, and all one with that which Heaven once was, may by the like art and cunning be refined and parted from all his distempered dross and foul drossiness, and brought into a Heavenly nature of the best and goodliest thing in Heaven: And yet

you must not take me as though I would have the mind and wit of man, which is but a spark of the divine great mind, (I spoke in my book called *Ventus Ingens*) to be able to reach the excellency of his work, and to make so great perfection; if he do but shadow it, and make a Counterfeit, that is, if he reach not so far as to make all things, but to mend a few by this his Heaven, all is well, it is as much as I can look for at the hand of any man that is not a Rosie Crucian. Now is the time to rest a little, and pray for the good use and practice of those that shall read our *Harmony of the World*.

Chapter XVI.

1.   Of the Rosie Crucian Sun. 2. Or Spiritual Oil. 3. Of the Divine Works of God not yet observed. 4. How to make Æther. 5. Examples of Medicines Rosie Crucian and Grecian. 6. Of Poison. 7. Of the Supernatural Miracles of the Rosie Crucians. 8. With obedience to Reason. 9. Another Medicine of Supernatural effect. 10. Of the power and secret skill of Nature. 11. How to dissolve Minerals. 12. And how to prepare for Men's Bodies.

1.   Eugenius Theodidactus has shown you this heaven, nay this Sun of ours, which is naught else, as I told you in one of my books of Astrology, *The Temple of Wisdom*, but an oil full of heavenly spirits, and yet in quality of his body just, even and natural, fine and piercing, close and lasting, able as well to rule this little World, as Mr. Thomas Heydon says, the great sun is able to govern the great World.

2.   But what is he that can see this Divine Art and Way, whereby God made his great and mighty work, viz. ברא as I showed in my Book, intitled, *Moses Speech to God*, upon the second chapter of Genesis? Or if he saw it, learn and match it by imitation? I answer, None but Rosie Crucians to whom I am a friend, and they God has enlightened and unsealed their eyes, they have found the way lying open in all places, and in all natural changes, they see them pass and travel, I say still, the course that Mr. Thomas Heydon calls soft and witty, that is, kindly separation: and if he be not swift and rash as many, such as Thomas Street, but will have sober patience, his own skill and labor will be but little if he please; for Nature herself very kindly will in her due time perform all, and even all that heavenly workmanship be easily performed: and yet I mean not so, but that Art must accompany and attend upon Nature (though with no great pains and skill) both forward and backward in this Journey (Doctor French knows my meaning, so doth Doctor Owen, if his angry Censure will suffer his Natural judgement) until he come to his wished rest, and to the top of all perfection.

3.   If you perceive not, consider the way whereby we made our Æther in our Book above named, and matched our own first moisture, a thing Ætherial, I say, and almost Temperate; mark what I say, there is a further end in the matter, hold on the same means, whereby you came so far through *The Wise Mans Crown*, and are gone so far in the *Harmony of the World*, which is that I spoke of, and you may reach It.

4.   Then you see the way to cure all diseases by the third way of Egyptian healing, which they do, and we may well call it the Egyptians Heaven, and yet it is a way far beneath the Rosie Crucian Art of Healing, as we shall show hereafter.

5.   But if they will not yield to reason, but mutter still Thomas Street—like, that these Heavenly Medicines of ours are very high for the reach of men's silly wits, here strowed[24] below upon the ground for other lesser and baser uses, and that no man since the first man, or if I will say Moses was the first that first found out these inventions, as they call them, after Adam; and that none but the Successors of Moses have been ever yet known to have found and wrought the same; I will not stand to beat Reason into such giddy-braid men, but go to the other two ways of healing, which the Egyptians found out and used, and called the first Mineral Medicines, and these Moses taught the Children of Israel in the valley of Mount Sinai, when he took the golden calf which he had made, and calcined it in the fire, and ground it to powder, and incorporated it with a solar V[25] eagle, and made the children of Israel drink aurum potable.

6.   And the next mysteries and secrets, as may appear by Riverius his speaking of Rosie Crucian secrets, we may fitly call this second kind, because that is too large a Name (if it be lawful for us as well as for all other learned men, where a fit word wants to make a new) we may do well, I say, to call it a cure it self, because it is by that way of healing, whereby every selfsame thing further broken may cure Itself; and this inward and hidden thing, as they say, the outward and apparent by the course of kind, whereby the stronger like eats up in trial and consumes the weaker.

7.   If this leave be once granted, we will borrow a little more for the other two likewise, because their names are not pertinent to our purpose, and call that heaven a cure all, for so it doth, and the next a cure the great, because the order of the Rosie Crucians is always to

---

[24] Spelled exactly as in the printed edition. -pnw
[25] Exactly as in the printed edition. -pnw

match the greater and more stubborn sort of diseases with the stout and mighty Minerals. And the rest, with those hidden, cure themselves, or at least in the lower rank of lighter diseases, with their likes; only raw, as the Grecians use them, without any curious dressing.

8. Let us draw nearer a conclusion of the matter; because Grecians themselves are able, and our English Physicians that learn of them, to cure the lighter sort of diseases, and to heal all but the four aforesaid, we will leave the rest for them, and so let this second kind of healing go, called our hidden cure themselves, and bend all our batteries against these four, which they call incurable, and see how by force of our Mineral Medicines they may be cured; we see the poisoned spirits and breaths of venomous things, with what force they work upon our bodies, things in Nature set against them, and how they consume them; if you do not see by imagination, reason with yourselves; if not, remember these above named, that killed with their sight; Hear one or two more that work the same by touch as violently. The hare-fish, a most cold and dry Creature (to omit that she makes a man's head ache by sight) if you touch her aloof only with a staff, that her venomous breath may go straight and round unto you, you die presently. The root Baazam in Palestine, as Pythagoras writes, kills the man that handles it, and therefore they used to make a Dog pull it up, who thereby died immediately. To come into the body; that costly poison that is in Nubia, and one grain kills a man out of hand, yet stay but a quarter of an hours working, and that one grain divided will overcome ten men; I hope you doubt not but these mighty poisons, if they were like in nature to the four great Diseases, and by little and little to be borne by Nature, and set upon. them, would be able easily, by their great strength, to devour & consume them; or else sure such heaps of poison as the physicians give us would not dwell so long within us,

but would put out life in a moment. Now what are these poisoned Vapors, but most cold and dry bodies, wrought and broken up by natural mingling, unto great fineness and subtleness, by this piercing swiftly all about, and by these contrary qualities overcoming? Then let us take the stoutest minerals, such as are called middle minerals by Rosie Crucians, or hard Juices (to leave the Metals for a better purpose) be they poisons, as G. Agricola says, but what they be I care not; and after we have by mere working cleansed them, and stripped them of their clogs and hinderances, broken and raised them to a fine substance, then match them with their likes, the hurtful things in our bodies, shall they not let all the rest alone, and straightway cleave to their fellows, as well as a purging Medicine, and so devour and draw them out by little and little? If there be no likes, I grant they will as well as that, fall upon their enemies, or good juices, and feed upon them.

9.   Then what do you doubt, is not a Mineral body far better? And therefore if it be raised to as great a fineness, much stronger in working then the gentle and loose temper of a Wight or Plant: wherefore these our Mineral Medicines, and some other aforementioned Medicines, and cure the great, as we call them, shall in any reason work more violently upon their likes, then the natural poisons of Wights and Plants do upon their contraries, both because the like does more easily yield then the contrary, and for that the lighter here is the stronger.

10.   But if you cannot see these things by the light of the mind, open your eyes, and cast them a little into the School of Alchemy, into the lesser and lower School, I mean of German, and you shall see the Scholars, especially the Masters, by striping the Minerals, and lifting up their properties, but a few degrees, to work wonders; as to name three or four, by quenching the Loadstone in the oil of Iron, his proper

food, they make his ten times stronger, able to pull a nail out of a poet, etc. And by this natural pattern they make Artificial drawers, not for Iron only, but for all other things, yea, and some so mighty, as they will lift up an Oxen from the ground, and rent the Arm of the Tree from the Body, as Mr. Comer doth witness, who reports again, that he saw a Flesh-drawer that pulled up one hundred weight of Flesh, and a Man's Eye out of his Head, and his Lights up into his Throat, and choked him. They make binders also to glen two pieces of iron together, as fast as the smith can Join them. To be short, they make eaters also, that will consume iron, stones, or any hard thing, to naught in a moment: They dissolve Gold into an Oil; they fix mercury with the smoke of brimstone, and make many rare devices of it: And all these wonders, and many more, they do by certain reason; I could tell you if I could stand about it. In the meantime, consider, if these or any other such like Minerals were raised higher, and led to the top of their fineness and subtleness, and matched with their like Companions, or with their Contraries, if you will, those great Diseases in our Bodies, what stirs they would make among them, how easily they would hew them, pierce, divide, waste, and consume them? But you must always have a special regard, that the Medicines be not like our natures, then the nature of the thing that hurts us, for then they would first fall upon us, and let the Diseases alone; which heed is easily taken in Minerals, things very far off our nature, says Des Cartes.

And with these Experiences of the wonderful virtue of the oil and water of Tobacco, Wise men I have known do Miracles with it, but the smoke of it is the abhorred at thing in the world.

11.   What is to be said more in these matters? I think nothing, unless through the countenance of an idle opinion that reigns among them, they dare fly to the last, and of all other the slenderest shelter, and

deny our ability to break, tame and handle as we list, such stout and stubborn bodies: (what) because you know not how to do it, will you fashion all men by your mold? Wise men would first look into the power and strength of skill and nature, and see what they can do, and measure it thereby, and not by their own weakness; there shall you understand, that there is nothing in nature so strong and stubborn, but it has its match at least, if not his overmatch in Nature, such is the nature of Man's body, of his Souls, of signatures of Plants, of Metals, and Minerals, and other things also.

12. But admit somewhat weaker, as herbs and plants, & etc. Yet this, if he get the help of a Wise mans Art unto him, shall quickly wax great, and mend in strength, and be able easily to overcome that other: Mark how the dregs of Vinegar, a thing sprung out from a weak beginning, and it self as weak as water, is able, if it be but once distilled, to sake stouter things then Minerals, even Metals themselves, all but Silver and Gold, to yield and melt down to his own waterish nature, nay which is more, then Mildew of heaven, wrought first by the BEE, that cunning Beast, and then twice or thrice by the Distillers distilled will do the same, you may judge with yourself, what not only these, but other fiercer and sharper things, as Salts, & etc. more like to do upon Minerals; and by the way consider, if such mild things as Wine and Honey, so meanly prepared, are able to subdue in that sort the most stiff and tough things in the World, so Minerals cheaper than aurum potable, in their highest degree of dignity would cure the stoutest Disease (being prepared fitly) that can grow, in our bodies. Now let us sit and take our rest a little, and then we will lead you the way to the golden treasures of Nature, and safe, easy and effectual Medicines.

Chapter XVII.

1. How the Rosie Crucians make a Chirurgeons Instrument. 2. That it shall pierce through any part of the whole body, without sense or feeling, and sound the depth of a Wound. 3. The difference of Common Physicians, raw, blunt, and herby Medicines, and Rosie Crucians: 4. What a Physician ought to be. 5. What they ought to learn, 6. And what they ought to practice.

1.   But I wear away time in vain, to speak so much about this matter; and yet since all are not of like Capacity, I will add one yet familiar example; when a Chirurgeon goes about to search a Wound that is deep, if he thrust at it with a Butchers prick he would move Laughter, let him take a Thorn, and it will pierce somewhat prettily; but to do it thoroughly, and at his pleasure he will use (though to the great grief of his Patient) a fine and long Instrument of metal. But a right Chirurgeon (the common ones are but Butchers) such a one as is a physician, and astrologer, nay a Rosie Crucian also, would touch his Instrument with a Loadstone, that is commonly found, to make it pierce throughout the body without all sense or feeling: Even so good PHYSICIANS, such as these, are hard to be found in this Government, where none of these can live without great envy. If one of these Rosie Crucians be to encounter with our greatest enemies, these four we speak of, he would not, I hope, if he were a true Rosie Crucian, be so mad as to thrust at them with the raw and blunt Herby Medicines, such as Doctor Scarborough prescribes, no nor although they be sharpened by Mr. Jacob Heydon, by plain distillations: neither would he, I think, for pity sting the poor Patient with Martyrdom of rude and rank Minerals, and unless they were made into a fine, clean, natural, and temperate quality, which would work mightily, and destroy either

of these four great diseases, leprosy, gout, dropsy, and falling-sickness: but feed, comfort, or at least not offend and hurt his Patient; they labor in vain that practice otherwise. These are the Medicines which I only use, and which a good and wise Physician ought only to seek and follow, and if he cannot find it, let him use the cure themselves. But such a thing as this, I say, brought to this equality and fineness of frame and temper, (were it at the first, Wight, plant, or mineral) was it which our father and founder Moses (the chief of the Rosie Cross) said, is like to Heaven, and the strength of all strengths, piercing and subduing all things.

2. This was it that warranted his Sons the Rosie Crucians to avow so stoutly, that Art was long, and Life short, and all Diseases curable; when Hippocrates, the father of common physicians, was driven by the infirmity and endless matter of his weak body and envious mind, tinctured with Covetousness and fickle Medicines, to cry to Rosie Crucians, but they would not hear such hard-hearted fellows, nor give him long life; he said therefore, that Art was long, and life was short. And whereas he and his offspring were fain to leave many diseased helpless, to the great shame of Art, and plague of Mankind, is it any marvel when as they prick at them (as I said) with a Butchers-prick? Nay, see what they do by their practice, they be so far from all help and comfort to the Patient in greatest danger, that they increase his misery many ways, except the great Easer of all pain, and their common Medicine death, be quickly administered: First, they make the Patient suffer the punishment due to their own slothful idleness, burdening his stomach with that labor of loosening and sundering the Fine from the Gross, which they should before have taken into their glasses: and then by doing these often, they clean tire his feeble Nature (as it would tire a Horse) when as by stripping the foul and gross stuff, that dulls the working, and retaining the Virtue in a narrow strong

body, they might do as much at one time as they do now in twenty; and because their Medicines applied are of smaller power and weaker than the things that hurt us, they feed, nourish and strengthen the Disease and sickness; but for all this, if some of this company and side of Leaches have been and are yet sometimes able to heal all Diseases in our body (though with much ado, as you have heard) save the four named remedies, yea and those as well in their spring as before their ripeness, as they themselves report. Is there any proportion in geometry? Let the College of Physicians lay measures why the Rosie Crucian mighty medicines, which I call cure the greats, passing these in power, as much as the ripeness of a disease is above the spring, shall not overmatch the ripe as well as the green Diseases: Wherefore there be no doubts left, but this plainly true, that albeit the Grecians are weak and halting in this kind of healing, yet is the Egyptian, or (as now they term it) the Paracelsian and Mineral skill sufficient to cure all diseases: Then I have paid the whole sum of my promise, touching the second means and helps to Happiness, Knowledge of all things past, present, and to come, long Life, Health, Youth, Riches, Wisdom and Virtue, how to change and amend all Diseases in young or old by Rosie Crucian Medicines, which is Life and Health.

3.   Before I close, I think it very meet, while the time and place very fitly serves, to do a good deed, and this shall be my intent, to admonish and exhort the Grecian Leaches, and their Scholars, the English, Spanish, and French Physicians, whom if they follow Hippocrates, Plato, Pythagoras, and his fellows, I love for their Learning, and pity for their misleading others, (although it be grievous) I know too old Scholars, wone[26] in a kind of Learning, to unlearn all, as it were, and begin again, for their own credit and virtue,

---

[26] Spelled exactly as in the printed edition. -pnw

yea, and profit also, if they esteem that best, to leave those gilded pills and sugared baits, and all other crafty snares, wherewith the World has been so long caught, and so long tormented, and to seek this only heavenly Society; as (to you that are learned) easily say temper your selves, and be acquainted with the ready, true, plain and certain way of healing Diseases. I think in former time they were not greatly to be blamed and accused but of dullness and weakness of understanding, in not applying and seeing this perfection, and supplying of all their wants; but since they have been so often warned, not with words only, but with examples of Learned men, Matheolus Fernelius, Severinus Danus, Philo Judaeus, Diodorus Siculus, and other such like which have and do revolt, and the away from them daily, yea and by the certain deeds of Paracelsus, it were impiety to sit still: Well, few words will serve to wise and virtuous Physicians, such as are of themselves forward.

4. But there is another, and I am afraid, the greater sort, less honest, more idle and covetous, full of windy pride and words, but empty of all good learning, and they are no friends to Rosie Crucians, nor they to them, and these no gentle warning to any, no though a Rosie Crucian himself should come and bring truth herself along with him in person, would prevail; who care not, it seems, if halt mankind should perish for want of help and succor, rather than lose their gains; and which not only speak foully, but write foolishly, against this over-flourishing virtue, but also like the giddy people of my time, where they catch the State, banish the men that hold and possess it; whereas if it were a good Commonwealth (quoting Aristotle) the matter would be so far from banishment or imprisonment, as they would esteem such a man as well as the laws (for he is himself a Law) exempt from all obedience and judge him worthy to be followed and obeyed as a perpetual king.

5.   This untowardness and crookedness in men, caused all our all healing ancestors the Rosie Crucians, from time to time, never to abide their Sentence, but to the great hurt and loss of mankind, go into willing banishment, you have established a kind of Government among you (to pursue the same, like a little ------[27]) wherein you rule alone over the weak and sorry subjects of men's Bodies, then their health and safety you ought to seek only, besides enough to maintain a contented estate also, which Plato allows his Governors, and not profit only (that were Tyranny) both for humanity and Religion sake; for to omit Religion, which they do lightly omit, if a Physician begin once to make a prey of men, he is not only no man, but a most fierce and cruel Beast, not fit to be compared and matched anywhere; if you seek all over the world, as with the misshapen Monster of India, which Aristotle describes, and calls martichora, which being by nature or custom, I know not whether, very greedy upon man's flesh, is with manifold and wonderful helps furnished and armed unto it.

6.   First with a face like a Man, a voice like a Trumpet, two fit things to allure and call him in, and then if he flies, with the swiftness of a Hart to overtake him; he darts like a Porcupine, to wound him afar off, and with the tail of a Scorpion, as it were, a poisoned shaft near-hand to sting him: Furthermore, lest all this might not serve, by occasion of Armor, he has feet like a Lion, fiercely and cruelly to tear him, and three rows of teeth on each chap for the devouring. Apply you and the Apothecaries the rest your selves, in secret, for my part, as I am not a Rosie Crucian, so I am as well as they sorry to see evil done. And I am loath to speak evil of it; and sure were not the great grief and envy I do bear, and always did, to see desert trodden down by such unworthiness, and some little hopes I have to hear of the amendment,

---

[27] Exactly as in the printed edition. -pnw

and so of the return of the Truth, and good Men out of banishment and imprisonment, you should have found me in Westminster Hall, as I have been an Attorney in Term time, and mean to continue my practice there so long as I live, except in the Vacation, which I intend to spend in chemical and Rosie Crucian medicines, for the good of honest plain meaning men: As you shall find in the fifth Book, after we have proved, the way to Happiness, the way to know all things past, present, and to come, the way to long Life, the way to Health, the way to wax Young, and to continue so, the way to Blessedness, the way to Wisdom and Virtue, the way to cure, alter and amend the state of the Body; the way to find out the Golden Treasures of Nature and Art, and the way to prepare Rosie Crucian medicines, their use and virtue; they being safe, easy, experienced, and effectual Rules and Receipts, and such, as whosoever puts in practice shall find true, to the Glory of God, delight of his Soul, and cure of his Body.

Chapter XVII.[28]

The Way to Wax Young.

1. Old Age, gray Hairs, dim Eyes, deaf Ears, rotten Teeth, and lame bones renewed to strength and youth; John Macklein's example and others. 2. The Reason. 3. What makes us young and flourishing. 4. Of cherishing life. 5. Why children and old folk are less Active. 6. The decay of the food of life. 7. Example, of renewed youth. 8. Why Princes are not long lived. 9. To preserve you. 10. Of Brachmans and Indian secrets. 11. The stay of the Law of kind. 12. The first moisture in Nature. 13. Motion. 14. Heat. 15. How to move the spirits. 16. Fruitfulness and Activity. 17. Of Frost. 18. Of youth. 19. Kinds of waxing young. 20. Various opinions. 21. Strange changes. 22. To spring to youth from Age. 23. Medea and Jason. 24. Of the Deeds of Nature. 25. Man restored. 26. To renew the skin, nails and hair. 27. Of order in youth. 28. That an old man may be taken as it were from the brink of the grave, withered, feeble and crooked, and led back to his former youth and lustiness. 29. Acts of Kinde. 30. An old woman turned into a man and of other things. 31. How to accomplish these things.

1.   Howbeit we live long and in health, if our bodies be weak and unwieldy as it is in age, it must needs let & clog us much in this happy race; wherefore the third step and help to health, that in youth was not idle, nay out of order is youth; what then is youth? They know best that have lost it; it is the most active, fruitful, and beautiful estate of the body: these be the marks and differences, whereby we may know it

---

[28] There are two (consecutive) Chapter XVII chapters in the 1662 edition. -pnw

from all things else. I mean activity, not in deeds of moving only, but of life and sense also, this is it which makes up the Nature of youth: the other two marks are taken in, not as need full helps, either to youth or health and such as may not be spared (especially beauty) but because they be very notable marks to know youth by: and that as we heard of true honor and pleasure above, so these will also perforce hang on and follow, though they be unlooked for and disregarded.

2. Then this is the matter under hand in this place, this we must prove possible to be kept and preserved to our lives end; yea and though it were lost before, that it may be gotten again and restored; and yet, first, as our natural heat is the cause of our being, so the cause of our best estate; and youth is the flower of it, that, is his chief strength and quickness. Then keep and recover this, and all is done.

3. But we had need be sure of this, that the flower of heat makes us young and flourishing and sure by proof and experience, the best assurance in the World: let us look all over and we shall find it so; for to begin with Plants, although their life is dark, and they be but lame and unperfect Wights, see my *Harmony of the World* (for Plato gives them sense) so clearly follow the quickness and dullness of their inbred heat, caused by the two seasons of summer and winter, as appears in India, where for the continual heat and moisture and summer of the Country, no plant feels age, or fall of the leaf, that word is idle in those parts, because by a strange property besides the rest, it has strange cooling above the rest, standing in water first, and then somewhat deeply from the Summer sun. Nay amongst us we see those Plants which are hot and dry, sound hardy, able to withstand the force of cold, to keep their leaves in winter, as holly, ivy, box, & etc.

4. Moreover, keep off the starving cold, and cherish the life within, and you say help and amend Nature, and make any Plant flourish and

bear fruit in winter; How is that? But an easy matter: Plant it in a stove, and cover the root with Horse dung, and the rest with chaff, and you shall see the proof, if not the profit worth the cost and travail. The same is seen in beasts, but let us leave the middle that we come not to the end too late.

5. Then why are Children and old folk less active, fruitful and beautiful, then the middle sort, but for want of heat? For let the summer first dry us; Galen says and that before the birth as I showed, be great there in store, bulk & quantity, because it waxes & wanes still with his food, our first moisture and this from thence decays daily; yet his quality, strength and activity, which makes him worthy of the name of heat, is then little, as over much drowned with overmuch foreign and strange wetness. (Like as we see in a green fagot) unable to work his will, and show himself, either to knot the sinews for strength, or concoct the blood for food, and Color before the foreign moisture be spent and gone, which is not in a long time: The cause I have proved in my Temple of Wisdom.

6. Now for old folk what is so clear as this, that by reason of the daily decay of the food of life, the fainting heat lets the Knot of strength and lustiness slack and loose again, and the concoction and Color of blood, which before made seed and beauty, to decline and grow to wateriness? In sick men and women for the same cause; and albeit women have their seed, yet is it not hot and quickening seed, but as dead stuff only fit to receive Life and fashion; and admit they be more fair and smooth then men, which are hotter, it comes by chance because the foul leavings, the blemish of beauty, by the force of manly heat are driven outwards, when the slackness of the heat of woman suffers to remain within, and turn into menstrus, a thing more grievous and noisome in truth then beauty is delightsome. And

therefore, Aristotle very well calls her a weak man; and he makes the male in all kinds to be that which is able to concoct the blood; and that which is not, the female. Then if it be cleared of all doubt, that the chief strength of heat is the cause of the flower of age and youth, and nothing else in the world; let us take a stick to the matter, and see how it may be maintained first, and then restored.

7. I will not urge the way of upholding heat in Plants above said, nor yet this witness of the German who has found out means for the same, both in Plants and Wights, as he teaches in his high opinions, nor yet make account of those examples, which by course of nature and good order of life have done well, and drawn near to this matter: as of Lucius the player, who pronounced upon the stage at Rome an hundred years together, nor of Cornelius who bare Saturnine the Consul after sixty two years; nor yet of King Masinissa, who about flinty got a child, and ever travailed both in frost and snow bare headed, and such other like, marked with signs of long continued life and lustiness. I will come to the point at once. Pliny (such an Author) reports that the whole nation of India lives long free from all diseases, well-nigh two hundred years without any grief of Body, not once touched with ache of head, teeth, or eyes, nor troubled with spitting, all the great companions (as we see) of age, that we may gather by likely guess, when they know not the companions, the thing itself as unknown unto them; but what needs any guessing when the same man for certain and in plain terms assures us, that in that part of India where the sun being their zenith, that is right over their heads, casting no shadow, the men are five cubits and two handfuls high, and live one hundred and thirty years, never waxing old, and being when they die, as in their middle age and chief strength & lustiness? What needs more words? If this report be true, as we may not easily doubt of such an Author, then sure this matter is not impossible, as they would have

it; but all men if they lived in such an sire, and took so good a care of life as I described (I must still fly to that succor) might preserve their youth, and never wax old until that term and stint of life appointed; or if this kind of teaching be now somewhat stale, yet bear with my meaning, and yet perhaps some other means may be found for the matter, in the storehouse of skill and cunning; let us see much more briefly then we have done before, because this part is already well-nigh dispatched; so straight is the link of all those helps, that one can scarce be loosened without the rest, and all must go together.

8.   Then what means may we find? What preserves this natural & heavenly heat of ours? That common people take hot meats & drinks, & think that these preserve heat & nature, as simply as if a man should put lime to the root of a tree which he loved: for as this hastens the fruit with heat, but kills the stock with drought, & soaks up the lively juice & moisture: so in them their hot meats out of kind, laid to the root of life, quicken and stir up the spirits, the fruit of life, for a season; but withal understand, drink it up & waste the first moisture, that is, the whole stock of nature; and so by softning thus the hardness of age; as it were Iron in the fire, they make it seem for a time youthful, and lively, yet it is but a vain and empty show and shadow; and as iron when it comes out of the fire, is the harder; so they make their age more unweildly, and draw it on the faster by that means, and that is the very cause, together with care and pleasure, why princes and nobles, by drying up their bodies in that sort, live not so long for the most part, nor in so good health as other folks, and depart especially at such time, (if the report be true) as those hushed stars called comits, appear; because whether it be a stedfast star, or an elemental flame, (I am not to dispute such questions here) it is not to be seen in a very fine and dry weather which consumes dry bodies, and sends them packing; and besides, (though it be besides my purpose,) turns good humours into

scum called choler, cause of Broyles and sedition; and so making, as we see, the bush Star, a plain sign of both those matters, but cause of neither: As you may read in our *Harmony of the World*, lib. 2.

9. What then preserveth heat? Learned men have brought in certain fine fat and airy meats, as butter, oil, and honey, and commended them for very great helps & means to preserve life and youth, (for both are done by one way, and under one) but especially one of them, that is honey, have they lifted up above the rest, for that the Bee, that little cold & bloodless beast, by reason it is both made of, and fed with the same, lives so long above the kind of parted Wights, even eight years as they report; and because Manne the famous nourisher unto man, is nothing else but a dew concocted in hot Countries, by the heat of Heaven instead of the Bee, and for such like causes too long to be told in so short a race of speech as I have throughout appointed. But these men are wide as well, though not as the former; for if you remember well, when we spoke of things that preserved life, (which is nothing, else as Thomas Heydon has said, but Heat, there were found only two belonging to the uses; viz, meat and exercise, and that to let pass exercise, although the finer breath of the outward AIRE of our meat may serve the Ætherial Spirits, which carries life as well upward as downwards; see my *Temple of Wisdom*; yet our Heavenly must have finer food, an Ætherial body which Is ready and at hand, nowhere in nature save in our first moisture, then this fat & airy meat of theirs, may help to lengthen life, and youth indeed; but not directly by feeding life and maintaining the first moisture, but by another byway procuring health and soundness, (for sickness and diseases bring age and death a pace)and this is, because for their great cleanness, whereunto they be wrought by nature and Art together, they neither breed (as other meats do) many drossy diseases, nor stop the lives and heats free passage.

10. Since then there is nothing in the world, within the Compass of our reach, able to maintain and nourish heat, but it meets needs faint and wane daily with our first moisture; how falls it out, say you, that those Indians so kept their youth, without waxing old, as we heard out of Pliny? I cannot tell, unless the sun, for that great and familiar acquaintance sake, has favored and blessed them above all people, and brought down Æther, and given it them to nourish them; for their Boyle and seat, because it lies right under the suns walk and travail, is not through extreme heat inhabitable, (as Thomas Street fondly supposes) but of other the best and most temperate, by reason that extreme heat of heaven is most equally answered, and justly tempered with cold and moisture of the ground proportionable, which thing they knew not, because their eyes were set to high to see the lower cause and course of nature, most plain & certain, For God when he meant to make our changeable world here below, by a wonderful foresighted wisdom, stinted the sun within those known bounds the north and south turns (which they call Tropics) least when he had run round about, he should have worn and wasted it everywhere alike, and made It smooth and even in all places; and so all either a dry ground or a standing pool, both unfit for the variety of change which he meant to see play before him; but now he is so curbed and restrained within those bounds aforesaid, he can wear the ground no further than his force can reach, nor any otherwise then as his force serves; so that the earth must needs be most worn and wasted, where it lies within the compass of his walk; and so rise by little and little, on both sides without the Turns, until it come to the top and highest pitch, where it is furthest off, that is under the two pins (which they call Poles) of the world. Then bear for the coldness, the earth Is fit to thicken the air and breed water, and for the bent and falling to send It down to the midst and lowest part; whereby the great strength of heat is drawn upon heaps and in great plenty; and for this cause and the

length of the nights, it cannot scatter abroad, and vanish away to naught, but thickens a pace, and falls again abundantly, raining three or four times a day. Whereby we may Judge, that this middle girdle wherein we inhabit, cannot be so broiled and insufferable as our STAR-men avow, but in all reason very mild and temperate; & think that as the sun meant to favor all parts as much as might be, so chiefly and above all that (as reason, yea and necessity bound him) with which he is best acquainted; and as this is certain (by report of all Authors) in all other things, yea and in men touching all other gifts and blessings, so we may guess this one which we have in hand, was not slipped and left out in so large a charter.

11. But for all this, and in good sadness (we have but argued hitherto, it is not good to seek to dispense against the loss of nature, and it were better to discredit Pliny the reporter, (though he be never so good an Author) then Nature herself the Author of all things: for this story is set against the whole course and drift of nature whose works as they be not woven and made up at once, so they decay and wear away by little and little: and therefore admit these men of India by special license from above, do bear age fresh and young a long time, in respect of other Nations: Yet we must in no wise think this is forever and not for death; as Pliny says, for then they should not die and depart as other men do, naturally, which is when age creeping on and changing by little and little, is as last made ripe and falling; but rather by some sudden force be taken, and as it were delivered by and by to I know not what hangman among the destinies, to be cut off and put to death by violence; but what force can that be? Nay I assure you further that if the streak of sickness and diseases were away (as says he, it is almost) they might live forever; another breath of the never broken laws of kind.

12. Wherefore let this story go, and let us hold this rule for certain, that by reason there is no other food for natural heat open in kind, but our first moisture, which because for want of supply, it likewise wastes daily, youth must need by nature fall away and cannot last forever. And yet we must also (to come to the purpose) remember how it was full often above proved, that such a free supply of due food for life were to be made by skill, and fetched out of the bottom of nature and all things by the Divine Art of Hermes. Wherefore to avoid the jar and ill sound of our often beating upon one thing, our Pantarva and heaven above declared, is it that feeds our heat, that holds and preserves youth; that is, it I say that doth the deed, for many causes set down before; I will send them that come not hither the right way back again to take all before them.

13. But there is another thing; motion I mean, and that helps to bear up the state of life and heat, which I scant touched in my discourse of Physick there, and yet it should be handled: because although it be not so needful as the former, yet it cannot in any case be wanting; for as Martial people like unto Mars (as we term it) and valor itself loses his glory and brightness In peace and quietness, as you may read at large in our idea of the law, government, and tyranny, the second Edition; so this heat that rules our body, though it be never so strong and lusty, yet it cannot so soon rest, as it decays and as it were rusts with Idleness, nay the body itself being as I showed above, an airy and fiery temperature, must needs have quick motion, as one of the two pillars of his state; and therefore Plato by the example of the great world, very well advises us still to move both body and mind, and that together, if we mean to have them long continue. And we find his Council good by daily pro ff, when we see those that move the memory most, as wise and learned men, do hold it longest, but because they do not for the most part exercise their bodies, to lose that quality: Whereas quite

contrary the common sort, by reason they move thus much, and that other little, are a great while in body lusty, when their memory is gone as quickly.

14. How moving increases heat, it appears in all places, first in the spring of all heat the sun above, which could in no wise serve to streach so far as to heat half the world at once, if those heaps of heavenly beans and spirits did not help him. See my *Harmony of the World*. Then they would be shut up fast as they be in stones, and metals and such like also, and hard lower lodgings, and not (as we see them) most free, quick, lively, and swiftly stirring; no more doth any fire below burn so fiercely as that which by a cold blast is driven up close and round together, and we see by those that move and stir most lively; to pass by the lightning, (as the weakest to the strongest in the world? And a number more such proofs; for what should I stand so long upon so plain a matter?) motion doth not only increase heat where it is, but begets and purchases it of nothing. And not only that way which every man sees, by rubbing two hard things together, but also by grating a hard thing against the soft and yielding mire, which is somewhat rare, and yet known to the Babylonians in times part, when they used to roast eggs by whirling them about in a sling in the same manner; and so these archers that have seen the leaden heads of their arrows, to melt in flying, so great a father of heat is motion that we may judge how able he is to keep it when it is once gotten: Read our *Temple of Wisdom*.

15. Now if this be sufficiently shown and proved, we need bestow the lesser labor to teach men, how to move their heat and spirits, because every child that can go, can do it; and it is enough to exhort them that love themselves to do it.

16. Then by these two means of like meat and motion, we have our youth still (that is) our chief color, fruitfulness and activity; is there anything else? These make up all the being and nature of youth; except you fear the loss of his hang-byes, and appurtenances, which are teeth, the sweetness of breath, the smoothness of skin, and of hair the color natural.

17. But it is no danger if you will let me run them over; for if our heat and moisture remain without decay, first the jaw bones wherein the teeth be mortized, will be full and moist, able to gripe and glue and so to bold the sane from falling, then all ill smell comes of rawness and want of heat to concoct it; wrinkles of cold which make the face to shrink, & gathers that together, which heat spreads a broad smoothly, and gray hairs from the same cause; for when our natural heat faints & fails, it withdraws itself from the outmost coldest parts soonest, and leaves the moisture raw, which for lack of inward heat to salt and keep it, lies open to the force of outward cold, whence comes all rottenness, and from this a white coat and hoariness. Therefore we see why sickness and sorrow bring gray hairs so fast, yea, sometimes presently: as to pass by the plainer, you shall hear by one strange example of a sorrowful young gentleman of Italy, that being fallen into the hands of pirates, and laid wrapt in a sail ready to be cast over bound, and within twenty four hours space, released and set at liberty, who by great grief and fears forcing his heat to retire to the heart her Castle, made his head white and aged in that space, & could never get it turn again all his life, which was a long time after. And so we have this point briefly and easily dispatched because it was a loose and easy matter; but the next, that is to recover young years spent and blown away, seems no such thing, nor to be used in that order; for as a new and strong building by due and daily reparation, is kept sound a long time; whereas if for lack of care, it be once fame to decay, it cannot

without great cost and time be recovered; even so it is with our body; as it is easy if it be taken in time with heed to preserve it; so if by negligence the weather have once beat in and made it rotten, it seems a marvelous work to repair it.

Although indeed it be much harder, not only then his fellow, but then all the rest that were before; yet we will not give it over now, and like our idle poet, fail the last act of life; wherefore let us go forward, and with all our endeavor strive to show, that youth long before lost (though not so easily) yet as well may be recovered as it was preserved.

19. There be so many kinds of waxing young again named in philosophy, and given to the nature of Wights, that it were good first to sort them out, to see which we mean in the place, least our labor fall into their hands that can quickly mistake; one of these ways is by name only, and not in deed, as when the soft and bare skinned beasts use by course of kind, twice a year, the spring and fall of the leaf, to cast off their upper coat and skin, they say they put off old age, and wax young again; when it is in truth the putting on of age rather and decay of Nature, as appears to them that know the cause, that even for very cold and drought, the true plain ear marks of age, their akin doth loosen and wither away.

20. There is another kind, as far in extremity as that other, and all together indeed, which Alconeon calls joining of ones end to his beginning, and which he says man cannot do, and therefore dyes; and this is and ever was, not the opinion of poets only, but of philosophers, and not of Greeks only, but of all Nations except our old Egyptians, and Rosie-Crucians , men always in all rare wisdom esteemed.

21. These men, as I said above, do not use to mark the steps of kind, and her most strange and unwonted changes, but also set and enter upon the like by skill; yea and to pass further, if any reason will carry them; and so at length they came, and I know not how, or whether by guess or knowledge to this ruled certain ground, of raising the dead, and whether it was possible for any man, put out by forcible and violent death by natural means to rise and quicken again, and so to be renewed, and as it were by a new birth restored.

22. But what be their new and marvelous means? Which way is this incredible course performed? After they saw not only some parts of other Wights (as the tails of lizards, the eyes of snakes and swallows) but also the whole bodies of cold and bloodless ones, clean razed and destroyed, naturally to spring a fresh, and to be restored, as a snake cut in pieces and rotten in dung, to quicken, and every piece to prove a snake again, they began to reach by divine knowledge and practice at some further matters; and to say some whole and bloody rights, that spring not out of nothing, but are breed by force of Seed and conjunction of Male and Female, and the like kindly corruption, to raise them up again and renew them (as a bird burnt alive in a close glass, and so potted, and then enclosed in a shell, to hatch it under a hen, and so restore the same) and other strange proofs they ceased not to make, until at last they durst be bold to think, that any right, even a man and all might by the same course wax young and be born again still, and live forever.

23. This is the second way of waxing young again, and as great an extreme as the other, and as far from any meaning, though there be divers reports and stories flown abroad, of men that took the same race in in themselves and others, and found both good and bad success (according as a man that favors it will think) as the work was intended

by them which were put in trust: Medea sped well say they, in proof, upon Jason's body, and made him young again, as Tully says, recoquendo; but Hermes, the poet Virgil, and the Spanish earl failed upon themselves, as some hold, they had good luck and came to their purpose. To know the whole Art, read the second Book of the Harmony of the World, and the Temple of Wisdom. What should a man say to this matter? Albeit I do not choose this kind of renewing, yet I will not condemn it without cause, and judge it for a thing Impossible; for I see no reason but that the story of the Snake may be full easily true, because it is bred by itself, and of more unfit stuff in the same manner; and for the rest all is one to nature, if the stuff and place be meet and convenient, having her general seed of begetting (which I said was all one in all things) in her bosom ever ready, and thereby making (yet as we heard before) all seeded Plants without seed somewhere, yea and perfect rights both water and Land ones; and at first when the stuff, and womb, and her own heat, and all served very fitly, having wrought man and all so.

24.   But now why is seed given up to things? Because nature for want of the former helps (as they could not last forever) so not able in all places to work the raw stuff of the beginnings so far, to such perfection, unless she find both the stuff well dressed and half made to her hand; and a whole womb like an artificial furnace, to help and set her forward; well then for this one matter and manner, of restoring man, let us call it to the account of reason, and consider what is that seed that makes man, and the place where he is made; what is all the work? Is it anything else but a part of man (except his mind) rotted in a continual, even, gentle, moist, and mighty heat? Is it not like that the whole body rotted in like manner, and in a womb agreeable, shall swim out at last, quicken and rise the same? I cannot tell, I will neither avow nor disavow the matter; nature is deep and wonderful in her

deeds, if they be searched and unwound to the bottom. I cannot tell, I say; mature may suffer this, but not Religion; and yet it is a dangerous trial as our men, and the Poet found by some men's sayings.

25.   They might more safely have made a proof upon a piece of themselves, which we call seed, ordered by that skillful kind of recoction[29] (which has been found true, a some report, and I think it certain) or perhaps more kindly and thoroughly, but sure more civilly and religiously in the due place appointed; for this also a kind of renewing of himself and waxing young again, when his child is (as Aristotle says well) another himself, only sundered and set apart from himself; but neither is this third kind enough for us; we must have the whole and unparted man restored.

26.   Then the fourth is it I mean, which is indeed a mean between all the rest, especially between that empty and dangerous deed aforesaid, performing more than the one in the outside, and less within then the other: for this way doth not only by a better race of refreshing it with heat and moisture, renew the akin, nails, teeth also, though these by the sane way of putting off the old ones; but for the inward chief and needful parts, how out of the seed at first by the natural workman, it shall neither make nor mar any, only change and alter, purge and place them all in their former state and soundness, youth and lustiness.

27.   Then let us see how we may be renewed and wax young in that order; beginning first with those idle and needless things (I cannot call them parts) of the body, which after were made up, finished, grew, and sprung out from the leaving of our meat and nourishment; the teeth, nails, and hair; as for the akin it is a part of the seed, or the crust that over cast the thing, when it was fully baked; then as these had no

---

[29] Spelled exactly as in the printed edition. -pnw

certain course and order of kind in coming; for (to omit hair that goes and comes upon every light occasion) some are born without nails, and some with teeth; when others again have none before they be old, and such like disorders; so they doubt by skill they may come and go again without any hurt or great change to the body. Pliny tells of one whose teeth came again after he was an hundred years old, and upward: and John Macelain an English Minister renewed his age and waxed young being very old as they say: and I know not well whether the soldiers in Germany by drinking of a spring, by the river Rhine, had their teeth shaken out, and loose and had them come again a new. But this is certain that there be waters in the world, which by a special quality make those beasts that drink thereof, cast their hair, horns and hoofs, and so renew them, as the Poet says.

28.   What needs many words? This part is easy and of small weight, and we may pass it over: but that am old feeble, withered, crooked, and barren man, should be taken from the brink of his grave, as it were and led back to his former youth, and lustiness, is a thing say they, both in truth false, and in reason incredibles; nay if two such men were set before us, it would seem in sense ridiculous: indeed it will seem so to such men as are either all sense and no reason, or else whose wits are all bestowed upon the search of such troubles as is not worth the searching.

29.   If it had been spent about the deep and hidden works of nature, there would some have appeared as great as this is, and stayed all childish words; for my part I am willing to supply the want, to unfold the greatest acts of kind, and set them before you but this work grows too fast and proves bigger than either I rat, or would; it is planted upon so good and fruitful a ground; yet have one or two of the fittest examples, and nearest and match them, and this together that you may

see it, at last, fall out to jest, and worthy laughter, (I am loth to fall into the mouths of jesters) about a solid and earnest matter, written by a young hand.

30.   Is it not as hard and wonderful a change thinks you, to see a woman suddenly prove a man, as to behold an old man by little and little wax young again? Compare this if you but doubt of the story. Pliny is my Author still, who reports of three such sundry chances, which he himself saw, (he named the place and party) and how performed upon her marriage day; Cardan doubted no whit of the truth, but ventures at a reason for it (which because it is both likely to be true, and unseemly to be told, I will let it go) and he vouched the same change again, but in another kind, and yet more strangely then the first, and whereof no man ever durst or could hold a reason: The same man again says, that the Beast hyena, every year changes her sex, being by course one year male, and another year female, never ceasing nor missing that strange and marvelous turning; is not this a much more harder & greater kind of change and alteration then that we speak of? Then we grant nature is able to do this if she be willing: but it seems not because she never doth it; she runs still if she be not letted, her appointed race; but if there be many dead chances able to hinder and let this course of nature, how much more can the wit of man, (which is a spark of the wind which gave by kind her Commission) doth it? As she doth often, if I might stand to show it, both stoup and lengthen, and yet she is never willing and consenting. Let us see then for this matter in hand, how skill is able to overrule kind by her own con consent, and make her willing to return and wax young again; first let us know that all philosophers, Rosie Crucians and physicians hold, the life and soul and natural heat to be always Of itself young and lusty, and never old, but to appear so by reason of her falling part her instruments, and that I have often showed in a kind of fire waxing and

waning according to her heat and motion; then here one good help to the great work of renewing; In like sort the parts of the body are not marred and lost, as they say of a rich man that he is decayed, when his money the life of the World, has left and forsaken him; even so when our natural heat the life of this little world, is faint and gone, the body shrinks up and is defaced; but bring again heat into the parts, and likewise money into the bankrouts Coffers, and they shall be both lusty and flourish again, as much as ever they did.

31.   But how may this heat be brought again? To make few words, even as she is kept and held by due meat and motion; for if she faint and fail for want of them only, then give her them, and she shall recover herself again; meat is the bait that draws her down; Motion comes forward like a gad—bee to prick her forward; but that work is performed in. this order; first this meat which is that fine and ÆtherEAL oil, often described by Eugenius Theodidactus in his book entitled the *Rota Mundi*, and in my Rosie Crucian *Axiomata* lib. 2. chap. 7. exceeding piercing swiftness, divides, scatters and scours away the gross and foul degrees and leavings, which for want of the tillage of heat, had overgrown in our bodies and which was cast like a blockish stay—fish in our way, to stay the free course of the ship of life, if she lifted to stir and run her wonted race, (which in some think it enough in this matter) but also scatters all about her dew and desired meat, and first moisture to draw her forwards. By which means our life having gotten her full strength and liveliness, and returned like the sun in summer into all our quarters, begins to work as fresh as she did at first, (for being the same upon the same she must needs do the same) knitting and binding, the weak and loose joints and sinews, watering and concocting all by good digestion, and then the idle parts like leaves shall in the hot Summer spring and grow fourth afresh, out

of this new and young temper of body, and all the whole face and show shall be young and flourishing.

This is quickly spoken, say you, if it were as soon done. It were happy medicines, nay that were a miracle; but I work no miracles; I only help, as I said, the willing race of kind, wherefore is long decaying and wearing away, or rather in making and waxing to his perfection; so in mending no doubt, he cannot return all at once, but must creep back by little and little, and so be restored; or else I would have told you at first dash of that spring in the isle Bonica which Master Edwards doth witness, will in few days restores a man quite (saving gray hairs and wrinkles) and make him young again. Nay if I had taken a course to delight woman and children, and to win credit among the common sort, I would have sought the Legend, and rifled all the goodly wonders in the world, and fitted many to my purpose. But as I serve Madam Beata my love, and wait upon a wiser mistral, yea and in the most inward and secret place among them, so I would by my will, speak nothing that should not be pleasing in her sight, and well sounded in the ears of wisdom; Wherefore let these few suffice for this matter being the truth, as I learned of Nature.

Chapter XVIII.

1. Of poverty and riches. 2. Of worldly wealth. 3. Of Arabian underground bodies. 4. Of Socrates underground secrets. 5. That the heat of the earth boils Rocks into Minerals. 6. Of Quicksilver. 7. The sulphureous and Mineral quality of the breath of man. 8. The Frozen Aire. 9. Of purging Quicksilver. 10. Dissolving of seed, and breaths of

metals. 11. Earthly Brimstone to make a perfect weight. 12. Sun and Moon make Man. 13. Of making Metals. 14. Instruction. 15. Earth in Power, rater in Quicksilver equal. 16. Nourishment in Minerals and Plants. 17. Of gross and fine bodies. 18. Mineral heat. 19. Heat perfects Minerals, cold covers the work. 20. Iron and Copper of the Nature of Quicksilver. 21. That Nature intended white and yellow Copper as Gold and Silver. 22. Silver and Gold in all Metals. 23. Degrees of Metals cleansed. 24. Nature change able. 25. Of mending Nature by Art. 26. The imagination of Birds. 27. Guides. 28. Antimony. 29. The Color of Gold fastened. 30. To die Metal. 31. To stay flying spirits. 32. To wash the hands in Molten lead unhurt, 33. To stand the force of a Bullet unhurt. 34. To keep a Cloth from burning. 35. To use unripe Gold and make it as good as the best. 36. How to make Aurum Potable. 37. That Quicksilver may be bound, colored and made Gold, and of strange things.

1.   Even this is the point at last which the golden world looked for at first, the way to pleasure, because it is indeed the last & lowest part, (being servants, and so to be used) and yet very needful and not to be spared in this blessed household; for although we have all helps to long life, health, and youth that may be, yet if we want the service of riches, poverty will besiege us, and keep us under and cut off and hinder many goodly deeds and works of wisdom and virtue; but what are riches? For the world by Rosie Crucians and philosophers agree not in this account; nor this within itself; the world reckons store of Gold and silver to be riches; Aristotle enough of needful things: the stock enough of earth and air. To begin here, these might be stretched and made larger enough, but that we know their straightness: would they have us live by breath alone, and never eat, according to the guise

which I set out in healing? Be it possible, as it seems; yet it is somewhat feeble, as I showed there, and so somewhat halting and unperfect (by lack of youth and lustiness) for our full and perfect life appointed, besides the means and hurts of poverty, which I right now touched. Aristotle is somewhat strait also for beasts, that I reason are riches as well: if he had put in enough of things needful for good life, wherefore we were made, he had said much better; yet not all, for so should all the bodily means and helps aforesaid be counted riches, a great deal too confusedly; now much less can we rate the golden wealth right and true riches, because a man may dye with hunger for all this; so he that sold a mouse for two hundred pence, dyed himself for lack of food, when the buyer lived, and this was done (to let go famed Midas) when Hannibal besieged Casiline; then true riches are enough of outward things needful. for good life, that is, for our physick above set.

But because that golden and worldly wealth is a ready and certain way & means to this (out bearing violence which no man can warrant) we will use the cause for the effect in this place & to strive to show how all men may get enough of Gold and Silver and that by weaker means then Hermes medicines as the place requires, although by the same way concerning the stuff we work on; that is, by turning base metals into silver and Gold, by the Art of preparing Rosie Crucian medicines named in my *Harmony of the World*, lib. 1. chap. 12. this is the hard matter which turns the edge of worldly wits, the brightness I say, of these healthful things dazzles the eye of the Common and bleared people, because it is in their account the best and highest, and most happy in the world; when indeed and truth as it is the least and lowest, and worst of all the helps unto physicks belonging, so it is in proof and trial the less hard & troublesome, both to art and nature, the most ready and easy to be gotten and performed.

3.   And to show this (we will make no long tarrying) it were first good to enter into the way and order which nature below keeps, in making the metals underground; if I thought I might not run into Socrates his accusation, for searching over deeply the underground matters: but I hope I shall not now by the mighty pains of the miners spades, and mattocks, the way is made so plain before me, or else sure as they be, indeed I would account them over deep and hard for my pen to dig in.

4.   Then all underground bodies, which the Arabians call minerals, are either stones or hard Juices, (which we name middle minerals) or else they be Metals; these as all other perfect things have all one stuff, earth and water, and one workman the heat of Heaven, as I said above, for their womb, because they be but dead things as they call them, the earth will serve. But for that nature meant to make most perfect things in that kind, which require long time to finish them; she chose a most sure and certain place even the dead and hard rock itself, not to the end the earth might hide them as hurtful things, and lean upon them with all her weight, as Seneca says very severely, or rather very finely, (for we know how he hunts after fineness) like an orator, to whom it is granted to lie a little in histories, that he may bring it in more prettily, as the Orator himself confesses,

5.   Then the manner of the work of Minerals is this, first the water piercing downwards, softens and breaks the rock, taking her course still that way where it is softest, to make the cross and crooked race, which we see of wombs, called veins or pipes of the Minerals; but as the water runs (to take the stuff as the next thing in order) it washes and shaves off small pieces of the rock, and when it stands and gathers together in one place by continual draining, cleaning and refining the same, until the mild heat of the earth, which is the heat of heaven,

come and by long boiling, makes it thicken and grow together into one body of many kinds, according to the difference of the stuff and heat, which they call hard Juices, as I say, or middle Minerals.

6.   This workman continuing and holding on his labor (this Agricola says, the cold and drought of the rock now lays upon the stuff, and by little and little, and at last binds it into that hard form of a metal; nay though Aristotle from the beginning gives the work to the same cause) out of the heart as it were, and best part of them, wrings out at last a clean, close, and heavy, raw, waterish, and running body called Quicksilver; here it stands in perfection of this Mineral work, except there chance (which chance happens often) by the means of the boiling, any contrary, whole and dry breath of the same kind, to be made with all in the same place, then the meeting with the raw, waterish, and unhappy lump, like a rennet with milk, or seed with menstrue, curdles, thickens, and fashions into the standing body of metal.

7.   This mineral breath of man, for his likeness in quality (though their substance do greatly differ) do use to call brimstone; now then this second and earthly heat is come into the work, the middle heat of heaven sets the stuff which stayed before to work again, and drives it forward, and these two together by continual boiling and mingling, alter and change, cleanse and refine it from degree to degree, until at last, after many years labor, it Comes to the top of perfection, in cleanness, fineness, closeness and Color, which they call Gold; these degrees if the heat be gentle and long-suffering (as they say) be first lead, then tin, thirdly silver, and so to Gold, but if it be strong and sudden, it turns the weak work out of the way quickly and burns it up quickly, and makes naught but IRON, or at the least if the heat by somewhat better, copper; yea and sometimes the fullness of the earthly

Brimstone alters the course of nature in this work, as also there is odds of Quicksilver; but indeed the cause of all the difference, is in the working—heat that makes and disposes the beginning, midst and end of all, thus and thus, according to her strength and continuance, and which is the main ground to this purpose, Quicksilver is the mother of all the metals.

8.   Now when the work is done, it lies yet as it did all the while in a thick flowing form, like the form of a molten metal, and then it is fit to make Telesmes for love, marriage, health, Long life, youth, gaming; for fortune in Merchandise and Trade; for war and all other things. But when the owner comes to enjoy it, bringing in the cold breath of the air upon it, like unto Coral, and other soft and growing Sea-plants, it freezes and hardens of a sudden fit for the turn and use of man in other things, wherefore it was made and ordained. These be the grounds of the most and best of our men Rosie Crucians, that is of men best seen and furthest traveled in such matters, where unto Cardan a man indifferent, and none of us, yet very learned, agrees jump as may be; but least these dime and little lights may seem to be darkened, with the brightness and fame of Aristotle, and his Scholar Theoperat, and the late renowned Agricola, holding hard the contrary, and the same sometime stiffly maintained; I will as much as in me lie, and my narrow bounds will suffer, endeavor to lay the reasons all down in order, which moved them to think thus, and staid them in the same opinion; that wise men at least may lay one reason with another, and judge which is the weightiest and worthy to bare the best price, with the vain regard of outward shows and authorities.

9.   First, That the Mineral stuff sprung out from the rock shavings aforesaid, all cunning Miners can tell you, who still by the nature and grill of the stone, though there be twenty several sorts (as there be

sometimes in the rock) are able certainly to say this or that vein follows. But to pass over lightly the lighter matters, and such as they grant as well as we; that Quicksilver is the nearest stuff or menstrue, or mother of metals, that is the thing in great strife and question, when it needed not in my opinion, if we mark the consent of all those men, in all Nations, that put the name upon things, which were not of the unwisest sort, flatly to allow this saying, when they by calling it in Greek, Latin, and all other tongues quick or Liquid Silver, in secret meaning plainly say, that if by the force of those two whole workmen aforesaid it were stayed and better purged, it were nothing else but Silver, for indeed Avicen and some other of the Learned side, leaving out the middle degrees, hold the very fine opinion, which I also think true, if the stuff and heats (as they are in hot Countries) be good and faultless, but the disquiet will account this kind of argument unskillful, and so cast it off, then remove the cold that at last came upon the metal and hardened it, and it appears to the eye nothing else but such an altered Quicksilver.

10. Or if the witness of sense be sometimes false and deceitful, enter our School and behold them by a more kindly and gentle way lead back to a true Quicksilver, both in cold and heat abiding, being a true rule in Rosie Crucian Physick and Philosophy, everything to be made of that whereunto it is loosened and dissolved. But if all this will not serve, pass a little further into the border and edge of secrets, and you shall see them by following the steps of kind underneath (which I marked out before) that is by Bowing the dissolving seeds and breaths of metals upon Quicksilver, to curdle and bring her in that form of metal which they will and wish for.

11. Now for that earthly Brimstone, nature doth make a perfect Wight, and is fain to break her first order, and to take the help of a

whole womb, & of another workman; even so to frame a perfect dead creature besides the help of Eugenius Theodidactus; both to fashion and to boil it to perfection.

12.   Then as Aristotle says, the Sun and Moon make a man: and the rest have two working and moving causes, the heat of heaven, and the breath of the male seed; so, in this work of metals, there is not only the great and general begetting breath of heaven; but also, the private and particular seed of the earth their father, that there lacks a little earth to stay Quicksilver. Aristotle himself shows, by a pretty like example (he says) that hearts blood flows still when it is cold, when as others stands, because it wants those earthly streams which others have, to make it grow together, as we may see by trial finding no blood which has them with a strainer taken away, to stand and cluster, but run continually. Even so take away the Earth and Brimstone of a metal (which our Art can do) and the water will not stand again but flow for ever; and this is general if we mark well, that nothing stands and leaves his running before earth, ruling binds and stays him.

13.   Whosoever allows not this way of making metals, besides other fails and errors, he shall never unfold the nature of Quicksilver, as we by Aristotle's and Agricola's struggling and striving against the stream about it, giving the cause of his flowing and flying from the fire unto abundance of mire in him, for then his lightness and feeding of the fire, two things far from his nature would as well as in all airy bodies, shine forth and appear unto us.

14.   But he that stands upon Eugenius Theodidactus grounds and rules laid down before may easily perceive his own raw, cold, and watery condition, to make him fly the fire his enemy; and this even proportion in power and will rule of earth and water in him, to be the cause of his running.

15. The first is plain, but that there is as much earth in power, as water in Quicksilver (albeit it seems all water) and no more of this then of that, surely mingled and put together, appears because it is the only dry water in the world; her earth haling one way makes her dry, and her water another causes her to flow; but this is a certain sign thereof, that when we find by reason all other things, if either earth or water rules over them, either to stand with cold and harden, or else to melt with fire and water; yet we see plainly this one dry water called Quicksilver to stoop, and yield to neither; but to our purpose.

The reasons why the heat of heaven is the workman in the Mine, are many; but hear a few and briefly delivered, if he worketh and mingles (as I proved above) all perfect mingled bodies, then that shall let and bar him from this labor also, the depth and hardness of the rock? No, for if those subtle bodies which we call spirits, are able in the opinion of all men, to pierce through stone walls without breach or sign of passage, how much more subtle and able to do it, is this heavenly soul? But all men grant the workmanship of livings to flow, from that only cause and fountain? Then tell us how it comes to pass that fish (by the witness of good Authors) are sometimes found in the deep and sound earth, where no water runs, nay which way do very toads get into certain stones in Germany, and millstone rocks in France, even so close that they cannot be spied, before they be felt in grinding and break themselves as George Agricola reports.

16. But if Minerals as well as Plants take their food and nourishment, wax and grow in size, all is clear I hope and void of doubt; this will I prove hereafter.

17. In the meantime, let us win it again by proof and trial, the strongest battery that nay be; cold binds and gathers in the stuff both like and unlike, gross and fine together, without any cleaning or

sundering; but metals especially are very finely and cleanly purged bodies. Again, if cold frozen and packed up Gold together, the force of heat (as we see the proof in all things) should cut the binds and unmake the work again, which is not. To this, what Color springs from cold, but his own waterish and earthly Color? That if a thing be dyed with other colors, we know straightway where it had them; besides cold leaves no smell behind it; but heat is the cause of all smells, then to omit the fiery smells of some stones, and sweet savor of others, and the variety of sent in Juices happened it that Silver found at Mary-Berg smelt like violets, as Agricola reports; that all men feel the unpleasant sent of copper and other base metals. But mark the practice of the plain men, when they devise and judge of a Mine below, they take their aim at no better mark, then if by grating two stones of the hill together, they feel a smell of Brimstone, because they take thus the leavings of the metals in their concoction. To be short do but cast with your selves, why there be no metals but in rocks and mountains, unless they unload end shut them down into the plain, and then wherefore foul metal in cold Countries, and fine silver and Gold, besides precious stones in hot Countries, and you shall find the cause of this to be the difference of the refining and purging heat, and the closeness of the place to keep in the heavenly heat, and barrenness with all, and emptiness of Plants to draw it forth and spend it.

18.   Some cannot conceive how heat should cause this matter, when they feel no heat in the Mine; I will not say to such that this heat is most mild and gentle everywhere, and there especially; but bid them bring a piece of Mineral earth, and lay it in the open air, and they shall feel if they will lay their hand upon it, no small but a burning heat, by the cold blast stirred up and caused, even as the lurking heat of Lime is stirred up with water.

19. Wherefore we may so safely sit down and build upon it that all Minerals are made with heat, and get thereby their being and perfection, albeit the outward shape and last cover as it were of the work, is put on by cold.

20. Now for the steps and degrees of metals, that they all except Iron and Copper, though some do not except them, arise from the steps and degrees of backing the self-same thing and stuff of Quicksilver, it appears in Lead Mines, where is always for the most part some Gold and Silver found by report of good Authors; and therefore albeit says, that Cunning Miners, use in such case to shut up the Mine again, for thirty or forty years, to bake the Lead better, and bake it on to putrefaction, and that thing to have been found true in his time in Secavonia.

21. But what do White and Yellow Copper is found in the ground signify unto us, but that Nature was travelling by way of Concoction unto the end of Silver and Gold. Again, how comes it to pass that plain Artificers can fetch out of every metal Borne Gold and Silver? And out of these some base metals, unless Gold and Silver were the heart and best part of the whole body, and of one self-same thing with the metals.

22. Nay Paracelsus avows that not only in these, but in Mines of minerals, things further off as you know, are never without silver and Gold; and therefore, he giveth Council to water them, as if they were plants with their own mine, & kindly water, assuring us that they will grow up to ripeness, and in few years prove as rich as any Gold and silver mine.

23. Then we see at last the strength of this Metal in ground unshaken, and standing sure for all the battery of the stoutest

Grecians, that all metals have but one Quicksilver, kind and nature, being all oneself same thing, differing in degrees of cleanness and fineness, closeness and Color, that is from Accidents springing out from the degrees of boiling and decoction: it is now time to go to build up this matter, and to show how these low and unclean metals may be mended and change into Silver and Gold, to make the way to attain Riches, if all metals are so near & like one another, especially some of them (which I set down before) wanting nothing, but continuance of cleansing and purging by concoction, then sure this change may seem no such hard impossible matter, nor to need perhaps to help the divine Art of Hermes Medicines, but a lesser and baser skill may serve the turn.

24. And as nature is not poor and needy, but full of store and change, so may skill if she will follow the steps of nature, find more ways than one to matter then which Is the lower way and lesser skill follows nature? We will fetch from that way you say nature take even now below the ground what is that? I will tell you shortly.

25. As nature in her work below used hot workmen, so will I, example in this place: and mention the five lesser and impure Metals, viz. and may be amended and changed into the greater and most perfect metals, viz, into and but this cannot be done without the Philosophers Pantarva ; and now let us return to our work in hand in the Course of Nature. And because we cannot tarry her leisure and long time, she taketh to that purpose, we will match and countervail her little heats with proportion answerable and for our time, that we may do that in forty days, that nature doth in so many years, and this proportion is not hard to be found, when we consider the odds and spate, that lie between the founders fire, and the gentle heat of heaven; and again the difference of such a scouring purge, as that Eater above

consuming Stones and Iron so quickly and the mild heat and lazy breath of a thickened quicksilver. And therefore as the miners do well in trying and purging the rude metal from the outward filth and leavings, besides a great outward fire, to put to the limp many holle[30] and piercing things to further the work of boiling, and so after they have done and made the metal clear and handsome, if we mean to cleanse them further from the inward filth and drowsiness, we must take the same Course, but with greater force and skill, even so much more, as it is more hard to part away the inwards, and inbred uncleanness, then the outward and strange **scurse** and foulness.

Although I did set before divers differences and marks upon the metals, yet indeed there are but two to be counted of; and there is no odds between them, and Gold, but in closeness and Color; the rest is cleanness, fineness, and steadfastness in the fire, follow all under closeness, for a thing is close, when much thing is packed up together in a narrow room which cannot be except the stuff be clean and fine before, and when this is so pact up, it must needs be weighty and steadfast also, heavy for the much stuff, but steadfast for two causes; both for that there is neither Entrance left for the fire to pierce and divide the stuff, (and by division all things are spoiled) nor yet any gross and greasy stuff the food of fire, remaining Quicksilver as I said was clean at first, & if it with a fine brimstone you stay & fasten it, which is often in hot countries) it straight way (I mean without any middle steps) prove Silver and then Gold: but if that curdling breath be foul and greasy (as it is most commonly) it turns Quicksilver into foul metals first, and the work must tarry a longer leisure to be made clean and perfect, that is until such time, as that foul brimstone, be clean purged out as it is only in Gold.

---

[30] Spelled as in the original printed edition. -pnw

26. That Nature doth in due time, and Art by imitation may part and drive away all the filthy Rennet, this is a sign because it is no part of the thing; how in that proved all the filthy rennet. This is a sign because it is no part of the thing; how is that proved? For that is the male seed that begets, makes, and fashions all, and naught begets itself, but is made by a strange and outward mover, which is like the Carpenter, or other workman towards the work he makes, that this is so, it is, plain by the male seed of Wights, which is not the material stuff seen with eyes (that is but a shell, given for the safe keeping) but an unseen hot breath of their bodies, whereby alone without the help of the shell, many Wights beget their Mates with young, as we may read in Aristotle and other good Authors; what makes it so plain, as the barren eggs which many birds fashion fully in themselves by conceit of lust, wanting only an outward quickening cause from the male?

27. Then how shall we purge out this foul and greasy workman, to make the work of any metal close and well colored? Nature would have done this in time by concoction, without any other help; but we must best to shorten the time fit for use, two devises, one to breed closeness, and the other to bring on good Color; the first is a binding shell, the next is a dying cunning, for the first, let Nature still be our guide and leader.

28. As she in all her easy changes, uses to consume and raze out the weaker with the stronger; like so we, if we mean to devour and consume all the greasy & gross stuff of the metal, that when all is clean and fine, the metal may draw it up close together; we must encounter it with a strong like; what was the brimstone or any other filth in Quicksilver, and of what stock think you? Did I not tell you it sprung out of a confused heap of middle minerals, and was a Mineral breath

and vapor? Then let us take the foul and sharp minerals, and in a strong fire set them upon the metals, and they shall sure by searching and sifting round about, quickly draw to them, eat and drink up all the water like dross of the metal, and leave the rest which is unlike clean and untouched; I need not stand any more about it; do we not see how Soap a filthy strong thing in battle, and working with a foul and filthy cloth, makes it clean and spotless? Nay, to come nearer, how doth antimony that fierce and foul mineral, where he is set on work with Gold to cleanse him, search and run over all the metal, take and consume his like meat, and the strange and unclean parts, leaving the rest as unlike and unmeet for him; to be short, if you mark well, you shall find it the plain ready and kindly way, not only in all purgings, but in every natural thing.

29. Then let this part go by, and since now the metal is as clean, fine, weighty again as Quicksilver, or close & steadfast as Silver, or rather more; let us take the next in point hand, and bring on the Color of Gold, this stands upon two points, it must have the fairness and lastingness of Gold.

30. But hear is all the cunning, to dye the metal all over, with an everlasting Color; to this purpose, it had need to be able to pierce the metal, and to abide at fire; that first is not hard again, but how shall this be done? Perhaps we need not strive before we lay the Color, to make it steadfast and binding; but like as Gold will, so fast embrace, and hold his flying, make Quicksilver, if she be a little cleansed and made fit to receive him, that no fire shall depart them; so the closeness of this one steadfast metal shall descend and save the Color; but suppose it will not, yet if Iron and Copper, nay, the middle minerals may be bound and made abiding in the fire (as our men hold and teach) then their colors may be staid and made steadfast also.

31.   What is remaining, if you be not yet content go to school and learn to fasten and stay flying spirits, as they call them, Cardan who denies it possible to make an open metal, close and steadfast, yet allows this matter easy, and since we are here, and he so ready, let us talk with him a little, I marvel much at him, a man so well learned (but indeed not skilled in the Art, the chief of all Learning) that although he had spoken well a great while, and allowed all metals to be made of one stuff, and to travel by one way of concoction unto one end, Gold: and to differ by one accident only and chance of those degrees of boiling, and thereby yielded that all the foul metals may be turned one into another, and Silver—like wise into Gold; because it is nothing else but imperfect Gold, and the worse part thereof, wanting naught but Color which easy and a little closeness, which by purging out the greasy food of fire may be given him; yet for all this, he denies it possible to change any of the lower metals into either sol or luna, because of our sudden heat (as I said) of Mars and Venus being burnt they cannot be brought to their old Mercurial clearness, nor yet be made abiding nor steadfast in the fire.

32.   This he would never have said if he had been brought up in this our trade of Learning, he should have seen us easily lead the metals back from whence they all came, and then, by means aforesaid stay them; for, he grants himself that all the cause of uncloseness, unsteadfastness and wasting in the fire, is that our fatty Brimstone, and that it may be cleansed out of Silver; why not out of the rest also? Will they not abide the violence? Not at first, but by little and little they will, as gentle and wise men know how to use them, there are others also as well as he, Erastus and such like, that deny this art of changing; if I thought those men needed any labor of reproof, who through ignorance of the points they handle, blunder and rush in the dark, cross, and reprove themselves, all about in such sort as they seem

rather to move pity to the standers by, then to make a challenge, and to call forth an adversary. Then such men I will exhort to be better advised, by the view of certain plain examples which I will lay down before them, and thereby with them, to stay their over swift and fore running judgements, until they come to the trial and battle itself, in that which shall follow—lead, as that workman know is one of the greatest spoilers of his fellows, the foul metal in the world, save them from the rage of him upon a shell of Ashes, which they call a Test, and he is counted safe, sure, and steadfast enough against all essays. A few years ago, when I was in Egypt, about April one thousand six hundred and fifty, Transilantis a learned man of Alexandria told me of a man at Chassalovia, which I know not how so anointed and armed himself, I mean his face and hands, as he could suffer to wash them in molten Lead.

33.   Another time I was in Arabia, and from thence sailed to the rock called Alexander, and further to Ansalerne and Christe upon Euphrates, in the year one thousand six hundred fifty and two; about May I set Hemet Obdeloh and the Alcades amongst many wonderful things, this I saw, one who durst oppose his naked body to the violence of a bullet shot from a musket; and this naturally his flesh was hardened by a charm. Why nay not then by the same example a tougher and harder Metal be more easily armed and fenced against all force and violence.

34.   In my Oriental travels, you may read of the events in nature experienced, and again to be experienced by myself; nay you shall see more wonders by the skill of nature easily performed, clear crystal saves the cloth that is wrapped about it from the rage of the fire, so does oil defend paper, in so much that you may boil fish therein, without either burning the paper, or the oil soaking through, and all

this is because the extreme and deadly feuds do save the middle thing by their working. Is it then a wonder, if iron or copper be by some pretty sleight, or kindly skill defended from all fire, and made sure and steadfast.

35. To draw nearer unto you, it is very well known that base and unripe Gold, fit only for Tolesmes when it is, see my Rosie Crucian *Infallible Axiomata*, lib. 1. in the Preface; as it were a mean between silver and Gold, wanting Color and closeness, wasting much away in time of proof and trial, may by some of the lesser and lower degrees of binding, be refined and made as good as the best Gold in the world; then is there any let in reason, why the rest, especially Silver, by strong and more forcible means say not be bound and colored and reach perfection.

36. Now before I travel further; it is first necessary to prevent the delusion of the Mountebank of London, and in other places; and let you know the truth that you be not deceived by those that pretend to have aurum potable, and those experienced Medicines Madam Beata taught me, as you may see in the Preface and Chap. 12. of my first book of the *Harmony of the World*. None has these true Philosophical Medicines, (but Eugenius Theodidactus, and Doctor Culpepper's widow) imagined to be attained by the studies of one Dr. Nich. Culpepper who learned of me; 'tis truth Nicholas Culpepper was a friend, but not a Master to the Golden experienced Medicines. But to undeceive you, the Post Doctors and pretenders err, and instead of Potable Gold, the panareae Pantarva which is the quintessence i.e. and tincture of Gold, etc. have given to men an impure Calx of Gold, not considering the difference and evil that follows upon it: Gold Calcined or powdered, if it be given to men, is gathered into one lump in the stomach, and does no good to the patient, it guilds the bowels and

stomach, and hinders the concoction, whence many and various sicknesses follow, and at length death itself; I were going at the finishing hereof into Italy which I left for health sake, etc. Take no Metallic Arcanum or Medicine into your body, unless it be first made volatile, and it be reduced into no metal. The beginning to prepare potable Gold is this; so, may such a volatile be afterwards dissolved in spirit of Wine, that both may ascend together and be made volatile inseparable, and as you prepare Gold, so may you also prepare potable Luna, Mercury, Venus, Mars, Jupiter, and Saturn; it is hard to learn without practice and a teacher. But to return to our purpose, and to conclude if we may by tracing and diligently persuing the footsteps of nature, which she treads daily, turn and plant of sight into a stone, and minerals into a metal, and Lead into Tin, nay Lead into Copper (as I will prove hereafter) with so great exchange, and increase of center and closeness; then tell me why by means fitted in proportion, Lead, or rather Copper may not be turned into Silver, or either, of these especially Silver into Gold.

37.   Therefore to make up all Paracelsus reports for certain, that in Corinthia they commonly turn Copper into Silver, and this into Gold in Hungary, though he names not the means whereby they made those exchanges, yet we may easily judge those ways, of binding and coloring set down before, that is, lesser ways then Hermes MEDICINE, and yet sufficient to serve our turn, and to raise that wealth appointed, as we may see by guess of their common practice, which else were empty, vain, and foolish, as also by the light change of middle minerals in respect of the return and gain of Gold. And if the praise of an enemy be lightly true and uncorrupt, let us hear what Poetus a denier of the Art of Hermes confesses upon his own experiences; that Quicksilver may by divers ways bound and colored and made perfect Gold and Silver; and on may when it is with Brimstone burnt and made cinaber

very gainfully (which thing Joannes Chrisipius found true) and further that in his due time and place Mercury by the smoke of brimstone within one month sill be turned into perfect Luna. I might press you with more as good proofs and trial of men of credit, but here is enough, I say to stay your judgement for a while, let us go forward.

Chapter XIX.

The two guards of safety, Wisdom, and Virtue, to the Soul and Body, with other wonderful truths experienced and published by good Authority.

1. Will and diligence. 2. Of the difference of Sapience and prudence. 3. Of the mind and Soul. 4. Earthly Judges. 5. Of the Servants of souls and spirits. 6. Messengers of spirits. 7. The power of spirits in receiving shapes. 8. Motion of the spirits and members. 9. That the Æther carries the soul, and all his beams down into the body. 10. The excellency of man. 11. The nature of Age, and youth in cold and hot Countries. 12. Of Stars and Prophets. 13. That a beast may put on manly nature. 14. Of a Mole. 15. Of the degrees of Nature. 16. Of the cause and cure of Kind. 17. To mend man in nine or ten offspring. 18. The nature of Parents. 19. Of diseases and Leprosy. 20. Wit and madness. 21. The cause of foolish bodies and the Mixture. 22. The cause of Virtue. 23. The cause of manners. 24. Of the will and mind of man and stars. 25. Of the place of the Sun, Air, and food. 26. Of the Poles of the world. 27. Pepper turned into Ivy. 28. The cause of distempers, 29. The cause of monstrous children. 30. The cause of Madness. 31. The cause

of joy or fear. 32. How to temper the heart and liver. 33. Of that proceeds from the heart and liver, etc.

1. You have seen now happiness, knowledge, long life, health, youth, pleasure, and are dispatched, and we have got such a goodly quire of helps, instruments, and means, to wisdom and virtue, that is to perfect health and happiness; what is wanting, but will and diligence, to bring all men unto it, unless there be some as there be many, so lude and fond by birth and nature, having their difference defaced, and being so far from their kind estranged unto the kind of beasts, that although they lack not these helps and furniture, no nor good will and endeavor to set them forward, yet all will not serve to mend them and bring them to wit and goodness.

Then let us seek the salves for these two sores, likewise that we may make it at last a whole and perfect pleasure and happiness; let us, I say, bend ourselves to show the means, how all foul and vitious persons may be cured and brought to health of mind (which is wit and goodness) no cure can be skillfully performed, without the cause to be first known and removed; the cause of wisdom and virtue, and so of their contraries (for one of these do bewray another) I opened heretofore when I brought into the bound and household of wise men or Rosie Crucians, that two other properties that is clearness and temperateness of body, but because we have no such bounds and beginnings, as the measures have given end granted, and it behooves if we mean to build anything ourselves, to lay all the foundation; let us take the matter in hand again, that those two are the very causes and makers of this health of mind, that is of wisdom and virtue, and then teach the way to apply the remedy.

2. To begin with wisdom (for that knowledge had a being before doing) and therein to let pass all the idle subtilities about the difference between sapience and prudence (if I may so term it for Once and use it not) as one of them to be seen in general and everlasting, the other in particular and changeable things, and because they ought ever more (as I skewed at first) to go together (even as our English tongue better than either Greek or Latin, has linked and shut them up both in one word together) I will take the common and true bounds of wisdom; that is, wit and knowledge of divine and human things; these containing in all minds and bodies, and affairs of private men, families and Commonwealths, it will be very hard indeed to bring the French fools to understand all these matters; but let us march, we have passed great dangers, etc.

3. And if in this discourse of the mind (as well, as in the former of the soul, and some other) I call in again the best philosophers, and make them abide the brunt, I hope you will not blame me in a course ever blameless and allowed in matters of such weight, both that the truth might be the better bolted out, and the man warded with a charm against the shot of envy.

4. Therefore letting pass these earthly judges as Arestoxenus Didarchus, Pliny and gallenests, who rating the mind as an earthly thing, do judge it to dye and to be clean razed out with the body, and all, other wrong opinions with the same mind; old PHILOSOPHY and Rosie Crucians (where it is best advised) hold and teach, that as the soul and life of all things is all one with itself, and all the odds springs from the divers tempera of the bodies, so the divine and immortal mind proper unto man, and author of wisdom and virtue, to be wise and a likewise, and one and the sane in all points, in all men, as God from whom it came, is one and wise) and to differ when it is divided,

and sent into sundry places, even as many rivers passing through many grounds, of sundry qualities do lightly every one take a sundry taint, snake and nature from the ground, though at first they all sprung and flowed from one fountain or head, or more fitly like as there are innumerable kinds of lights in the world, differing to the seats and houses that receive them, when the light of the Sun from whence they all receive light, is of itself all one and the same in all places. Then as the Sun (think not much if I be still driven to likes, because it is the lightsome way of delivering divine things, wherein you see me plunged; for as the eye can behold all things but herself and the Sun, and those it cannot see but, in another thing, fit to represent the figure, even so the mind cannot understand herself, nor yet other divine matters, so well as in a like and comparison) as the Sun, I say, of himself ever shines, and sees all things, if his beams be not stopped with a cloud or some other thick impairment, even so, the mind alone, and before she fall, into the cloud of the body, is ever busy and likewise knows all things, as unto so divine a thing belongs, but now she is so entangled and darkened in this manner, she is sometimes idle, and never seeks all things yea naught at all, without the leave and help of the body.

5. This course therefore she now taketh, since she may not herself step forth and range abroad, to see things, she craves and takes the help of the soul and his servants, which they call beams or spirits; first she uses the outward spirits that sit in the edge and border of the body for messengers to receive (by means of their instruments, the parts where they lodge) and bring in tidings, that is shows and shapes of things, and then the inward beam sitting in the brain, takes the same tidings and represent them, as it were in a glass; before her, that she may cast her light (which they call the suffering or receiving mind) upon then and see them to skip over the known fine; inward wits

which we have (not unfitly) compared with glass, are divided into sundry and several seats and offices. First one sort called, thought inhabiting the forepart of the brain, takes, holds, and represents the shapes, let in at the windows of the fine outward senses; then another crew which we call remembrance, keep the hinder parts of the head, receives still those shapes in great plenty, and lays them up as it were in a storehouse, until first the third company of the souls and spirits called common sense, and sitting in the middle of the brain (as becomes a judge) calleth fat them to examine them and determine of them (though this lower judge heareth present matters in thought also) and then at last the great chief Justice called understanding, by laying the things together and gathering one of another, judges all. But which is the seat of the chief Judge, that is, the question among the learned: when I take it to be no question, if they all grant that the soul, by the pattern of her fire the Sun in the great world, dwelleth in the heart, the middle of the body; that by casting her beams all about, and equally to all parts, she might give life and light equally to all, as equal distant from all: and in the midst of the heat, as the only moveable and therefore to move others the only fit part of the body; for then sure the mind being in the inward kernel, as Plato said of the other two, the soul and the spirit, must needs rest and be rooted there also.

6. Seeing the mind earth and knows nothing but by means of the soul & his inward wits and spirits, not these but by the help of the outward ones, called the five wits or messengers, nor neither of both, without the parts where they lodge and rest; and even as the parts of the body stand affected and disposed, so doth the mind understand.

Let us go down more particularly to the matter, and see what condition or disposition of the body helps, or hinders the work of understanding.

7.	After that the five wits and messengers have thus received and delivered up the tidings to the threefold glass within the brain, this by stirring and running up and down, presents and musters them before the mind, and she by casting her light and view judges and determine, that we may easily and quickly gather two things needful to wisdom and good understanding; first such a glass, or such inward spirits, as are able to receive and hold many shapes imprinted that is, very clean and clear spirits by the example of an eye that kindly glass, or of an artificial one, or of a garment, all which will easily take and show, in that case, every little spot shape and fashion set upon them; whereas when they are dark, foul and uneven, they can take nothing, nor yet represent them, if they had them.

Secondly, these spirits had need be quick and lively, that is whole, to be able by their swift running to and fro, to represent and show them all apace, and easily; for the mind doth all by matching and laying things together.

8.	That heat is the cause of quickness and stirring of the spirits, appears in sickness, age, sound sleep, especially in age and sickness, more clearly than meets any light of teaching. But how in sleep? When the heat of the spirits serving wit, is either loaded with the clogging fumes and breaths of the stomach, or spent with labor, or with sweat, and still beholding, (for rest abates heat, as I ever said) or else lent for a time, unto his fellow servants, the spirits of life, for the digestion sake; then the spirits of the brain be still and quiet, and outward and inward senses, wit and understanding all cease at once: But if the meat (to omit the expense of heat) was neither much, nor of an heavy and clogging kind, and so neither breathing out leading stuff, nor needing foreign help to digest it, then our perceiving spirits begin to take their own and natural again unto then, and to move a little before the mind, whereby she holds some old shapes and shows of things in their passing, which is called dreaming, But in case they recover all that heat, they bestir themselves a pace, running to the outside of the body, and bringing back new tidings to the mind, which when she perceives, is called waking.

Then the cause of wisdom is clear at least as we see, to wit a clear and stirring glass, and of folly when the same is foul and still. If the glass be fouled all over, it causes natural or willing folly, as in fools, children, or drunkards; but if it be but here and there besmeared, and drawn as it were with dark strokes and lines of foul humours, the shapes appear to the mind, even as the forms appears in a broken glass to the eye by half and confusedly, and it makes madness.

9. But how come the spirits of this inward glass so foul and slow, when they are of themselves (as becometh the beams of a heavenly soul) both very clean, clear, quick and lively? But we need say no more, but clear and foul above, when these two qualities make or mar the whole work of perceiving; for if the spirits be clear, it is a sign they are in their own nature, and so whole & quick withal, but if they be foul, it is a token their whole condition and property is lost and gone, and so that stillness is come upon them also, neither is that Ætherial thing which is called by the name of a spirit, that carries the soul and all his beams down in the body and breaks (as I said above) between them, foul or still of itself; (for spirits are not as some leaches think, made of but seed with the breaths of our meat) but very fine, clean, and lively, as all men grant of Æther; how then? Must it not follow that all the cause of fail and want in this case, springs from the body, and from that part especially, where the wits inhabit?

10. If the waked reason brought in by Theodidactus above will not serve to content this matter, let us lead him forth clad with proof of

eyes, light and experience, the plainest, greatest, most filling and most satisfying reason in the world; if man alone doth pass all other Wights in wits, for his tory and Fiery temper above them, as we heard before, then if one man goes before another in wit, it must needs follow, from the same cause; now as Air and Fire are clear and quick, when Earth and Water are foul and slow, so are the might. where they bear the sway, affected both in wit & body, as appears with difference between the Rat and the Toad; and all other wholesome and noisome Wights; to go further, why are the men so gross & rude under the two pins of the world, in the frozen Countries? And so civil and wise in the whole as Aristotle well notes; but for that the outward heat cleanses as it is a cleanser, and dryer, and so clears their bodies? Whereas cold on the other side binds and thickens. And so likewise by stopping the flying out of the gross, foul and waterish humours and leavings, makes all not only dark and cloudy, but whole and moist also, as it were drunken by boiling together, as Aristotle terms it.

11.   But me thinks (I must favor them a little because they are our neighbors) he might have done well to have resembled those broiled people to old men otherwhere, and the aged men in frozen countries to the youth in hot soils, because the odds of wisdom between age & youth flows from the sane cause of drought and moisture, that is cleanness and foulness of the bodies; and therefore Plato was not ill advised, when he said, that at such time as the eye of the body failed, the eye of the understanding begins to see sharply; because when his waterish instrument dries up with the rest of the body, though it put out the sight of sense, yet it is a token that the light of wit increases; for drouth as I said, breeds clearness, if it be not mixed with coldness; for then it brings in earthiness the most foul and sluggish Element of all, and therefore those that are very old and cold, are very doting and childish again: but if that drought be seasoned with heat (the more the

better) they make the man very wise, and full of understanding, as it has been always observed: Caesar is described so; but more strongly before him, Alexander whose body by his great heat and drought was not only most sweet in his life time but able lying dead above ground in a whole soil and season without any balming, alone to keep itself fresh and sweet without all taint and corruption for many days together.

12.   But I am too long; wherefore Prophets are said to be wiser than men: and their spirits wiser then they, and the stars most wise of all; for the odds and degrees in the heat, drought, and clearness of their bodies.

Now then we know the cause of this hurt and disease, let us apply the Medicine; let us clear the Idiots body, in many kinds of foolishness, as in childhood, drunkenness, sleep and doting diseases: Nature herself is the same to disperse in her due time and season, and scour out all the foul and cloggy cold and gross humours which overwhelm the spirits, and make them unclean and quiet; or at least in the ranker sort of them, as in doting diseases she may be opened easily, and enabled by little skill to do it; that we may judge, if great and strong and mighty means of Art chanced once to join with nature, the rankest of all, and deepest rooted, That is, Natural folly itself may be rooted out and dispatched.

13.   But you may reply as some do, that the rest, which sprung out from outward light, and hang by causes, may be cured, when this being so rooted in nature, and first mixture of the seed (a mixture as ill as a beastly mixture) can never be mended, unless we grant that a beast nay be holpen also and put on manly nature. I had need send you back to the degrees of kind, allotted and bounded out above, by the Counsel of Philosophers, whereby you may see if you consider

well, that a beast standing in a lower sleep and kind of mixture, can in no case be bettered and made man, unless that his temper be marred first & made a new, and so his life being put out and razed, when as a foolish man has no such cause and reason, being both for his divine mind (though it be eclipsed, with the shadow of an earthly body) and in respect of his temper a degree above a beast, and in the state and condition of mankind, fire abounding in him as his shape declares, as well as in other men though not so much, and in the same point and measure.

14.   And what is the cause? Not because, Nature meant it so, but reason or nature was let and hindered by some cross thing lay in her way, within the stuff, whereby she was driven to stray and miss and come short of her purpose: like as the Mole, as Aristotle says for all her blindness, is in the same kind with all other whole and perfect Wights, which should have all their wits and senses.

15.   Because having all the parts of an eye whole and perfect, it is a sign that Nature went to have gone forward, and was let with the bar of a gross and thick skin. Now then we see the fail and errors of kind by skill daily corrected; yea and Bone hold opinion that the blemish in the Mole may be washed out and amended also: That we may hold it possible to do the like in this fault of folly, nay we may think it more easy then some of them, because there is no several purpose, which seems so in the work of the Mole; But some odd and rare examples, as it were monsters in kind, or more fitly diseases left by nature, decent, and inheritance, spring out from some ill temper of the Parents.

16.   But how may this disease be cured? All things in kind by the course of kind, have both their highest and deepest pitch and end, and as it were their South and North turns, from whence they still return and go back again, to avoid infinity.

17. So these natural and last diseases, have their race which they run and spend by little and little; when it is all run, and the stock of Corruption spent, (which is within nine or ten offspring) then they mend and return to health again; such is the race of wisdom also, and of health of body; for the health of the mind, is enclosed within the other, as we see by the children, which wise men beget, and so forth; the cause is plain and easy.

18. Then we see in this matter, how nature inclines, and is ready to hold herself; and if Art would lend her hand, we may think the cure would be much more speedy, and many parts of the time cut off and abated; and as we find in sores and other lighter inward hurts, this done by slight means of slender skill, so we may deem that by more mighty means, more great and mighty deeds may be performed; but what do .1 fetch about the matter, when it is above as I think sufficiently proved, that all left Leprosies, and other natural diseases of the body, by those heavenly and Mineral Medicines (which I call the Cure-alls, and Cure-the-great) may be quite cleansed and driven away; and this among the number of least and natural diseases, all sprung out from an ill tamper of the seeds of parents?

19. And to omit the rest, if the Leprosy flowing from the foulness of the blood of all the body, say be cured; much more this which proceeds from the frame of one part only, that is, frog a muddy brain; ox if that disease nay be said to come from one part alone, that is the Liver, because it is the maker of all blood, yet that one is a most dangerous part if it be ill-affected, because by need of nature, it sends to all places, and so reaches through all, and strikes all by contagion, whereas the brain as other more keep themselves within their bounds, and stretch no further.

20. But let us go further, if good fine temper through all diet and passions of the soul, has often failed from a good wit, to a kind of madness, scarce to be descried from the state of an Idiot; then sure through the contrary cause, a foul frame may be cleared and rase wisdom, by as good reason as the Art of reason has any, especially if those contrary passions and diet be helped and set forward by meet Medicines, which the Grecians know and teach, and wherewith they make great changes in men's bodies; but without all doubt and question, if that our most fine, clear, and whole Egyptian Cure-all come in place to help the matter; for if the mightier enemy shall in fight overcome the weaker (as you all grant, and thereon stands your Physick) then shall passing fineness and clearness, when it arises in the body like the Sun in the morning, scatter and put to flight all mists and darkness, clearing and scouring mightily by his matchless heat, strength and swiftness, every part of the body.

21. Neither shall you say, life will suffer no such violent and forcible dealing, when as life itself shall do it; for what is that which made and mingled at first the foolish body, but a beam of heavenly fire carried on a Couch of Æther? And what is this our heavenly Medicine but the same as is above shown at large; then let us put same to same, strength to strength; and if one before was too weak to break as it would and mingle the fond body finely; now both together, one helping another, and still with fresh supply renewing the battle, shall be I think able to overcome the work, and at last to bring it to the wished end, pass and perfection. If you fly to the last hole and shift, and say that time is now past, and occasion of place and stuff now lost, and slipped away, being too hard for nature upon so hard a stuff and place to work such exchanges; if you look to her ordinary race in all things, you shall see that she is able, and doth daily frame, rule and square very gross and unmeet stuff in most unfit places to our thinking, yea much more then

these in this work; and not only the thick and sturdy stuff of Miners, cleansing the works (yet in unseen places) down to the bowels of the ground; and the gross and rude gear in the bottom of the sea to make shellfish, but also living, moving, and perceiving land nights, in the close rocks (as you heard before) and in the cold Snow and burning fire, as those worms and flies in Aristotle. To close up and end this matter at once; if you remember how this our heavenly Cure-all, when he was sent into the body to work long life, health, and lustiness, did not only strike and kill, and put out of being all foul and gross distempers, his own and our enemies, but also cherish, nourish and feed our bodies, and bring it towards our own nature (even as far as we would by disposing of the quality) you may easily conceive the plain and certain way of this great exchange when you know his most clean, fine, clear, and lightsome Nature.

22.   Now I have dispatched the first part of Physick, not as some have pretended, but even to God himself, let us go to the second; and because we have not done it before, though we talked much thereof, we will now begin to bound the matter, and make virtue (as Aristotle and truth teaches us) a mean in inward deeds and dealings with other men; or a reason in manners and conditions, as Plato terms it, all is one; the cause of virtue is likewise set forth in the beginning, to wit, a temperate body; but I left the proof unto this place, which is all. the hardness in this cure of lewdness; for if it be once known that temperateness is the cause of virtue, we shall easily by that temperate medicine, so notable in the speech going before, purchase and procure the same; and why that is so, it has so often won before, that we may quite cast it off and leave it, being naught in this place to prove that a temperate state of the body is the cause and way to virtue.

23. But first let us see whether all manners flow from the body or not, and then from what state or condition of the body, among them that have searched the reasons and nature of things, the cause of manners is laid upon the disposition, either of Stars, or of men's bodies, or of their wills, thus or thus framed, either by the bent of nature, or by use of custom, let us scan the matter and that briefly.

24. They cannot flow from the will of the mind of man, least all men should perforce be good against our daily proof and experience; because the mind of itself is coming from goodness, is good and alike good in all men, as I said before; and sure no custom can alter and turn to divine and right a will to lewdness, but by great force of necessity, which force cannot be Bent and laid upon it by the Stars, as I know not how, the weather spiers, by long watching and besieging the Sky, the high and strong City, with empty and bootless labor, have observed; for whether the Stars be nights or no, they are all (as I showed before) of one good strain and quality, or if they were not; or whatsoever they be in either substance or quality, they cannot touch the mind immediately, but must needs be let in by the loops of the body, and so change and dispose the body first; and by means of this affect the mind; for if the mind itself, a finer thing then the Stars, cannot pierce out of the body, as we heard before, then much less shall they make way to get in by themselves, without the help to our mind allotted; and as these are all bodily (I mean the first helps) so the nearest cause of manners must needs flow from the body; and if the inward spirits and wits likewise do naught without the instruments of the body, and follow the affection and disposition of the same, the appetite of the unreasonable soul, common between us and beasts (upon which Aristotle and his heirs do lay the cause of manners) is dispatched also, and all the whole stream must needs clearly run from the body.

25. But let some old danger come & shake these old Grounds, which you saw the Philosophers lay so long ago, and so this building might fall and tumble, I will shore it up with experience a thing most fit to fill and please the sense of them which have nothing else but Sense.

As all diseases, so all manners spring, either from the natural and inherited, or from the purchased temper of the body; to keep the first till anon; this we have either from the air and soil where we live, or from the meat which we take; the Air follows either the place of the Sun or the nature of the ground; but this is somewhat too hard and thorny a kind of teaching; let us enlarge ourselves, and unfold and prove how (though I showed the manner at large before) the air and meat alters and changes and sakes to differ, the bodies first, and to the manners rude and fierce.

26. All Starmen and Philosophers (no otherwise then we see by proof) hold opinion, that where the Sun is either too near the people, or right over them, or too far off, as under the two pitches of the world, there the bodies are big and strong and the manners rude and fierce; whereas within the two temperate girdles of the earth, they keep a mean and hit the midst, as they say, both in body and manners.

To come down to the ground(for I must be short) we see that a fat and foggy land sakes the blood and spirits thick and gross and thereby dull and slow, and so the men fond in wit and rude and simple, faithful, chaste, honest, and still in that strain of manners; whereas a barren and dry ground, if the sun be temperate therewithal (as at Rome and Athens) makes the same thin and clear and lively, subtle and deceitful men, valiant, unchaste, and so forth of all other proper ties appertaining. For meat, manners in man are like the virtues and proper titles kind, swift and healthful for it; and enough such like examples might be brought, if time would suffer; to come to our bodies left us by our parents, if we see manners ingrafted and inbred in stocks, kindred, and children and nephews still down, to take one after another a long time by kind and nature, as that cursed father bearing kindred, set down in Aristotle, and other pilfering stocks, which though they have no need, must needs steal; to let pass lechery, valor, and other good & bad qualities, which we see daily descend and rain on kindreds, whence are these? Not from the parent's minds and offspring, nor can be left and ingressed, but must return straight and whole, & all at once when they flit out of this life to that heavenly place from whence they came: Neither are all their wits alike framed by use and custom, but brought up sometimes contrary.

30. Therefore to cut off the giddy reeling drunken opinion as a string too much discording, those manners spring out from the parents seeds, which is a part of their bodies, purchased by meat and nourishment, which bodies if they use good and temperate diet, are ever like the first; otherwise they follow the Nature of the meats, and of their distempers, as Cardan in a few of the worst diets, has most notably marked, that drunken, or over studious or too great fasting, or large onion eating parents do beget and bring forth, for the most part, mad and frantic children.

31. To close up all this first part, with this one little proof at once: If we find ourselves to do many things against our will, as when a fair thing is offered, our hearts pant and fail with fear; when a fair lust and his part will arise, whether we will or no, and all incontinency springs from that root, then sure the body must lay this force upon us.

But how is this? And which way does the body so violently overrule and carry away the will and mind after her? When any shape appears in the thought of man, the doing and mind takes it strength (we must wear these words with use, and make them softer) laying it with good or bad, and matching and comparing all things, degrees and determines; and then her will and reason which Plato places in the head, follows and desires: But at the same time steps in another double will and appetite sent from that unreasonable and perceiving soul, which is common between us & beasts, and sitting one part in the heart, and desiring outward goods of the Body; and look which of these is stronger, that is, which has the stronger house, either by descent or purchase, (or else baser mold, be still the weaker, and obey the better) that prevails and moves the spirits unto it, and those the fineness; and those again by other middle means, the whole body or part thereof, as is the pleasure of that Commander.

32.   Wherefore to come to the point more fully, we shall never be good and follow virtue, that is mean and reason in our desires and doings, before these two parts, the heart and the liver, be first by kind, and then by diet in order, square and temper, apt to obey the laws and rules of reason; for to begin with the root, if the heart be very hot and moist, the man is courageous and liberal, desiring honor and great outward things: If hot and dry, cruel, angry, deceitful: but if it keep a mean, obeys reason in that kind of manner; for the liver if it be hot and moist, likewise it follows venery and gluttony; if hot and dry, it doth the same, but crookedly & out of course; but if it be cold and dry, the man is very chaste and abstinent; and if cold and moist, somewhat chaste & abstinent, but outwardly: Whereas a temperate liver holds a mean in both, and following the race of kind desires to live soberly in company and honesty in marriage, a life as far from Monks and Hermits as gluttons and lechers.

33.   Wherefore we see that all manners proceed from the temper of those two parts (nay perhaps understanding also, if it varies still

according to the divers heats and moisture in the body) so that all good manners and all virtue bud forth from the good, equal & middle temper and mixture of the same parts; and all our Labor and travel (if we seek virtue) must be to bring those twain into square and temper, that is equality as near as may be of the four qualities; not only by the Philosophical salve of use and custom, (though Plato hits it right in his time, as when he will have no man lade by his will, and therefore not to be blamed but through his by—use or nature ill disposed) but rather by good diet, and by right Physick especially.

And thus, we have at last finished these parts, wherein we mean to prepare the mind both of the common and learned people, and to make the way to the truth of Hermes Medicines.

Thus, have we proved our way to happiness, knowledge, long life, health, youth, blessedness, wise and virtue, plain and easy; let us next passes unto the Golden treasures of nature, and the method of the *Holy Guide*.

The
Holy Guide:
Leading the Way to
The Golden Treasures of Nature.

How all may be happy in this world; Enoch and Elias knowledge of the Mind and Soul. Eugenius Theodidactus his discovery of the manner and matter of the Philosophers Pantarva, or anontagius, and the manner of working Canonically and orderly made manifest in the secrets of nature and art, by which philosophy is restored.

That anontagius will transmute Tin, Iron, or Copper into Silver or Gold, with what advantage you will.

The Rosie Crucian Seraphical Speculations and Ganathes, and how to extract the Soul of Gold, and put it to another body.

That Gold nay be wrought into a fine Oil, and transmuted into Gold again. Row to make the Bracemans Medicine that cures all diseases.

A manual experiment, discovered and communicated to the World.

By John Heydon Gent., A Servant of God, and a Secretary of Nature.

Si non ego mihi, quis mihi? & cum ego mihimet ipso, quid ego? & si non modo, quaudo, 1657.

John Heydon

To my honored friend
Mr. ROBERT RICHARDSON Citizen and
Merchant Adventurer of
LONDON,
All Celestial and Terrestrial happiness be
wished.

Sir; Following the Path of the Rosie Crucians, it is my ambition to let the world know why it is that I do especially honor men; it is not Sir as they are high born heirs of the great Potentates, for which most honor them (and upon which account I also shall not deny them their due) but as they excel in honesty, and are friends to The Fraternity of R.C. That poor Philosophers should take no delight in Riches, and Rich men should take great delight in Philosophy, is to me an argument that there is more delight, honor, and satisfaction in the one then in the other. have you not heard of a Nobleman's Porter that let in all that were richly appareled, but excluded a poor Philosopher? but I should if I had been in his place, have rather let in the Philosopher without the gay clothes, then the gay clothes without the Philosopher. as long as I have sense and reason, I shall improve them to the honor of arts. in the perfection thereof there are long life, health, youth, riches, honor, pleasure, wisdom and virtue; by art Artephius lived a thousand years; Des Cartes knew all things past, present or to come: by art Elias raised the dead, Joshua made the Sun stand still, and Moses with Aurum Potable healed the people. By these arts you may command lead into Gold, dying plants into fruitfulness, the sick into health, old age into youth, darkness into light; a month would fail me to give you an account of their power; but you may read them in The Rosie Crucian Infallible Axiomatia, and in our book called *The Harmony of the World*, and in our *Temple of Wisdom.* Now for the effecting of this, let me advise you to read well all my books; there you will find my mistress, she is a virgin, and a mother of children; court the mother, and you will win the daughter; prevail with nature, and the fair Beata is at your service; it is pity there is such great encouragement for many empty and unprofitable arts, and none for these and such like ingenuities, which if promoted would render an university far more flourishing than any in the world; but I never expect to see such days in this kingdom, till shadows vanish, and substances flourish, truth prevail, and The Fraternity of the Rosie Cross discover themselves to us, which time I hope is at hand, and desired by all true artists, and to my knowledge especially by yourself, upon which account I truly honor you. now to yourself therefore I crave leave to adumbrate this part of the art which I know you will be willing to promote for the public good. I dedicate this treatise to you, not that it is worthy your acceptance, but that it may receive worth by your accepting of it. I present it to you (as men bring Lead to the Philosophers to be tinged into Gold) to receive the stamp of your favor and approbation, that it may pass current with an acceptance amongst the Rosie Crucians, Astrologers, Geomancera, Astronomers, Philosophers and Physicians; whereby you will oblige, Sir, Your most affectionate Friend and Servant;

John Heydon.

London, March 15, 1662.

Book IV. Chapter I.

How to Change, alter, Cure and amend the State of Man's Body, when nature makes it deformed.

1. Of the Compositions of man. 2. The door of Light. 3. Order of speech. 4. Of Hermes Medicines and other things. 5. That an ounce of Gold in a year will make a Medicine as the Philosophers stone. 6. Of the Son of Gold. 7. Of the heavenly virtue of Wights. 8. Of eating Mice and other things. 9. Of the beams of Heaven. 10. Of Celestial spirits in Minerals. 11. The force of Heavenly spirits. 12. Envious Leeches. 13. Of stones, Trochises, Pills, Electuaries, (?)le Water and other things. 14. Of the virtue of calcined Metals in Physick and Chirurgery. 15. Of the secret Virtue of minerals. 16. Of dissolved Gold and raw Gold, and other things. 17. Of our first Nature. 18. Of the perfecting of the mind and body. 19. Paracelsus Opinion, of Poison.

1. Do you see how we have shown heretofore in the *Axiomata*, lib. 3. divers ways to our guide to happiness, etc. and sundry means whereby the whole kind of men may come to the knowledge of the Composition of man, and of the infusing of the soul, and how the supernatural things being the secrets of God alone, are artificially made helpful to mankind, and of the power of the soul being separated from the body, at the command of the spirit, and how it becomes like the heavens, and of the virtues of the mind and soul and how God wonderfully works effects in the imagination, and what is the first matter of all things. Yet in truth they are all by long and cumbersome ways, fit rather to put them in mind of a better way (which was the

drift of that purpose) then to be gone and travelled by lovers of wisdom and virtue: wherefore I would not wish them to arrive their councils in many of those places, but to seek to the haven of Hermes or Rosie Crucians, and of their sons the wise Philosophers as to the only one, ready and easy way to our guide to happiness, then we are come at last to that which was the first intent and meaning of all this labor, that Hermes and the Philosophers medicines are the true and ready way to eternal happiness in Physick.

2. But how shall we prove this unless we unlock the door of secrets, and let in light to those matters which have been ever most closely kept and hid in darkness? We must I say first open what is Hermes medicine, except we would put on a vizard[31], and make a long buzz and empty sound of words, about that which no man understands. We are like now to be driven into a marvelous straight either to fly the field, or to venture upon the curse and displeasure of many wise and Godly men, yea and of God himself as we heard at the beginning. If Plato thought he had cause when he took in hand that mighty piece of work, the world, first to make his prayer; how much more may we in such a world of doubts and dangers? And to desire of God that we say prove our question, not only with sufficient evidence, but with such discretion also, that those men which can use it, and are worthy of it, may see the truth, and the rest may be blinded.

3. Then both to direct my speech, which must have some ground to stand upon, and their steps which crave a little light to guide them, I think it beat to come to the entrance of this way, and to point afar off unto the end, leaving the right into their own wit and labor, for I may not be their guide, least the rest should espy us and follow us as fast.

---

[31] A mask or disguise. -pnw

4. Hermes Medicines and the Rosie Crucians Medicines lie among them, even in Gold; and the end of this Journey where happiness begins, is the son of him; albeit that I am not ignorant that father Hermes and the rest of his wise foster children, hold and teach that out of any Plant, Wight or mineral, may be fetched a medicine for all diseases of men and metals, as good as this which we have described; neither do we, as though we had drunk the water of Lethe, forgot the reason of it above declared, because all things are in all things, and the same and one thing, as having all one stuff and soul; if their stuffs had the like and not divers singlings, and for that all things it they were wrought to the top and highest of perfection (as they may be) show a like with all the virtues of heaven and earth, soul, body, life and qualities; but those ways are long, cumbersome and costly, as well as the rest, and I know you seek the most ready, near and easy, which is Gold far above all other things in the world.

5. The reason is because nature has powered herself wholly upon his, and infused his of far more and greater gifts both of soul and body then all the rest, having given his not only greater store of the heat of heaven, but also the most fine, temperate and lasting body, whereby, but especially by reason of his exceedingly tough and lasting body, wherein he wonderfully passes all things, we have him half ready treat to our hands, and brought very near the Journeys end, quickly to be led forward and finished with little labor; when as the rest are left in a very hard way very many miles behind him; it is strange I am persuaded that a thousand ounces of a plant, or night (as for Minerals, they be much better) cannot with great labor, cost, skill, and time, be brought to that goodness and nearness to perfection, as an ounce of Gold has already given him by nature; and I durst warrant you, that out of an ounce of Gold in less than one year's apace, with a few pounds charge may be gotten a Medicine as good as the Philosophers

Pantarva, of plant or night, that taketh a thousand ounces of stuff, many hundred pounds of charge, three years' time, and the wearing out of many man's bodies, that we may think, although the wise Philosophers in Egypt saw and showed the depth of Nature, and these works, yet they were not so mad and fond as to put them in practice; and therefore Hain Geberin says, it is possible out of Plants to make the Medicines, and yet almost impossible also, because thy life would first fail thee; wherefore we may be content also to know the secret, but let us use no other way but this, and so dispatch not only plants and nights as foul earthly things, but also middle Minerals, which are like the standing lights of heaven in this comparison. Nay, neither hold we his fellow plants to be his equal, no though they be Quicksilver or Silver themselves, the best and nearest of all the rest, especially Silver the wife of Gold, but even let her pack away with the rest; for as her fire above glister and sakes a fair skew until she cone in presence of her husband (as the want of bad woman is) so this our earthly Moon be she never so bright and excellent in another's company, yet in sight & regard of Gold her husband, she appears as nothing; if you marvel why? It is because she wants much of the heat of heaven, temperateness and toughness of body, but in fineness a hundred-fold; these things are high and lofty, and soar above the common sight, we will fetch them down anon and make them plain and easy.

6.	Then let us fall to the matter, that the son of Gold may be found, the ready way to Gold, the perfect Medicine both of man and metals; and first as it is meet, let us regard ourselves and cure our own bodies before we help a stranger. There is no gift property or virtue but it springs either from the soul or the body; the best gift of the soul is most store thereof; as we showed before; and of the body first temperateness thereof in the first qualities, and then fineness and closeness, which causes lastingness, in the second; let us see how Gold excels in all those virtues and overgoes all other things, first by the gift of nature, and then by a divine Science; but it were not good in such an heap of matters to be disposed and dispatched at once, to regard those that be clear and received; so then let the fineness of Gold go his ways, as clear in all **sens** eyes, and his temperateness, which all Leaches grant, and take the rest as things both more in doubt and of greater worth.

7.	Those that are longest a ripening and growing to perfection, are both the most tough and lasting, and fullest of heavenly virtues; whereas on the other side, soon ripe; soon rotten, as they say, and an ill Weed grows a pace and so forth; the cause of this in bodies, is because the first moisture, if it be fast and close, that is full, proceeds and spreads slowly and is hardly consumed and eaten up with the fire of life, when thin and waterish moisture spreads apace, and spends as fast; and for that heavenly virtue, when the stuff has long lain open under the hands of the spirits of heaven, it must needs receive greater store of them, and hold them surely with his strength & roughness; what reason can show this more plainly, except you will call me to examples? Then bend your ears a while and mark the Elephant, two years in making in his mother's womb, and a long time in growing to his best estate and lustiness, to reach the highest and best pitch in mortality (for man is mortal) and not only by strength and long life which you heard before, but through a kind of wit and good conditions also, drawing near to the nature of mankind.

8.	Consider again of Mice, those little vermin, how soon they be bred, as sometimes the earth creates them, sometimes the mother without the male by licking salt, and other whiles (for a wonder in nature) they conceive and are big with young in their mothers' belly. Consider I say, how soon again they be swept away, even with a shower of rain, as Aristotle reports, who tells of a one-day fly bred in a leaf in the forenoon, at midday fledged, and ever dying at night with the setting of the Sun.

9. Again, Pliny writes of a child that within three years space grew three cubits, and was now grown to mans state (which they call pubertateum) but haste made waste as they say, and within three years after his limbs shrunk up again, and he died. Nay he says, that the whole kind of woman among the Catingians, conceive at five years of age, and live but eight.

To cut off living nights, and come to plants, are not trees the longer-lived the better in use, for the long growth and ripening? And among trees, doth not the Oak, after his long growth to perfection, stand to our great profit even for ever almost? It is strange that I say, and yet Josephus writes of one that stood from Abrahams time, to the razing of Jerusalem, two thousand years at least; and God knows how long after that time it lasted; to be short, the best tree of all the earth that brings forth the Coccus of India, in one man's age, scarce begins to bear any fruit, and lasts after that almost past all ages.

Wherefore the minerals by the course of reason and custom, being by the grant of all men, longest in making and perfecting, must needs of all other both be best in virtue, and last the longest, and among them Gold above all, because it is the end of all, and so far in that point passes the most part of them, that as some men think, a thousand years are spent before it come to perfection; for his long lasting we see plainly he is everlasting; and if we doubt of his heavenly virtue, let us weigh the place and womb where he is fashioned, and we shall see it a common gulf of all the beams of heaven, as the Sea is the receipt of all rivers that run.

10. How is this? Albeit the beams of heaven set forth from a round and wide compass, and likewise leave a circle after they have traveled a great wide way one from another, do meet at last together, Jump in the Navel of the Earth, yea and with great force and strength, above all other spirits in their places, not only by the reason and the length of their Journey (for all natural things the further they go, the more they mend their pace) but chiefly because meeting in such a strait, with such abundance, they violently thrust, and throw one another on heaps together, as we see the force of wind and water meeting in that order, or rather as the Sunbeams, falling upon the stone Hephaestus's, or the Steeple fashioned burning-glass, thereby skews such strange and unwonted force to burn dry things, melt metals and such-like, because the beams that light upon it, do meet all in heaps and apace, in one narrow point of the Middle.

11. Wherefore the Minerals, because they be bred and brought up about that place, first receive great plenty of those heavenly spirits, and then those very surely set on by the swiftness of the stroke, and as fast held and kept, for the sound and close bodies that take the Printing; when as plants and nights, dwelling in one place, and outside of the ground aloft, where those beams and breaths of heaven, are more scarce, slack and weak, must needs have not only less store, faintly put upon them, but also those which they have, for their loose and soft stuff, quickly lost and forgot again.

12. But if the edge of some mens wits, be too blunt and dull to cut to deeply into the earth, to find this matter, let then cast their eyes and behold the daily experience, how these heavenly spirits in Minerals, for all they shut up and bound so fast in the prison of the hard and sturdy stuff, yet are able to show their force, as such and work as

mightily, as the free breaths of other things enlarged in their soft and gentle bodies.

13.   It would not be amiss to bring in a few and set before us, because for the sloth of times past, and spite of the latter leaches, these things have lain for the most part, buried as they be, and hid from the light and common knowledge.

14.   Then to pass by the Pearl, that helps swooning and withstands the Plague of poison; the smarage and jacinte likewise, which keep off the plague, and heal the Wounds, Venomous Stings, and many more such rare and worthy virtues, which they themselves grant and give to precious stones in their writings, nay in their Trochises, Pills, and Electuaries, let us come to hard Juices and middle minerals, the water of Nile, which makes the women of Egypt so quick of conceit, and so fruitful, as to bear seamen at a birth, as Phroates writes, is known to be a Saltpeter water; it is found by common proof that the same Saltpeter, or common Salt, or Copperas Matter, made a water, kills the poison of the Toadstool, and Juice of Poppy: that a Plaster of Salt and brimstone heals the hurt of venom in stings; That Amber which is no stone but a hard. Chany Juice, called Bitumen, eases the labor of women and the falling sickness of Children; it is known likewise that all wholesome baths, both wet and dry, of water or its vapors, which are without number in this world, but especially that famous Rot-house in Italy, called Salviati, for the space of three miles compass wrought and hewn out of the ground very daintily, deserves to be named and delivered to the memory of men to Come) flow from a brimatony ground, and draw from thence, all their nature, quality, force, and virtue, except a few of copperas water, as appears by their dying property, whereby they give any white metal their own yellow and Copper Color; Now for Metals; if it be true that precious stones in that hard and ungentle

fashion, show such virtue and power of healing, why should we mark the German for a liar, when he awards great praise to the mixtures of all the Metals, made in the conversion of their own Planets which he calls Electrum, saying it will cure the Cramp, benumbing Palsy, falling sickness, if It be worn on the hearty finger: and give signs besides if the body ails of anything, by spots and sweating; and bewray poison, if it be made in Plate by the same tokens? For all that Pliny will have poisons so described by the natural Electrum and Mass of Gold and Silver, and not by the artificial mixture to be made of silver and Gold and Copper, *adulteranda adulteria naturae*, as he more finely then constantly says, when he allotted so chaste virtue before unto her.

15. But suppose this virtue in the hard form of metals nor apparent; yet no man shall deny the daily proofs of them openly by rude skill, and set a little at liberty, as the great use of burnt brass, Iron-saffron, Metal-smoke (and this by Gallens own witness) and marvelous help in Chirurgery; nay the mighty power, both within and without Antimony, which is unripe Lead, and of Quicksilver, very raw and running silver, so after tried before their eyes, has amassed and daunted the better Leeches, though Gallen himself in times past has termed this rank poison, set straight against our nature, and the least part thereof taken inward, to hurt and annoy is, to the great Laughter of the Country wits, which even Children, a dangerous time to take Physic in, take, without any hurt at all, nay which they use to drink it against worms in great quantity; but Gallen did but rove by guess at the matter, when as in another place (forgetting himself as be doth often) he says he never had tried its force neither within nor without the body.

16. But if those stones, Juices and metals were by great skill more finely dressed, and freely set at liberty (as they be by the German)

what wonder where they like to work in the art of healing? Neither let us think (as Gallen and his herd think of all things) those great and rare mineral virtues, could issue out, and come front the gross and foul body, but from a heavenly gift of a mighty soul, which cannot be kept in awe, and held so straight, with those earthly bounds, as it shall not be able in some sort to stir and break through and show it force and power.

17.   Wherefore to return to my purpose, if nature has bestowed upon these three sorts and suits of minerals, so large gifts and virtues, when she has given to Gold the end and perfection of them all, so receiving and holding the virtues of them all at once; What says the Leaches to this matter? They are loath to say anything, albeit their deeds speak enough, when they lay raw Gold to the right side of the head to heal his ache; right against the heart to comfort his sadness end trembling: and when in such sore they apply it to such purposes; again why do they boil it in their coulisse, mix it in their pills and their electuaries, bid the Lepers swallow it? Do they not seem to smell its great and matchless power against diseases and marvelous Comfort and wholesomeness to our nature? But like rude and unskillful Cooks they know not how to dress it? But if they know the skill, they should see it rise in power and virtue, according to his degrees in freedom, & when it comes to the top, which I call the son of Gold, to prove almighty, I mean within our compass; for consider Gold is now good and friendly above all unto us, for his exceeding store of comfortable heat of heaven, shining through the midst of a most fine and temperate body. Then what would it be if the properties of the body were by great, mingling and breaking of the stuff, refined and raised in their kind, a hundred degrees at least? (which our Art possesses) and those lively and piercing helps of comfortable spirits, freed and set at full liberty, and all these seated, upon a mighty body subduing all things?

Is there anything in the world to be compared to the mighty and marvelous work which he would sake in our bodies? Could any of these very violent and mortal poisons, which I brought in above, so easily and roundly destroy us, as this would help and save us? But to come to the point.

18.   If that our old close fine and Æthereal oil which they call a first nature, was able alone, for the reasons set down in their places, to breed and beget all those blessed bodily gifts and properties, that is, health and youth, and the two springs of wisdom and virtue, clearness and temperateness; how much more shall this son of Gold the medicine and stone of Hermes, and his offspring be sufficient and furnished for it?

19.   For first, when his soul and heat of heaven is such more great and mighty, and his body a more fine and fast Oil, that is a more like and lasting food of life, it both upholds and strengthens life, and natural heat better, and so proves the better cause of long life and youth; then being temperate, and that quality carried upon, and that quality and finer and tougher that is a stronger body, it Is able with more ease and speed to subdue his and our enemies, the distempered diseases, and to cleanse and clear fashion and bring into good order and temper the whole frame of our body & to procure health, wisdom and virtue better sort, and in more full and heaped measure; for you must not think that a fit nature of wine or such like that I brought in above, and which many men do make for their bodies, is so good by twenty degrees as the Philosophers stone: I mean the same measure of both; when besides that it is not temperate and near unto Heaven, (though the name be never so near) for it wants twenty parts of the soul, and as much of that fine stuff, closely and finely tied up together; and therefore one part thereof will last longer, and spread further with

all his virtues, and so do more good in our bodies, then twenty times as much as the former; deliver to mind what I say; it is worth marking: I shall not need to stand to show you the reasons why, and manner how this great Medicine of Hermes, shall be able to get and purchase those pleasures of mind and body, because it is already done at large elsewhere, and it may suffer in this place to win by force of reason (which has been done as such needs) that this medicine is such better, and more able then an Æther, Heaven of first nature.

20. Then those men may see (I mean Paracelsus and such as know whereof they speak, let the rest go) how rash and unadvised they prove themselves when they are content to let the name of Poison into this happy medicine, and to avow that it worketh all those wonders in our bodies, by that way of curing which I skewed, by stronger-like poisons: for then it would be at most but a general medicine, and cure-all against diseases, and fit for health alone, but no Physical or joyful way to long life, youth, wisdom, and virtue; which grant as well as the other, both he and all the rest do give unto him; for it might not be taken and used in a second body, no more then a purging medicine, except it were of the Viperous kind aforesaid; for he then would battle with our nature, spoil and overthrow the first moisture, and the whole frame of the body; so far it would be from nourishing the natural heat and moisture, from clearing and tempering the body to cause long life, youth wisdom and virtue.

And the reason of this reproof is, because every poison is very barren and empty of the heat of heaven, and very distempered cold and dry in the body, set straight against our hot and moist nature (as appears by flying the fire, and oil his enemies) the Philosophers stone was temperate in respect, at first, and is now exactly so and a very fine oil, and full of heavenly spirits; and so for these three causes, not only most friendly and like to nature, but also a very deadly enemy and most crass contrary to all poison.

Chapter II.

Hermes and Paracelsus Medicines.

1. Of the four Complexions of the body. 2. Of Malice and Ignorance. 3. Of Diseases, Age and Death. 4. How to make minerals grow. 5. How to make Lead grow. 6. That Gold has life. 7. The unwinding of Secrets. 8. Authors Opinions. 9. How to order the seed of Gold. 10. Experienced truths. 11. Comparisons. 12. Of turning wood into Iron and Stone. 13. Of turning Iron into Copper. 14. Abraham Judaeus experiment. 15. Irish eaters and other things. 16. Of Geber and Agricola. 17. Of Salt Gesm. 18. Iron may be made to cut steel as fast, as steel cuts wood. 19. Of Silvery and Golden Copperas. 20. The virtue of Copperas Water. 21. Of Art and Natural changes. 22. Natures Medicines. 23. Of the food of Gold and other things.

1. Over this we have lightly run, being the former part of long life, health, youth, clearness, and temperateness, which make up all good gifts of body needful; let us now come to the outward help of riches, and borrow so much leave again, as to use the cause for the effect, and take Gold for riches, and strive to show that the son of Gold is able to

turn any metal into Gold; and not so sparingly and hardly as we did before, by those bastard kinds of binding and coloring: (though a little of it were without mispence of time & travail would serve our turn) but as fully and plentifully as any of our men allow to that amazement of the world; they set not down nor stint, which I will do, because I have to do with thirsting ears, and because again I love not to run at random, but to make a certain mark whereat to aim and level all my speeches.

Then let us say, by this great skill of Hermes, & a little labor and cost we may spend with the greatest Monarch in the world, & reach the Turks revenue, yea though it be fifteen millions Sterling, as I find it credibly reported; yea let us be bold, and not as Socrates did when he spoke of love, hid his face from the matter, the truth is vouched before God and man, and will bear itself out at last, though it be my luck still to be crossed by men of our own coat, Hermes foster Children: But why do I call them so? Albeit Paracelsus of whom we deal of late, was plainly so; Tet his Scholars Physick, which now comes in place, is out of this account as clearly; this man says to excuse his own Ignorance, has learned a new trick, in unfolding Hermes Riddle, that neither Hermes nor any of his followers, in saying they turn the sour soul of Metals, Lead, Iron, Tin, Copper, into Silver and Gold, mean plainly according unto common speech, but still riddle and double the matter, understanding the four Metals in so good form and temper changed: And these to be silver and Gold which they make at any time, and that by this token, because they fetch their medicine as you heard even now out of all things; then he flies out and lifts up his Master, with high praises, for finding first, and untying the knot and riddle; whereas there is nothing so plain both in Paracelsus and all other of his hidden science, as their opinion as touching this matter: Nay see the worthy memory of the man himself, in construing the words of his Master concerning the same matter, makes it as well as he and the rest, a plain division of this matter, and yields in open terms, that our Medicine serves both for men and metals.

2.   This noble Dr. Elias Avert when I was a novice and firstling in this study, as he mislead me in other things which he took upon him to unfold, so he amazed me in this, before he himself knew the least of them: But after I went forward and began to consider earnestly, and weigh the things by their own weight (and not by the weight of words and authorities) the only way to knowledge, I quickly saw the falsehood of the new onion, and more plain reason and cause of belief, for this point then for all the rest, which he allows, and which I showed before: then let us not stay, for him nor for anything else, but let us march forward with all speed and courage, and if it be never good in discourse of speech to heap and huddle up all together, but for light sake to Join the matter and cut it in divers pieces, let us do so too, and prove that the Son of Gold is able to turn metals that are base into Gold, then that he can change so such, as to make up the sum I left as needful. Heydon is to turn metals two ways; first, as a seed if a man lift to sow him upon them; and then after his birth, by nourishment, and turning them into his own Nature, and this is either into his fathers, which is his own after a sort, or into his new being, and selfsame nature; of these I will treat severally; and first of seed which cannot be denied unto Gold, if all things have life, and life have three powers and abilities, to be nourished and to wax, and to beget his like also; the second part is clear and granted among all Philosophers; and that all things have life, it has been often showed before by their feeding and divers other arguments.

But because it is a thing whereon almost all the frame of my speech leans, and yet such in doubt, and hardly believed among the learned; let us take it again, and prove it by name in Minerals, because they be both farthest from belief, and nearest our drift and purpose.

3. Those things that have diseases, age, and death, cannot but live, and we see plainly the diseases, age and death of precious stones; but most clearly in the precious Loadstone (though he be foul in sight) which is kept, fed and nourished in the filings of Iron, his proper and like food, when quicksilver, or Garlic quite destroys him, and puts out all his life, and virtue.

4. But how if the Minerals by feeding wax and grow as well as plants or Wights? As miners have good experience of that, when they see them, by those due and constant fits, so dangerously void their leaning. Agricola says, that Saltpeter, after that by draining it has lost his taste & virtue, if it be laid open in the weather, will within five or six years space, grow and ripen, and recover his power and strength again; the same man tells of one lead Mine, and two other of Iron, which after they be dug and Emptied, within few years space, ripen and grow to be full again, and one of these every tenth year.

5. But admit these by the flight and canvases of a crafty wit may be shifted off; yet they shall never read the next that follows of Lead, after he has been taken out of his proper womb, where he was bred and nourished, and fashioned into his form for our use requisite; yet if he be laid in a moist place underground, it will wax and grow both in weight and bigness by many good Authors, yea and by Gallen his own witness, which although it be light otherwise, yet is of weight in this matter, because it makes so much against his own cause; nay mark what Agricola reports that the same has been found true on the top of houses, and shows where and how the proof was taken; but to come to

the very point, Paracelsus says, that Gold buried in good soil that lies East, and cherished well with Pigeons Dung and Urine, will do the same, and sure I dare not condemn his witness in this matter, because the rest that went before, see me to say as much in effect, and to vow the truth of this story.

6.   Then if it be so certain, that Gold has life, there is no help, but it shall beget his like also; if Philosophy and Common proof be received, but they will say that nothing does so, that want seed, as many nights and plants do, and all Minerals, no man says so, that knows what seed is; seed is no gross thing, that may be seen with the eye, but a fine and hot heavenly breath, which we call life and soul, wherewith not only the common rule of the world, but also nights, yea and perfect Wights sometimes beget without the company and sense of that frothy stuff and shell, as I said above: but yet more commonly nature takes the help, & guard of that body called seed, that was proved not only to be a branch and part, slipped from the whole body, but the whole itself sometimes, as, by kind in the four beginnings, and in Minerals, and in seedless plants and nights, and by skill in all.

Therefore, minerals and all have their seed, and their whole body is their seed.

7.   Then as by nature they are wholly, sown, and die, and (or else under-Moon things would prove Metals) rise again the same increased according to the wont of nature, even so they will above ground if we can by skill use them kindly, which we may as well as nature, if we could espy her footing, not impossible to be seen as I could show you quickly, if I might a little unwind the bottom of secrets, and lay them open; but I must take heed.

Then as the seeds of plants and nights rise again, much increased in store and bigness, because it draws unto it, and turns into it in his own nature, much of the kindly stuff and ground that lies about it to corrupt it; even so if you make the metals and ground fit to receive and corrupt the seed of Gold, it will after his due time rise again, turning them, or much of them into his own nature.

8.   Now Doctor Freeman or Moore may see if they be not blinded, that this is no riddled matter, but a plain and certain truth, grounded upon the open and daily race of nature, which not I espied out first (as they spied out the subtill falsehood) but the same tell the troop of the wise Egyptians say, and taught before me, yea and some of them that set in darkness as those worthy Leaches, whose aid we took before, Puine, Fernet, and Cardin, especially the two first, because they bear good will to the truth of this science: But Cardin as a mate that neither knew nor loved it, halts a little; for when he had all about held for certain, that minerals and all had life and were nourished, and grew, and waxed, yet he buried the third point with silence.

9.   But let us not urge this so much in this place, because it is not the right Son of Gold, and stone of Hermes, but a lesser skill and lower way to riches, fit to have been followed in the second Book. Then how doth the Philosophers stone, and the natural Son of Gold, turn base Metals into Gold? For that was the second thing to be handled in this place: When this child is borne, keep him in his heat, which is his life, and given him his due and natural food of Metals; and he must needs, if he be quick and abled to be nourished, digest, change and turn them into his own Nature, much more easily than lead, and he in a cold place, and rude, and hard fashion, was able before to turn strange meats and digest it, as I showed above the change of natural things when they meet in Combat, to be either throughout or half way; that is

either by consuming to raze one another quite out, and turn him into his own nature, or when by mixture, both their forces are broken and dulled equally; Even so in this great skillful change, we may so order the matter, and match the two Combatants, that is the meat and feeder, stuff or doer, with such proportion that one shall either get the victory, and eat up the other quite, or both maimed alike and weakened.

10. To be plain, if we give this mighty child and son of Gold, but a little food (the quantity I leave to discretion) he will be able to turn it thoroughly into his own self-same nature, and thereby to mend himself and increase his own heap and quantity; but if you will make Gold which is your last end and purpose, match your Medicine with a great deal and hundred times as much, or so (your eyes shall teach you) and both shall work alike upon each other, and neither shall be changed thoroughly, but make one mean thing between both, which may be Gold if you will, or what you will, according to your proportion.

11. And if you perceive not, mark how (the comparison is somewhat base, but fit and often used by our men) they sake a sharp and strong Medicine, called leaven of the best wrought flower which is dough; and such another of milk well mingled in the calves bag, called Rennet; and how by matching them with Just proportion of flower and milk, they turn them into the middle natures of dough and curds, nothing so fit; mark it well; nay since you begin to call me to examples, I will play and load you with them, and yet I will lay no strange burdens upon you, no not the quick nature of the Scottish Sea, turning Wood into Geese; nor yet the Eagles feathers that lying among Goose quills, eat them up, two more marvelous changes, then all these that are professed in the Art of changing; yet I leave them, I say for things too strange, and far off my purpose.

12.   There are many waters and earths, which I am credibly informed by G. Agricola, and others as good Authors, are endued with the properties to turn any plant, night, or metal into stone. Cardan tells of a lake in Iceland, wherein a stake stuck down, will turn in one year's space, so much as sticks in the mud into stone, and so much as stands in the water into Iron, the rest remaining Wood still.

13.   There is an old mine pit in the hill Carpart in Hungary, wherein the people daily steep their Iron and make it Copper; the reasons of these things are plain, that which I brought for our great and golden change, and likened to Rennet and leaven here before.

14.   The waters and earth which astonish things in that order; are ever more infected and mixed with some very strange stony Juice, as Agricola says, and reason agrees plainly in the matters, when they no sooner rest from running then they go into stone; nay Pliny says the stony sticks in, Arcadia goes into stone running. J. Hetham meant thereby to try such a thing upon his Lord the great Grecian Monarch, when he gave it to him to drink, it killed him.

15.   The Irish water is without doubt mineral, and as I gather by the description tempered and dyed with the Iron Juice which is called Ferrugo; but every man knows for certain that that the matter of Carpat is Copperas water; now Carpat is as near the nature as the name of Copper, which the Greeks set out most clearly, calling Copper Chalcum, and that other Chalcanthus, and the stone pyrites or Marcasite (as is termed in Arabia) that breeds them both, it is like leaven to dough made of Copper, and raised to a sharp quality which when it is loosened into water, and by draining and by distilling up and down in that hill, refined, it becomes yet more sharp and strong,

able easily to overcome Iron, a like and near weaker thing (for what is near to Iron as Copper?) and turn him into his own soil mean and middle nature.

But how shall we show that Copperas comes of Copper in that order? First the proof of our men makes clear, when they turn that into this, and this into that so commonly.

16. Then the authorities of Geber and Agricola (the best skilled in mineral matters of all that ever wrote) the one after that he had observed it long in Mines, setting it down for a rule, and Geber calling it the Gum, as it were droppings of Copper; but chiefly the workman's daily practice who by following the steps of nature, softening and dissolving the brazen stone pyrites, do commonly make Copper; let us now see what art has done by counter—felting these patterns by Nature set so plainly before her; if she has not done as much and more, surely she was but a rude and untoward child; let us see what is done.

17. She has likewise, and as well as Nature; by a sharp stony water, called Salt Gemme water, turned wood into stone, yea and metals also into precious stones, not by any counterfeit way which Glassmakers use, but Philosophically and naturally, by a marvelous clear and strong water of Quicksilver, leading them back to the middle nature of fine stones.

18. To let pass middle minerals which by the same course we easily change one into another, she turns Antimony into Lead, and this into Tin easily, because as that is unripe Lead, so this is unripe Tin also. These things Agricola reports and tells the way of the first by concoction only, but not of the second, which Paracelsus supplies, by purging him our way of binding with Sal Armoniack. I could set down a way to turn Iron into such Steel as will cut Iron as fast as this will cut

wood, and bare out all small shot, but that they are both but one kind, one better purged then the other, as indeed so are all the metals, though not so nearly allied.

19. Even so 1 esteem of the Silvery and Golden Copperas, which nature sometimes yields underground, and Art counterfeits by our binding, and coloring rules above set, as Agricola tells and teaches; neither think these bastard wits cut quite out of rule, but so follow the same reasons of nature; and as the rest take the finer like part, and leave the gross unlike, so do these feed upon their like, the fouler parts, and leave the better as unlike their Nature.

20. But to proceed to turn Iron into Copper by Copperas-water, is somewhat more ordinary then the rest; Agricola says an old parting water which is made thereof (as we know) will do it, but the workmen in the bill Kuttenberg in Germany, do more nearly follow nature in that hill of Carpat, for they drain a strong Lee from the brazen stone, that is, they make Copperas—water strongly and kindly, and by steeping their Iron in it make very good Copper; nay further, Paracelsus says again, that in Casten they turn Lead also into Copper, and though he names not the means in that place, yet otherwhere he does, and reacheth how by Copperas sundry ways sharpened, to turn both Lead and Iron into Copper, in which place he delivers another pretty feat to unloose both Iron and Copper into Lead again, and this into Quicksilver, by the force of a sharp melting dust which Miners use, and this our common rule still of stranger likes; for this dust being of the same nature still, which exalted Lead and Quicksilver, two great softeners and looseners of hard bodies, is able to make the stubborn metals, retire and yield into the middle place of Lead, and this is Quicksilver.

21. Now then we see that Art has reached and overtaken all the Natural changes of Minerals; why may not she by the pattern devise more of herself, as the grief of good workmen is, and go beyond nature, and turn the foul metals into fine Silver and Gold? She has a great advantage of nature; first for patterns, and then her helps in working; and lastly the help and instruction of a divine wit and understanding, whereby no marvel if all wisemen have said, she passes nature.

22. Albeit it is uncertain whether nature have such a Golden Medicine in. her bosom hid, or no, as well as those of Copper, Stones, and such, yet this is sure, that by the bastard way of binding (as we have heard before) she turns Lead and Tin, and perhaps Copper too, but surely by Quicksilver and Silver into Gold.

Then I say it is a sign of a weak and shallow wit, if Art cannot by these patterns aforesaid, devise further to turn other metals into Silver and Gold; is it any more then to raise and exalt Silver into Gold? But this will serve for both into very sharp, strange qualities, able like the rest, to devour and turn their own like meat into their own middle nature from whence they sprung. Certainly, the reason is so plain and ready, that I must needs deem him, less than a child that cannot conceit in it; nay bend your ears and minds.

23. By reason, if the workman be very strong over the stuff, he will turn in trial, things unlike and contrary as well, though not so easily as like and friendly.

And for the proof of stony Juices, turning all sorts of things, even metals themselves into stone, as has been found by the stamp remaining; of Antimony and Copperas turned into Lead and Copper, of the ripening of the Mineral Mines of Lead and Gold eating dung and urine, and such exchanges set down before, I am led to think that a very lusty and strong Medicine would be able to change other things as well as metals, especially Minerals into Gold; some of our men say no, because their wants in the rest the ground of Quicksilver, the knot of friendship and unity. I grant it were hard in respect of the right way, and yet I hold it possible.

And thus you have seen the ability of Hermes medicine, to turn base metals into Gold by three sundry ways; first as he is sown and rises again to be made medicine, which I call begetting; and then by changing the little food that is given him into his own nature; to make him wax and grow in heap and bigness, which I term nourishment; and lastly by changing the great store of stuff, wherewith we march half way into the middle nature of Gold, which is the best change and drift of our purpose. And this I may do well to call mixtion, though Sir Christopher Heydon and Fernel name it begetting also, as it is a kind indeed, but because it goes not the kind way, let it go and us keep our order with our brethren.

Chapter III.

The Rosie Crucian Medicines.

1. Of seed. 2. Of increasing Gold. 3. The quality of Gold. 4. Of nature in concoction. 5. Changeable stuff. 6. How nature made Quicksilver and turned it into Gold. 7. Of purging. 8. Of lightnings. 9. Of fire—flies. 10. Of the Star—fish, and other things. 11. Of the nature of fires. 12. Of

Hellen star and cause of lightning and thunder. 13. The power and virtue of Rosie Crucian Medicines. 14. The first matter of Gold. 15. Of hot spirits. 16. Of the fiery quality of Gold and its power. 17. Of the pernicious quality of cold frozen countries. 18. Of the understanding spirits of the air, and the lively spirits of heaven. 19. Of the spirit of metals. 20. Of a natural stone that consumes all the flesh and bones of a dead man in forty days, and of other things. 21. Why Copper—water parts silver from Gold. 22. Hot stomachs. 23. Directions to Philosophers. 24. Examples. 25. How Gold got its high red Color.

1. Now, how shall our son of Gold be able to subdue and turn so much of base metals with so little change and travail, and so great return again as we have promised? It is for three causes; first, for the bitterness and readiness of the stuff to be changed; and then for the great store and strength of the changing workman, to send away the lightest still first and foremost; and lastly for his increase in store and quantity, which may be made by sowing and nourishing the son of Gold without number; for Bowing first, There be sundry sorts of sowing and making this our medicine; one is an excellent way, but a bare and naked and lone way; because if Gold can be made fit and open to be wrought, as behooves a seed, he has all both stuff and workman, male and female, seed within himself; and the less contagion there is of unclean stuff, the more excellent and mighty will he rise again; this way by deep and painful wit, has been sometimes taken, but very seldom, because it is very hard, long, and irksome, and therefore we will leave it also; but chiefly because it crosses my purpose abovesaid; for if it be sown alone, he cannot rise increased; whereas we desire to augment his quantity, then there are two kinds of grounds, and yet both one kind, which we may put unto him to

corrupt him easily and raise him again with great increase , and quantity; one nearer his nature then another, so much is enough for that.

Now for the store of ground fit to be laid about him, there is a choice better or worse also; but that is no great matter, so you keep the measure and discretion which a common seeds—man can keep, neither to overlay and drown him, nor to leave him dry and barren; then to our purpose; cast in yourself what increase in store one grain of corn will yield, within few times sowing; when I bad a little leisure I did once cast what one grain, by the increase of fifty (which happens often) would arise to in seven times sowing, and I wearied myself with an endless matter.

A greater sum then any man would think, I have forgotten it, cast you that have leisure: Now a grain, I mean an ounce, of our seed, though it rises not with such advantage (for If it were so sown, It would be quite drowned, or at least not worth the tarrying) yet it rewards it another way, with speed in working; For albeit, the first time be much alike, about forty weeks or such a matter; yet the second is run much sooner, both because now he is softer then the first seed, and easer to be loosened, and also mightier, and more able to turn the work over, so that we keep our selves within the number of ten, as some do set the bounds, yet I think the midst between, duplum and decuplum a notable mean, although that be as it happened, yet by this great haste & speed, we may quickly overtake infinity.

2.   But if you think this too slow a course, let us run to the next increase by nourishment, whose great speed and readiness will easily supply all and fill the biggest desire in the world; after the Son of Gold has been once sown, and raised again, he is now able to work mightily, and not before, and to turn one hundred parts of his due meat, into a

third middle thing, Gold his father's nature; this now will show hereafter; then if he be able to turn an hundred times as much half way, he can surely as easily and quickly one part, that is no more than himself, quite through into his own self-same nature, especially if that food be silver or Gold, which is best of all to the purpose: then he is now twice as big and as strong as he was before, able to devour as much again: and so forever, for this strength shall never be abated, when after his feeding he is left the same still, or even as one Candle lights another still or more strangely, though not so largely, like unto the Loadstone, which as Plato reports, after it has drawn one ring of Iron, it giveth power to draw another, and thus unto the next until you make a long row and link of rings, close and fast, one hanging upon another.

3. Then since we may so soon heap up so great a quantity of this Golden medicine, it may chance we shall not need any great help of the readiness of the stuff and strength of the workman; and if but ten parts of the Gold might be made at once, between a weak workman and a stuff, yet perhaps it would serve the turn to raise the sum appointed: But suppose it comes short ten parts of the way, yet if through the means of the nearness of the stuff, and force of the door, one part may come to turn a hundred, then we shall supply and overtake the want and hinderance: Let us see.

4. And first again of the stuff, because it is the shorter and easier matter; a thing fit and easy to be changed when it is like the nature of the workman, & nearer the ways end.

The straight affinity and nearness of the Metals one to another we have opened above, when we found them all to be one thing, differing only by certain hang-byes of clearness, closeness, and Color springing out from the odds of concoction, and that if the same concoction hold, they will come at length to their Journeys end, which they strive unto, the perfection of Gold, except perhaps Iron and Copper; by over sudden heat or some other foul means, have been led out of the way, yet they may be led back again and cleansed as we heard before, and yet they were all made at first of quicksilver, a foul and greasy thing in respect, and then were grimed and bespotted greatly again, with the foul earthly Brimstone which afterwards came upon them, whereby they were all gross and ill colored, open and subject to fire, and other spoiling enemies, before by long, gentle and kindly concoction, all the foul and gross stuff was cleansed and refined, and so made apt to take good Color, (as we see in plants and all things) and to gather itself up close together, and likeness to be weighty, for the much fine stuff in a narrow room, when lead and Quicksilver, heaviness follows from the rawness and lastly, to be steadfast and safe from the fire, and all other enemies, because there was never any way of entrances in so great closeness, lest, to make division and dissolution, that is destruction, nor yet any greasy stuff the food of fire remaining.

5. Wherefore we see the near neighborhood of metals, and easiness to be changed one into another, (especially if we work upon Silver, which is half Gold already) when they want nothing of Gold, but either long or gentle concoction, or instead thereof (because we cannot tarry) as strong and fierce one answerable unto it, first to cleanse out all the gross and greasy stuff, and then to bring Color upon it.

So that I cannot but wonder at those men if they be learned, who, in reproof of this Art unknown, vouch, unfitness of the stuff to be changed, saying that Metals being of sundry kinds and natures, cannot be turned before they be brought into that stuff, whereof they were first made and fashioned, which we do not when we melt them only, and which is not easily to be done. It is a sign that either they never knew, or at that time remembered not that nature of a Metal, or of the first stuff, for If they mean the Grecian supposed first empty and naked stuff without shape, but apt to receive all, even that which is the middle state of a thing lasting but a moment, when by the way of making and marring (which our men with Hippocrates call changing) it is passing from one to another, then if yielded and quickly granted with Geber, Arnold, Lully and many more learned men, on our side, that in that very violent work of changing the Metal being so far altered and broken, even into dust of another fashion, I think I must drive them to blow the seed, as they say, and they know not what to answer.

6.   But if they mean as they seemed to do, we should not melt our Metal, but bring him back unto his nearest beginning and stuff Quicksilver, and then put on our shape and form upon him, according to the kindly sowing of Gold, upon his base ground above said, they are deceived not knowing the nature of Metals; for they be not of sundry kinds and beings (as they say) but all one thing differing by degree of baking, like divers loaves of one paste, that it were madness if any of them lacked baking to lead him back, or mar or spoil him of his fashion, but in the same form and being to bake him better, and so did nature in the Ground, in baking quicksilver, or lead into Gold, she went forward and not backward with the matter: Nay why go I so far with them? They never marked the nature of their own words, which they use in their own Philosophy, where changing is fitting only, and shift of those hang-byes called accidents, the form; kind and being of the thing remaining.

Then if the stuff be so fit, let us see what the work is, not in store which is done already, but in force and powers, his strength and power is seen in two things, purging end coloring: First he must mightily she, himself in purging and driving out all the gross greasiness of the stuff, and then when all is fine, clear, and close, he ought to stretch himself at large, and to spread far forth in Color upon it; for albeit long & gentle heat purging by concoction, of itself breeds and brings good Color, yet this over-short, and violent heat proportioned doth not so (as I showed above in the discourse of binding and coloring) but needs bring Color with him already coined.

7. So that when he purges the stuff understands; he draws not out the foul and gross stuff, and departs away from the work withal, as the foul purging hinder did; but being a clean and fine thing like the nature of a Wight, he purges by digestion and expulsion, driving out the foul and unlike parts as leavings, taking and embodying with himself the fine and clear for food and nourishment.

Then let us see how this work of purging is performed, for that is all, and the Color hangs upon the same, and is done all under one, as we shall hear in going out of this treatise, if nothing purges but heat through concoction, and this ever to be measured according to the need and behoof of the work underhand; and we must scour an hundred times as much stuff in one or two or three hours space at most (for that is their task) when we had need of a marvelous fiery Medicine, besides the great outward heat, to prick him forward, scarce to be found within the compass of the world and nature; it must show itself an hundred times fiercer then a binder, which was scant able in longer time and stronger heat to scour and purge one part; and as much of the same stuff.

8. This is a marvelous hard point: I had need what my thoughts and memory, and all the weapons of wit unto this matter; if we search all about and rifle the corners of kinds, we shall find no fire in the world so hot and fierce, and the lightning able to kill plants and Wights &

melt metals, and to perform other such like marvelous things in a moment.

As (to let pass plants not so strong) I have read of eight Lepers in the Isle of Lemnos which as they sat at meat under an Oak, were all suddenly stricken stark dead therewith, setting still in the same guise of living and eating creatures, again that it has sometimes passed through a purse at a man's side, and melted the Coin without hurting the leather, because such a suitable and speedy fire found that resting stay to work on, in the Metal which it wanted in the open and yielding leather, and many more such strange deeds we may find done by that most violent fire, then our fiery work; man if he be tasked as he is to work as great wonders at these be, had need to be fierce and vehement, as the fire of lightning, as it is sometimes termed in our Philosophy.

Let us match these two together, and see how they can agree, that all things are laid, and as it were stricken together, the light of truth may at last appear, and shine forth of the comparison; let us as till! faith, at the first setting out, launch and row a little easy before we hoist up sail.

Gold of itself in Philosophy is a fire that if it be raised and Increased one hundred degrees in quality it may well seem to prove the greatest fire in the world.

9. But our men as they speak all things darkly, so this perhaps in regard of other metals, or rather because like the Salamander, not like the fire flies (for though the Salamander can as well as Serpents eggs, by his extreme coldness, quench a little fire, yet a strong fire consumes him and puts him out of being) because I say, like the fire fly he doth

live and furnish in the fire, when as indeed Gold, as all other metals, is cold and waterish far from the fire.

10. And yet it is not the outward show of the body alone that makes a fiery nature, but sometimes the inward quality doth the deed of fire, (if we speak at large as the common custom is) and so the Star Fish in the Sea burns all she touches, and a cold spring in Selavonia sets on fire any cloth spread over upon it, and to come near by such fiery force doth the water fix in Thessaly pierce through in any vessel save an horse hoof.

11. But now we are come unto the deep, let us hoist up sail and speak more properly and philosophically, and more near the purpose; let us I say hear the nature of fire and how it combined fire, as they bound it, and we shall find it if we mark this offspring, as a very hot and dry substance; the first cause of fire is motion, a gathering and driving much dry stuff into a narrow straight, which by stirring and striving for his life and being, is still made more close, fine, and hot, that its nature will bear and suffer; and so it breaks out at last, and is turned into another larger, and thinner, dryer and hotter nature, called fire: hence the great underground fires, in Aetna Hecla and many other places, grow and spring at first, when the cold drives a heap of hot earthly breaths and vapors, either round up and close together, or along through the narrow and rough places, rubbing and wringing out fire, which the natural fatness of the ground feeds forever.

12. So the Star called Hellen-Star, that lights a sign so dangerous upon the table of the ship, and falling melts Copper vessels, and comes of an heap of such vapors, carried up by violent cross winds, so that by rubbing Millstones, flints and such like, we see fire arise after the same manner; and this is the manner of the spring of all fire, others flow from this, one still sowing as it were one another; but if the stuff of this

fire be tough and hard, and then when it is wrought into fire, if it be moved again apace, it proves for these two causes a marvelous hot and violent fire, whence springs all the force of Lightnings; for it is nothing else but a heap of thick and brimstony Vapors (as some hold with reason) by the coldness of the cloud; beaten up close in that order, and now being turned of a sudden into a larger and thinner Element then it was before, when it was earth and water, his own place will not hold him, and so by the force of nature, striving for room and liberty, he rents the clouds in that manner which we hear in thunder, and bursts out at last, a great and swift pace, as we see in lightnings; much swiftness together with the toughness of the stuff, finely wrought, makes up his violence above all fires In the world.

13. Now for the Son of Gold and Hermes his Medicine, what kind of fire is he, when he can be no such Element, extreme hot and dry fire; for he is temperate, and has all, the qualities equal, & none working above another, and yet indeed by reason of the fine and tough (and therefore mighty body) whereon they be seated, they work in equality together, much more forcibly, that the extremely distempered cold and dry poisons can work alone and as fast and faster than they devour and destroy distempered bodies; these do overthrow the contrary: Then what fire he is I showed before, how full stuffed with heavenly spirits above all things, and so be is an heavenly fire, which is much more effectual in power, and mightier in Action then that other: by reason of his exceeding subtleness, able to pierce through rocks, all things, where that other small quickly stays.

14.   Admit it say you, if that heavenly fire were quick, free and full of liberty: but it is fast bound up in a hard body; then I will give you all the reason, bend your wits unto it; Gold at first was fully fraughted with the most piercing fire in the world, and then came and wrought it

into a most fine flowing oil, and so unbound it and set it at full liberty: not so freely indeed as in heaven, but as it can be in an earthly body, closely crowded up together, (which help) heats as in a burning-glass, upon a most strong and mighty body far above all things in the world; and lastly with a violent outward fire, she sent all these apart away to work together.

15. Judge then you that have Judgement, whether it were not like to bestir itself as lightning; Copper, the heat of the hot spirits, is as great; and if it were not, yet their passing subtleness would require that matter easily, and make him even; yea and perhaps when they be drawn and carried up close together, make some odds and differences between them; but surely the exceeding toughness of the body(as we see in Iron and the rest) augments heat greatly, and carries him far beyond it.

16. Now for the pace, it is much swifter, and driven by a much stronger mover, even so much as a founders fire passes in strength, the top of a thick cloud; for this is he that sends that lightning which else would have flown upwards; therefore because the fire is Stronger, and has the helps of body and motion far more favorable, the fire of the Son of Gold must needs pass the lightnings in power, and wonderful working. Then bethink yourself, with what ease and speed, such a fiery medicine was like to pierce and break through, sift and search about, and so scour and dense a great mass of foul metals? How many times more than a weak and gross mineral binder? Fasten and bend your minds upon it: we see how a weak waterish or earthly breath ma narrow place, within a cloud, the ground, or a Gunn, (all is but thunder) because he is so suddenly turned into a large Element, and lacking room, bestirs himself, and worketh marvelous deeds; what may we think then of the heaps of those false reports of heaven, and of

that most strong Golden body, closely couched up together in a little room, when they be in a narrow vessel driven out, and spred abroad at large by a mighty fire, and thereby still pricked and egged forward (for as long as the fire holds, they cannot be still, nor draw in themselves again), what thing in the sturdiest Metal can be able to withstand? How easily shall they cast **doin** all that comes in their way, brake and bruise all to powder. May not we all say plainly that which the Poet by borrowed speech avouched, that Gold loves to pass through the midst of the Guards, yea and to pass through to rocks, being mightier than the stroke of lightning, it is so fit, as if it had been made for the matter.

17. I have heard that the extreme cold weather in Lapia and Finland (which are under the Poles girdle of the world) pierces and freezes, and cracks the rocks, yea and Metalline vessels; again, that the poisoned Cockatrice by his violent, cold, and dry breath, doth the same on the rock where she treads; then what may we judge of the force of our fiery medicine upon the metals, by these comparisons? How fiercely and quickly were it like to divide and break them, having an extreme fire, the greatest spoiler of all things, to over match the cold and dry quality? And a much stronger body then these vapors which carried the former qualities, and both these sent with far greater speed and swiftness, as appears in the difference of the movers?

18. Lift up your ears & mark what I say, a deaf Judge had not need hear these matters; who has not seen how Quicksilver enters, cuts and rents the metals, though many doubts and differ about the cause thereof? Cardan thinks that, like as we find of the cold weather in those frozen countries, so this marvelous cold metalline water, entering the metals freezes their moisture within them, and make then crack and fall asunder, and therefore Gold soonest of all other, because his moisture is finest, even as sodden water for his fineness freezes

sooner than cold. Surely very wittily Paracelsus deems this done by the spiritual subtilty of the body, even as the understanding spirits of the air, and the lively spirits of heaven use to pierce through stone walls and rocks, by the same strength, without the force of qualities; but I think it is rather for his stronger like qualities, seeking to devour them; else he would pierce your hand and leather, and such like easy things which he leaves untouched as unlikes and strangers; as for the qualities of Quicksilver, it is a question what they are, and which excels; some judge her very cold, some again marvelous hot (as Paracelsus for One) some moist, other dry, but as she has them all apparently, so I deem her temperate, like Tin that sprung from her, and almost like unto her. Gold I mean, though perhaps the qualities be not all in her, as in him, so equally balanced.

19. But let the case be what it will (I love not to settle upon uncertain matters) the great spirit of metals after she is first wrought into Gold, and then into his son our medicine, shall be in any reason both for body and soul an hundred times stronger and more able to do it; nay Antimony and Lead are much grosser then Quicksilver, and yet you see how they rend, tear, and consume base metals even to nothing; but what say we to plants? There is a great difference in sharpness and ability to pierce and enter between a thorn and a needle, and yet you heard above the gentle plants of the vine, and the middle dew of heaven yield stuff to an eating water, able within three or four distillings to devour and dissolve metals, then what shall not only sharp mineral eaters, but this our almighty Gold medicine show upon them, which besides that wonderful passing, sharp, and piercing body has the great help (which they want) of that heavenly fire, and of her swiftness, stirred up by a mighty mover? These things are enough to suffice any reasonable man (if they will not atop their ears against the

sound of reason) touching the power, might, and strength of our Medicines.

20.   What is then behind, *The Holy Guide* has taught us all things; yet I bear them whisper, that albeit these medicines of ours have such thundering power, yet they may not force so our purpose of consuming all the metals (as the guise and forcible use of such fiery things as) without regard or choice of any part or portion; but it is not always I hope the guise of violent things; I need not go far; there is a natural stone in Hazo, which by a mighty and strange property uses, in forty days space to consume and make away all the flesh and bones of a dead man's body, saving the teeth, which he leave ever safe and whole; and therefore they called it in times past flesh eater, and made tombs thereof for dead, and boots for Gouty men; I could clog a world of readers with like examples, if I might be suffered, but weigh this one and our artificial Pantarva together. Why may not it is as well have its choice and same, a part of this great waste and spoiling? They know not why, and how then? There are many deep, hidden, and causeless properties in the bosom of kind and nature, which no man's wit is able to reach and see into, the world is full of them, when Art is open, and all his ways known. Indeed the world is full, of late of such senseless and blind Philosophers (which like as the Poets when the Stoic a little calls on JOVE by many names, to help to shore up the fall of a verse, or stop the gap in the number) so they when their eyes are dazzled upon the view of a deep matter, fly, to nature as fast, and to hide her Unsearchable secrets, to cover the shame of ignorance, as though God moved all with his finger (as they say) without any middle means and instruments. There is nothing done without a middle cause forerunning, if it were known, as I think it is to some, though never so dark and hid from others; and therefore to come to the purpose, as the reason of the natural eating alone, was clear to Agricola though

unknown to Pliny, and many more the reporters) and found to be for the loose and light temperatures, and Copperas water, fit to eat the flesh and softer bones, and yet unable to do a thing above his strength, that is, to overcome the harder; even so you may think the reason in this like property of the R. C. Physick, Pantarva, etc. is seen to some: for certain, & however it was my luck to see it, I cannot tell, it has been sure unfolded twenty times at least, in the speech going before, if you remember well, it follows the high and common way of all nature, I mean that eating nature; for all things eat, and that the cause of things done below; then there is nothing eats and devours all the stuff which it overcomes, but so much as is like and turnable, the rest he leaves as strange and untouchable; so did all the foul binders purge above; nay so and no other ways, doth the lightning end all fire eat and consume the stuff subdued, turning the air. and water into fire, and leaving the earth and ashes; even so doth our medicine, after it has driven out and scattered all uncleanness, it takes and strikes unto the fine part, like unto itself, and makes it like himself, as far as his strength will carry.

What need I pray? I there need any of any more examples? Is it not clear enough that all things seek their like and shun their contraries? Yet because these mineral melters have been evermore very strange and unacquainted with the Grecians, I will set down one or two of the clearer examples.

21. Why does copperas-water part and draw away silver from Gold? But copperas is like to Copper, and this to silver; for as Lead is to Gold, so is this to Silver; cast in plates of Lead and Copper, and that will cleave to the Gold, and this to the Silver. But Silver is liker to Lead then Copper, therefore to part silver from Copper, the Miners use to season a lump of Lead with a little Silver that softens the work and makes it ready, then one Silver draws the other part unto her, nay raw

Quicksilver as she is strong in all things, so in this very wonderful, Quicksilver I say the grandmother of our medicine, and the spring of all her goodness, will quickly receive and swallow, either in heat or cold, her near friend, or very like clean, temperate, and very fine body of Gold (and therefore as the one is termed unripe Gold, so the other ripe Quicksilver) when the rest she refuses, and bears a loft as foul, gross, and unlike her nature; and this secret the miners also by their practice have opened unto us when they so part Gold from the rest, mashed altogether in a dust heap; wherefore when this fine and clean body Quicksilver, is made by nature, and Art yet much finer and cleaner, and again as much more piercing and spiritual, and able to perform it, how much more deadly will she run to her like and devour it, the clean, fine, and spiritual, that is the Quicksilvery part of the metal, and if she devour it, then it cannot be lost, but must needs go into a better mature, even to the nature which we desire.

What then is to be said more? I have not yet bounded the matter, as I promised, and chewed how the golden stone should turn a hundred times as much into Gold, I have shot a large compass, but all at random; now it is time enough everything has its due time and place.

22.   You have heard I am sure of the hot stomach of the Elephant, Lizard, and Sea-calf, able to digest and consume stone, yea and to come to the point, the Struchio (Estridge) that marvelous beast, Iron also; if the stomach of a Wight be able in a short space to divide, expel, and turn the fine part of a metal into his own selfsame nature. How much, and how soon may the stomach of our medicine turn into Gold? Not only an hundred times more than the beast, because it is an hundred times more fitter and able to do it; first for the likeness and nearness of the stuff, and then for the two great heats I speak of; and thirdly for the wonderful, subtle, strong, piercing and cutting

workman, but especially because he goes not quite through with the work, as the beasts did, but half way to the middle Nature of his father: consider and weigh the matter, but if he be somewhat far off the mark, see how woade & other things of like strong gifts and qualities are easily able to overcome and change, with whom they meet, even without this great mingling and boiling: why shall it then be hard for our medicine, with great concoction, to do the like upon his own subjects, for proportion of strength, for strength will follow him, as able to overcome the stubborn Metals, as these two the weaker water.

23. To close up all, remember what I said, and what is most true and certain, that Gold is closest and most full of fine large spreading stuff, of anything else in the world, passing the wonderful gift of Silver, in this point an hundred fold, in so much that one ounce of Gold, by the blunt skill of the hammer, may be drawn out and made to stretch over, above two Acres of ground: Consider well this one point, all shall be plain and easy; I mean to them that are learned, for these be no matters for dull and amazed wits to think on; then after this spreading Metal, is made a fine flowing oil; and drawn out at length; and laid out a broad most thinly, by a vehement heat of fire upon, how much will it spread; may you think in reason? But such a view may quickly dazzle the eye of the understanding; let us picture out the matter as Plato uses.

24. Think the difference in fineness, in Color between the Son of Gold and Silver (if you will take him to turn as I bade you) to be like the odds between very fine Scarlet, and course white sack—cloth; let this be closely shut up together in a Walnut shell, this packed up as hard in a very round pot of a quart, or of that bigness, which will take the measure of an hundred Walnuts; you see the bulk of both; and so,

if you weigh them, one will prove an hundred times as much in weight as the other: but draw them out, and spread them one upon the other, & one shall overtake, match and fit another on all sides: Now owne[32] is very course and big, and the other is very fine and small, as appears by their threads, yet the small may be full as strong as the big, as we see in a little gall, poisons, etc. it is common.

Then these two encountering (as we must suppose) shall of force, hurt and change each other equally, and so the exceeding fine and gross mingled, make a middle thread, and the extreme red and white colors carried with their bodies, take a yellow mean also: even so you must think when an hundred ounces of silver, and one ounce of our Medicine, are both by the fire beaten, and driven out at length and to the furthest thinness, every part overtakes, fits, and reaches other, and the small part being as strong as the bigger, in striving one overcomes, consumes, and turns the other, that neither shall be quite raised, but both equally changed and mingled unto a third mean thing, both in fineness and Color, and all other properties whatsoever.

25.   And so you see the Color also dispatched which I kept in their place, and which seemed a wonder in some men's sights, so I hope you will not ask me how Gold got his high red and unkindly Color unless you be ignorant how all such hang-byes flit and change up and down, without hurt to the thing that carries them; and except you know not, that by a kindly course (whereby all soft & alterable things, gently and soft boiled, wax first black, then white, next yellow, and lastly red, where they stop in the top of Color) we see changed and drawn up our seeds of Gold unto this new unwonted Color; of this I have spoken largely in the nature and dignity of Angela.

---

[32] Exact spelling found in the printed edition. -pnw

And thus, you have at last, all the reason which I show, or at least thought good to deliver in writing; for the truth of Hermes or the Philosophers stone and Medicine, why is it the ready way to bring all men to all Rosie Crucian—happiness in the world? That is to long Life, Health, Youth, Riches, Wisdom, and Virtue: It is now time to sit down and take our rest.

Chapter IV.

What the Pantarva is: The true matter in Nature and Art: The manner of working: Canonically and orderly made manifest in this Book.

1. The place for working. 2. Heaven unchangeable, all beginnings even and of other things. 3. Of end and everlastingness. 4. Heaven and Earth. 5. Of God and Man. 6. Of blood. 7. Of Making and perishing. 8. Of the four seeds of strife in the world. 9. The dissolver and destroyer of Gold. 10. The ray of making and working the thing sough after. 11. Of the body, fire and blood of our matter. 12. The due of Starry blood and womb for seed. 13. Influences of Heaven. 14. Of Instructions. 15. The Quality of Countries. 16. The Pantarva. 17. Dr. More and Dr. Freeman convinced, and all the Art made manifest.

1. Eugenius Theodidactus hears them mutter among themselves, that there is never a reason given as yet, no not one, because all stands upon a famed and supposed ground, which being nothing, all that is built upon it must needs come to nothing: For even as Paracelsus in his

supposed paradise, in the end of high opinions, concludes, that if it were possible to be made, by any labor or wisdom, it would prove no doubt, a notable place for long life and Health; even so nay be thought of this stone of Gold, if any Art or skill were able to contrive it, that it would without doubt work these wonders aforesaid; but as his Paradise (if he mean plainly as he says, and of the Philosophers stone whereto it may be wrested) is impossible to be made, unless he would include himself in a place free, first from a the contagion and force of outward Earth, Water and Weather, yea and therefore of the fire of Heaven, and light also; and secondly where all their beginnings were in their pure and naked Nature, which they call the fifth nature, which is nowhere save in heaven, and which were a miracle to be conceived; and lastly except he would live without meat and his leavings, which both learned and unlearned hold ridiculous to think.

Even so it is as hard in opinion and unlike, that Gold may be spoiled and brought to nothing, as he must be first, and then restored and raised to such dignity; because as heaven is ever one and unchangeable: for that in it all the beginnings are weighed so even, and surely tied together, in a full consent, unable even to jar and be loosened; in the like manner Gold is so close and fast, for his sure and equal mixture of his fine earth and water, that no force of nature, neither of Earth, Lire, or Water, no nor fire, although he be holpen with Lead, Antimony, or any such like fierce or hot stomach, easily consuming all other things, will ever touch him; nay which is strange, the greatest spoilers in the world, fire and his helps, are so far from touching him; that they mend him and make him still better, and better; what is to be said to this? Albeit I confess that to be the main ground and state of all the work and building, yet I suppose It not nor took it as granted, as if I had been in geometry, but lest it to be proved in the fitter place; as for that supposed paradise it is hard to Judge, because he did but glance at it, and so leaves It unlawful to be told; albeit a man may devise in thought as well as he, (for I think he had not tried it) what may be done and what mature will suffer.

Then what if a man enclosed himself in a little Chamber, free from outward influence which is easy; overcast for light sake, if need be with such Marble as Pero made his Temple, shining in darkness, with all floured thick with *terra lemnia,* or the earth of a fish nature (which is better, but much harder to be gotten) and had such water within the lodging, as that not long since found under ground in Italy between two silver Cups; then if he could ever live quite without meat, (which I showed not Impossible) or preserved himself with a first Nature, which breeds no leavings: What thank you of the matter leavings; but think what you will: if it jar and sound not well in the ears of any man, let it be among other his Incredible and impossible monsters, yet our cause shall not be the worse for it, but easily possible, as I will open unto you, as far as my leave will suffer me, which has been large indeed, and must be, because I made a large promise at first perhaps too rashly; but for the good meaning) which must be payed and performed to my brethren of the R. Cross.

3. Aristotle says like a wise Philosopher, that nature makes her creatures and subjects apt to move and rest, that is, changeable; and again, that a body that is bounded cannot be without end and everlasting; and therefore that when heaven ever moves, and earth ever rests, it is beyond the Compass of nature, and springs from a more divine cause; if this rule be true, as it Is most certain, then Gold a thing not unbounded, nor yet an extraordinary and divine work, but made by the ordinary hand of kind, as we heard above, must needs decay and perish again, and cannot last forever; and if nature can dissolve him, much more shall she with the help of Art perform It; and

that which was said of fire and his helpers, is nothing; for why doth fire better Gold, but by removing his enemies, which nature secretly laid above him to destroy him? And so, every stick as I said above, may be saved from decay; but let nature have her swing underground, or skill above, they shall cause his enemies in time to spoil and consume him.

We cannot tell (say they Country-like) it may be a divine and no natural work, for we see it everlasting.

4. Go to, be it so, I will overtake them that way too; for as we know that which Aristotle knew not, that both heaven and earth by the same divine cause that made them both, may be, and once must be marred and changed; so we may think that Gold, although it were a divine work, yet by the like skill follows the divine pattern, might fall to decay and perish.

5. But what Is the divine pattern? And how shall men be like unto God? Even by the goodness of God, who has, as I said above, left his pattern open in all places, and easie to be seen to them that seek to be like the main pattern wherefore we are all made; and this as Hermes says, gentle and witty separation, wherewith he avows both the great, and our little work made and woven, and so to be married and unwoven again, to figure unto us privily that there Is no great and cunning work performed by such rude and Smith-like violence as you speak of (*vis consilri expers mole ruit sua*) but by this gentle skill and counsel, as we say see very plainly and fitly, by a thing in virtue and price, I mean in the worldly estimation most near unto Gold, the noble and untamed Diamond, which when he comes into the Smiths hands, will neither yield to Fire nor Hammer, but will break this rather then he will break, and not so much as be hot (as Pliny says) but not be hurt

(as they all grant) by that other, and yet by gentle means of Lyon or Goat's blood.

6. Though they be hot bloods (that by kind, and this by a disease of a continual Ague) you may so soften and bring under this stout and noble stone, as he will yield to be handled at your pleasure, nay by the flowing tears of Molten Lead (a thing not so hot as may be) he will quite relent and melt withal. Even so we may judge of Gold; that albeit the more roughly he be handled, the less he stoops, as the Nature of stone things is, yet there is a gentle and heavenly skill and way to soften him, and sake him willingly yield and go to corruption, though this as well as that be not common and known abroad, as no reason it should.

7. But what need we fly with Aristotle to any divine shelter? As Gold was made by a common course of kind, and must dye and perish the same way; so, this skill of our needs not be fetched from any hid and divine secret (whatsoever our men say, to keep off the unworthy) but from a plain Art following the daily and ordinary steps of nature in all her kindly works and Changes; then mark and chew my words well, and I will open the whole Art unto you.

8. God, because he would have none of these lower creatures eternal (as is aforesaid), first sowed the four seeds of strife in the world, one to fight and destroy the other; end if it would not serve as it will not here, he made those that sprung from them of that same nature; and there is nothing in the world that has not his match, either like or contrary, able to combat with him and destroy him.

9. But the like eats up and consumes the like, with more ease and kindlier then the contrary, for their nearness and agreement; then if nature mean to spoil Gold and make him perish, because it is so strong

a thing, she takes the nearest and most kindly way, she sets a strong like upon him to eat up and Consume him. What should I say more or more plainly? You know the thing most like and nearest unto this, is in all mens sight corrupt, and subject to decay, and then when is loosened, very strong and fierce; it is ever more wrapped about him, and so by contagion it strikes and enters, and so pulls him after, and all in their own nature, heat and furnace rot together, and in due time rise again and the sane; for being all one in effect, as the seeds of male and female, it boots nothing whether overcome in the end, and a new thing like the old must needs arise, if some occasion in the place (as I said of heat and Brimstone) come not between and turn the course.

10. You have heard of nature, let us come to Art; if she cannot follow the steps of nature, she is but a rude skill; nay she must pass them far, if she means to take profit by the work; for albeit I deny not that ail things may fall out so luckily, that our son of Gold say start up underground (though never found, for who would know it?) yet nature may so easily fail in the choice of corrupting ground, but chiefly in tempering the degrees of her kindly heat (without which the work will never see end) and again the lets are so many and so casual, that perhaps we would be worn before the work be finished.

Then how should Art, her counterfeit, pass this kindly pattern? Very easily, by the understanding skill of a divine mind, which I said doth pass nature in her own works; first in choosing the best ground, and best proportioned for generation, which nature in this respect cannot, as aiming at destruction only, then in removing all lets to come between.

But especially in well ordering that gentle and witty fire of Hermes, wherewith all the work is sundered, that is turned, altered and mingled.

But what is this witty fire? For here is all the hardness, here all the world is blinded, all the rest is easy; bend your minds, I say, I will tell you all the Art; Enclose the seed of Gold in a Comora, yet a kindly place; Lo here is all the Art, all the rest is written to blind and shadow this; so far as I may do good and avoid hurt, I will unfold this short, hid and dark matter, and yet Hermetically and Philosophically. As the Sun is the father of all things, and the Moon his wife the mother (for he sends not down those begetting beams immediately but through the belly of the Moon) and this double seed is carried in a wind and spirit into the earth, to be made up and nourished; so our Sun has his wife and Moon, though not in sundry Circles, but Adam-like, and both these are carried in a spirit also, and put into a kindly furnace.

11.   To be more plain, this seed of Gold Is his whole body loosened and softened with his own water (I care not how, but best for his beloved for ease in working) there is all the stuff and preparation, a very contested strife; here is the fire, this belly is full of blood of a strange nature; it is earthly and yet watery, airy and very fiery; it is a bath, it is a dunghill, and it is ashes also, and yet these are not common ones, but heavenly and Philosophical, as it becomes Philosophers to deal with nothing but heavenly matters or things; search then this rare kind of heat, for there is all the cunning; this is the key of all; this makes the seeds, and brings them forth; search wisely and where it is, in the midst of heaven and earth; for it is in the midst of both these places, and yet but one indeed; you may think I cross myself and know not what I say, but compare and look about, and you shall. find nothing prosper but in his own place.

12.   Let the dew of his starry blood beat about the womb, and the seed shall joy and prosper, yet so much the better and so near also, if that blood be whole and sound, and standing of all his parts;

wherefore no marvel though the world misses this happy stone, when they think to make it above the ground; I say they must either climb up to heaven, or go down deep within the earth; for there and nowhere else is this kindly heat.

13.   Wights are heat with blood, and plants with earth, but Minerals with a heavenly breath; to be short, because men are too heavy to mount up to heaven, you must go down to the midst of the earth, and put the seed in the sine again, that he says take that influence of heaven equally round about his again.

14.   Muse and conjecture well upon my words, you that are fit and skilled in Nature; for this is a very natural heat, and yet here all. the world is blinded. Nay indeed if a man could read little and think much upon the ways of nature, he might easily hit this Art, and before that never.

What now remains? We have all the way to mar and spoil the Gold, and that was all the doubt; I answer, for if he be once down so kindly, he will rise again sure, or else all nature will fail and lose her custom; and if he rise, he shall rise ever in virtue tenfold increased; I mean if it be not embossed as the seeds of Wights and plants are, and as the seeds of Gold was by that base way abovesaid, with the ground that corrupted it. So, if a poisoned plant or Wight be rotted in a glass, she will rise again a most venomous beast, and perhaps a Cockatrice, for that is the offspring; corrupt in like sort a good plant, and it will prove a worm or such like, with much increased virtue; what is the reason? Because the same temper and measure of qualities, still rises in power as the body Is refined, and the gross stuff that hinders the working, stripped off and removed.

15. Wherefore Gold is now temperate; loosen and refine him often over by corruption, that is, strip off the lets of the body, and all the qualities shall be raised equally, and shall work mightily, devour, and draw things to their own nature, more than anything else; because they be not only free, and in their clean and naked nature, but also seated upon a most subtle and tough body, able to pierce, divide, and subdue all things. Again, both metals and stones, the more heat they have (as in hot countries) the finer and better; and therefore, the oftener they be brought back to their first matter, and baked with temperate heat, the more they increase in goodness.

16. And if he be brought to such a temperate fineness, that Is, to such a heavenly nature, then he keeps no longer the nature of metal in respect of any quality save the lastingness of the body, nor of any other gross meat nor medicine, and therefore he cannot be an enemy to our nature, nor yet any ordinary digestion in our body, but straight way flies out, as I said before, and by extraordinary means and passages as well as nature herself, and so joined with our first moisture, and doth all other good deeds belonging to this Rosie Crucian *Infallible Axiomata* of long life, health, youth, riches, wisdom and virtue in such sort and better then I have shown you of a fifth nature in that book abovesaid; and so Appollonius, Philostratus and Erastus, and all other slanderous mouths may now begin again; for there is not a word spoken to any purpose, because all runs upon a false and unknown ground; a wise man would first have known the nature of the thing he speaks of, if he mean not to move laughter to them that hear him and know the matter.

17. But indeed Van Helmont, Glauber and Behemon the Cobler, and other railers, are safe enough, because these things are so hid and unknown to the world, that no man, but one of this our household can

espy them or control them; therefore I took in hand this hard and dangerous labor, which all other of our ancestors to this day have refused, both that they might be ashamed of their wrongful slanders, and the wise and well-disposed see and take profit by the truth of so great a blessing freely bestowed upon them. If they find it, let them thank God, and use it; no doubt they will do good unto good men. If I have slipped in words, or abounded in truth of matter, or failed hitherto, mark well the subsequent discourse; although you think I speak strangely, yet assuredly you shall find something that was never revealed to any, but of our laudable order; if in this or that Chapter you find anything amiss, think how common it is among men, especially of my age; I may be excused; and weigh the good and bad together, or else Homer himself an old man in his time, when he skips now and then could never escape it, and yet he was in an easy matter (a man may find I think howsoever) and he had Orpheus and Migaeus, I think, before him; but you see the hardness of this shift, although my pattern you do not see, because it is not to my knowledge in the world to be seen, but what care I; these men whom I regard, will take all things in good part, and then the rest I passed by long since unregarded; now let us sit down and rest a while, having perused the way to happiness, knowledge of all things, past, present and to come, Long life, health, youth, blessedness, wisdom and virtue; how to alter, cure, change, and mend the state of the body in young or old; and showed you the golden treasures of Nature, and the Fountain of Physick and Medicines; and this being all possible to be obtained, we shall. next lead you the way to prepare the Medicines which are experienced to be safe and effectual for all bodies, and you shall find their wonderful, incredible, extraordinary virtues, if you practice and use them as you are taught in the fifth Book; but you must remember to know the name of your patient, and the number of his name, Genius and Planet, and choose a fit time as you are taught in the second Book;

then prepare the Medicines as follows in the fifth Book; and now having guided all men to happiness, knowledge of all things past, present, and to come, long life, health, youth, blessedness, wisdom and virtue; and to alter, cure, change, and mend all diseases in young or old, I have proved these mysterious truths practical, and therefore next we will teach you the receipts, their virtues and use in the fifth Book. The Theory being sufficiently cleared from all objections, and the mysteries of nature made plain and easy, both in the structure of man's body, mind , soul, and spirit, of the nature of Stones, Herbs, and Plants, Minerals and Metals; then I having proved the power of nature, and the temper and order of happiness what it is, and how all may obtain it, vim, knowledge of the time when to give Physick, when the party will recover.

And thus, having passed the Theory and Practique part of Art and Nature, I shall proceed to the practique part of Physick.

And first you must observe the nature of your patients, their Ages, what number Governs each name, and what Genius attends that name; what Physick is proper for that person, when it is good to give it; this you will find in the second Book. If the number be in the Lawrel, it is good, if the number be in the Serpent, it is evil. Again the number of your question, name, Planet, and the day of the week must be added together, and divided by thirty, and what remains you shall find in that Figure; and if it be in the Lawrel, your question or what you desire shall be obtained, and your patient shall be cured; if it be long life, it is good, for you shall live long; and if the number be in the Serpent, it is evil, and the patient will die. And thus, may you do of any other question whereof you would be resolved; you must note the numbers in the Figure exceed not thirty, as you are taught in the second Book in the Rules of the Holy Guide. The young man that sits upon the

Mountain of Diamonds, is the servant and child of the holy Guide; he receives his knowledge from Mercury in Virgo, and his completeness of body from Caput Draconis in Gemini; Saturn and Venus in Libra direct him to the light of Nature; Fortuna Major & populous Figures of Geomancy give him health, and they receive it from the Sun and Moon; the Angel defends him from the Dragon, and the spiteful Dragon bites his tail in Sagittarius in anger, because he cannot destroy the youth; Jupiter in Capricorn with two Ideas of Geomancy conspire against him; but he receive Medicines and treasures from the Sun, and Jewels from the Moon, and gives them to Mars in Cancer, and Jupiter in Capricorn, who rewards him evil for his good will; the numbers in the Lawrel are heavenly and defended by an Angel; they grant you your requests; and the numbers in the fold of the Serpent destroy all your hopes, being earthly and evil. And now the Medicines follow; practice them to the Glory of God, and help your diseased neighbor. And so, we end our fourth look.

BOOK V.

The Holy Guide
LEADING THE WAY TO THE WONDER OF THE WORLD

A complete Physician teaching the knowledge of all things, Past, Present and yet to Come, viz, of Pleasure, Long Life, Health, Youth, Blessedness, Wisdom and Virtue, and to Cure, Change and Remedy all Diseases in both Young and Old.

To the Learned
Jeremiah Mount, Esq;
Celestial and Terrestrial Bliss and
happiness be wished.

Sir,

Your own worth and their attendants have in ways of Civility, to whom I hold my self-obliged for your Favor to me, forced this public Action, which perhaps you may think strange, that a Person so wholly a stranger as I, should tender you such a piece as this: Yet will, I doubt not, acquit me of rudeness and incivility in so doing, when you consider the present discourse, as there is no humor at all in it, so I hope there is less hazard of Censure: For here's no Lavish mirth, no Satirical sharpness, no writing or distorting the Genuine Frame and Composure of mine own mind, to set out the deformity of Another's; no Rapture, no Poetry, no Enthusiasm, no, no more then there is in Euclid's *Elements*, or Hippocrates his *Aphorisms*, but though I have been so bold as to recite what there is not in this Book: yet I had rather leave it to your wisdom to judge what there is, then be put upon so much modesty myself as to speak any thing that may seem to give it any pre-excellency above what is already extant in the world about Philosophy and Physick:

Only I may say thus much, that I did on purpose abstain from reading any Treatises concerning this Subject, that I might the more undisturbedly write the easy emanations of my own Mind, and experienced Medicines; and not be carried off from what I knew to be true, which should naturally fall from myself, by prepossessing my thoughts by the inventions of others: I have writ therefore after no Copy but the Eternal Characters of the mind, and the safe, easy and effectual Medicines for all diseases in the known Phenomena of Nature. And all men Consulting with these that endeavor to write sense in these Matters, though it may be not done alike by all men, it could not happen but I should touch upon the same heads that others have, that have wrote before me, who though they merit very high commendations for their learned achievements; yet I hope my endeavors have been such, that though they may not be Corrivals or Partners in their praise and credit, yet I do not distrust but they may do their share towards that public good, under your protection and patronage I aim at.

For that which did embolden me to publish this present Treatise; and dedicate it to you, was not as I said before, because I flattered myself in a Conceit, that it was better or more plausible, then what is already ii the hands of men: but that it was of a different sort, and has its peculiar serviceableness and advantages apart and distinct from others, whose proper preeminence's it may aloof off admire, but dare not in any wise compare with. So that there is no Tautology committed in recommending what I have written to the public view, nor any lessening the Labors of others by thus offering the fruit of mine own, for consideration there are such several complexions and tempers of men in the world, I do not distrust but that as what Dr. Culpepper and others have done, has been very acceptable and profitable to many, so this of mine may be useful to some or other, and so seem not to have been writ in vain. Such as it is, I shall leave it here under your Patronage: and submit it to your judgement, if you shall think it worth the while to take cognizance of it, whether to peruse and consider the truth of it, (which by Reason of your good accomplishments in these, as well as in other parts of Learning, you are well able to do) or to lay it by for those that will: as being unwilling by any importune solicitation to trespass upon your Leisure, or divert your thoughts from matters of more Concernment, to consider of such things as these, desiring mean while your Favor so far as to give me leave to honor you, and (though I have not hitherto had the honor to be well known to you) to subscribe myself,

June 11, 1662.

    Sir,

        Your most humble Servant,

            John Heydon.

Book V. Chapter I.

Of Projection and preparing Rosie Crucian Medicines.

1. Of the Original of Gold. 2. Of Sperm. 3. Of the first matter of Metals. 4. Of the difference of Gold. 5. Of the difference of Climes. 6. What Salt, Sulphur, and Mercury, are. 7. Of the virtue of Sulphur of Metals. 8. Of the Nature of Mercury. 9. Of Salt. 10. Of Gold. 11. Of Silver. 12. Of the Preparation of Gold. 13. Of Aurum potable, and Oil of Gold. 14. How to make them. 15. The second process. 16. Etc. 17. The third process. 18. The true oil of Gold. 19. The Child of Gold. 20. The Sun of Gold. 21. The Moon of Gold. 22. The Star of Gold. 23 The Rainbow. 24. How to make Aurum Fluminans.

1. I shall now endeavor to show whence Gold had its original, and what the matter thereof is. As Nature (says Sendivogius) is in the will of God, and God created her: so, nature made for herself a seed, (i.) her will in the elements. Now she indeed is one, yet she brings forth divers things: but she operates nothing without Sperm: whatsoever the Sperm will, nature operates; for she is as it were the instrument of any artificers. The Sperm therefore of everything is better, and more profitable then nature herself: for thou shalt from nature without a Sperm, do as much as a goldsmith without fire, or a husbandman without grain or seed. Now the Sperm of any thing is the Elixir, the balsam of sulphur, and the same as *humidum radicale* is in metals: but to proceed to what concerns our purpose, Four elements generates a Sperm, by the will of God, and imagination of nature: For as the sperm of a man has its center, or the vessel of its seed in the kidneys: so the four elements by their indeficient notion (every one according to its quality) cast forth a Sperm into the center of the earth, where it is

digested, and by motion is sent abroad. Now the center of the earth is a certain empty place, where nothing can rest: and the four elements send forth their qualities into the circumference of the center. As a male sends forth his seed into the womb of the female, which after it has received a due portion casts out the rest; so, it happens in the center of the earth, that the magnetic power of a part of any place attracts something convenient to itself for the bringing forth of something, and the rest is cast forth into stones and other excrements. For everything has its original from this fountain, and there is nothing in the world produced but by this fountain: as for example, set upon an even table a vessel of water, which may be placed in the middle thereof, and round about it set divers things, and divers colors, also salt, etc. everything by itself: then pour the water into the middle, and you shall see the water to run every way, and when any stream touches the red Color, it will be made red by it; if the salt, it will contract the taste of salt from it, and so of the rest: Now the water doth not change the places, but the diversity of places changes the rater. In like manner the seed or Sperm being cast forth by the four elements from the center of the earth unto the superfluities thereof, passes through various places, and according to the nature of the place is anything produced: if it come to a pure place of earth and water, a pure thing is made.

2. The Seed and Sperm of all things is but one, and yet it generates divers things, as it appears by the former example. The Sperm whilst it is in the center, is indifferent to all forms; but when it is come into any determinate place, it changes no more its form. The Sperm whilst it is in the center, can as easily produce a tree, as a metal, and an herb as a stone, and one more precious than another according to the purity of the place. Now this Sperm is produced of elements thus. These four elements are never quiet, but by reason of their contrariety mutually

act one upon another, and every one of its self sends forth its own subtility, and they agree in the center. Now in this center is the Archaeus, the servant of nature, which mixing those Sperms together sends them abroad, and by distillation sublimes them by the heat of a continual motion unto the superficies of the earth: For the earth is porous, and this vapor (or wind, as the Philosophers call it) is by distilling through the pores of the earth resolved into water, of which all things are produced. Let therefore as I said before, all sons of Art know that the Sperm of metals is not different from the Sperm of all things being, viz, a humid vapor. Therefore, in vain do Artists endeavor the reduction of metals into their first matter, which is only a vapor. Now says Bernard Trevisan, when Philosophers speak of a first matter, they did not mean this vapor, but the second matter which is an unctuous water, which to us is the first, because we never find the former. Now the specification of this vapor into distinct metals is thus. This vapor passes in its distillation through the earth, through places either cold, or hot; if through hot, and pure, where the fatness of sulphur sticks to the sides thereof, then that vapor (which Philosophers call the Mercury of Philosophers) mixes and joins itself unto that fatness, which afterwards it sublimes with itself, and then it becomes, leaving the name of a vapor, unctuosity, which afterwards coming by sublimation into other places, which the antecedent vapor did purge, where the earth is subtle, pure, and humid, fills the pores thereof, and is joined to it, and so it becomes Gold: and where it is hot, and something impure, silver. But if that fatness come to impure places, which are cold, it is made lead: and if that place be pure and mixed with sulphur, it becomes copper: for by how much the purer and warmer the place is, so much the more excellent does it make the metals.

3. Now this matter of metals is a humid, viscous, incombustible, subtle substance, incorporated with an earthly subtilty, being equally and strongly mixed *per minima* in the caverns of the earth. But as in many things there is a twofold unctuosity (whereof one is as it were internal, retained in the center of the thing, lest it should be destroyed by fire, which cannot be without the destruction of the substance itself wherein it is: the other as it were external, feculent and combustible) so in all metals except Gold, there is a twofold unctuosity: the one which is external, sulphureous, and inflammable, which is joined to it by accident, and doth not belong to the total union with the terrestrial parts of the thing: the other is internal, and very subtle, incombustible, because it is of the substantial composition of Argent Vive, and therefore cannot be destroyed by fire, unless with the destruction of the whole substance, whence it appears what the cause is that metals are more or less durable in the fire: For those which abound with that internal unctuosity, are less consumed, as it appears in silver, and especially in Gold. Hence Rosarius says, the Philosophers could never by any means find out anything that could endure the fire, but that unctuous humidity only which is perfect, and incombustible. Geber also asserts the same, when he says that imperfect bodies have superfluous humidifies, and sulphureous, generating a combustible blackness in them, and corrupting them; they have also an impure, feculent & combustible terrestriety, so gross as that it hinders ingression, and fusion: but a perfect metal, as Gold, has neither this sulphureous or terrestrial impurity; I mean when it is fully maturated and melted; for whilst it is in concoction, it has both joined to it, as you may see in the golden Ore; but when they do not adhere to it so, but that it may be purified from them, which other metals cannot, but are both destroyed together if you attempt to separate the one from another:

Besides, Gold has so little of these corruptible principles mixed with it, that the inward sulphur or metalline spirit doth sometimes and in some places overcome them of itself, as we may see in the Gold which is found very pure sometimes in the superficies of the earth, and in the sea sands, and is many tines as pure as any refined Gold.

Now this Gold which is found in sands, and rivers, is not generated there, as says Gregorius Agricola in his third book *de re metallica*, but is washed down from the mountains with fountains that run from thence. There is also a flaming Gold found (as Paracelsus says) in the tops of mountains, which is indeed separated of itself from all impurities, and is as pure as any refined Gold whatsoever. So that you see, that Gold although it had an extrinsical sulphur and earth mixed with it, yet it is sometimes separated from it of itself, viz., by that fiery spirit that is in it. Now this pure Gold (as Sendivogius says) nature would have perfected into an elixir; but was hindered by the crude air, which crude air is indeed nothing else but that extrinsic sulphur which it meets with and is joined to it in the earth, and which fills with its violence the pores thereof, and hinders the activity of the Spirit thereof; and this is that prison which the Sulphur (as the aforesaid author says) is locked up in, so that it cannot act upon its body, viz., Mercury, and concoct it into the seed of Gold, as otherwise it would do: and this is that dark body (as says Penotus) that is interposed betwixt the philosophical Sun and Moon, and keeps off the influences of the one from the other. Now if any skillful Philosopher could wittily separate this adventitious impurity from Gold whilst it is yet living, he would set sulphur at liberty, and for this his service he would be gratified with three kingdoms, viz. Vegetable, Animal, and Mineral; I mean he could remove that great obstruction which hinders Gold from being digested into the Elixir, For, as says Sendivogius, the Elixir or Tincture of Philosophers, is nothing else but Gold digested Into the highest degree: for the Gold of the vulgar is as an herb without seed; but when Gold (I) living Gold (for common Gold never can by reason that the Spirits are bound up, and indeed as good as dead and not possibly to be reduced to that activity which is required for the producing of the sperm of Gold) is ripened, it gives a seed, which multiplies even *ad infinitum*. Now the reason of this barrenness of Gold that it produces not a seed, is the aforesaid crude air, viz, impurities: You may see this illustrated by this example.

5.   We see that Orange-trees in Polonia do grow like other trees, also in Italy, and elsewhere, where their native soil is, and yield fruit, because they have sufficient heat; but in these colder countries they are barren and never yield any fruit, because they are oppressed with cold: but if at any time nature be wittily and sweetly helped, then Art can perfect what nature could not. After the same manner it is in metals; for Gold would yield fruit, and seed in which it might multiply itself, if it were helped by the industry of the skillful artist, who knew how to promote nature, (i.) to separate these sulphureous and earthly impurities from Gold. For there is a sufficient heat in living Gold, if it were stirred up by extrinsical heat, to digest it into a seed by extrinsical heat I do not mean the heat of the celestial Sun, but that heat which is in the earth and stirs up seed (i.) the living spirit that is in all subterranean sperms to multiply, and indeed makes Gold become Gold. low this is a heat of putrefaction occasioned by acid spirits fermenting in the earth, as you may see by this example related by Albertus Magnus, but to which the reason was given by Sendivogius. There was, says the former author, certain grains of Gold found betwixt the teeth of a dead man in the grave: wherefore he conceived there was a power in the body of man to make and fix Gold: but the reason is far otherwise, as says the latter author: for says he, argent vive was by some physician conveyed into the body of this man when

he was alive, either by unction or by turbith, or some such way, as the custom was; and it is the nature of Mercury to ascend to the mouth of the patient, and through the excoriation of the mouth to be avoided with the flegme. Now then if in such a cure the sick man died, that Mercury not having passage out, remained betwixt the teeth in the mouth, and that carcass became the natural vessel of Mercury, and so for a long time being shut up was congealed by its proper sulphur into Gold by the natural heat of putrefaction, being purified by the corrosive flegme of the carcass; but if the mineral Mercury had not been brought in thither, Gold had never been produced there: And this is a most true example that as Mercury is by the proper sulphur that is in itself, being stirred up and helped by an extrinsical heat, coagulated into Gold, unless it be hindered by any accident, or have not a requisite extrinsical heat, or a convenient place, so also that nature doth in the bowels of the earth produce of Mercury only Gold and silver, and other metals according to the disposition of the place, and matrix; which assertion is further cleared by the rule of reduction; for if it be true that all things consist of that which they may be reduced into, then Gold consists of Mercury, because (as most grant, & Avenrois affirms, and many at this day profess they can do) and say be reduced into it. There is a way by which the tincture of Gold which is the soul thereof, and fixing it, say be so fully extracted that the remaining substance will be sublimed like arsenic, and say be as easily reduced into Mercury as Sublimate. If so, and if all Mercury may be reduced into a transparent water, as it may (according to the process set down before, and I know another better end easier way to turn a pound of Mercury of itself into a clear water in half an hour, which is one of the greatest secrets I know, or care to know, together with what may be produced thence and shall crave leave to be silent in) why may not that water in some sense, if it be well rectified, be called a kind of living Gold out of which you may perhaps make a medicine; and a menstruum unfit for the vulgar to know? It appears now from what is premised, that the immediate matter of Gold is probably Mercury, and not certain salts, and I know not what as many dream of, and that the extrinsical. heat is from within the earth, and not the heat of the sun, as some imagine (because in the hottest countries there is all, or almost all Gold generated) who if they considered that in cold countries also are, and as in Scotland were, Gold mines in King James his time, would be of another mind then to think that the celestial sun could penetrate so as to heat the earth so deep as most Gold lies.

6. Now having in some measure discovered what the intrinsical, and extrinsical heat, and the matter of Gold is, I shall next endeavor to explain what those three principles are, viz. Salt, Sulphur, and Mercury, of which argent vive, and Gold consist: Know therefore that after Nature had received from the most High God the privilege of all things upon the Monarchy of this world, she began to distribute places and provinces to everything, according to its dignity; and in the first place did constitute the four Elements to be the Princes of the World, and that the will of the most High (in whose will Nature is placed) might be fulfilled, ordained that they should act upon one another incessantly. The fire therefore began to act upon the Air, and produced Sulphur? The Air also began to act upon the Water, and produced Mercury: The Water also began to act upon the Earth, and produced Salt. Now the Earth not having whereon to act, produced nothing, but became the subject of what was produced. So, then there were produced three principles; but our ancient Philosophers not so strictly considering the matter, described only two acts of the Elements, and so named but two Principles, viz., Sulphur and Mercury; or else they were willing to be silent in the other, speaking only to the sons of Art.

7. The Sulphur therefore of Philosophers (which indeed is the Sulphur of Metals, and of all things) Is not, as many think, that common Combustible Sulphur which Is sold In shops, but Is another thing far differing from that, and is incombustible, not burning, nor heating, but preserving, and restoring all things which it is in, and it is the *Calidum Innatum* of everything, the fire of Nature, the created Light, and of the nature of the Sun, and Is called the Sun; so that whatsoever in any thing is fiery and airy, is Sulphur, not that anything is wholly sulphureous, but what in it is most thin and subtle, having the essence of the natural Fire, and the nature of the created Light, which indeed is that Sulphur which wise Philosophers have in all ages with great diligence endeavored to extract, and with its proper Mercury to fix, and so to perfect the great Magistery of Nature, Now of all things in the world there is nothing has more of this Sulphur in it then Gold and Silver, but especially Gold, insomuch that oftentimes it is called sulphur, (i.e.) because Sulphur is the most predominant and excellent principle in it, and being in it more then all things besides.

8. Mercury is not here taken for common argent vive; but it is the *humidum radicale* of everything, that pure aqueous, unctuous, and viscous humidity of the matter, and it is of the nature of the Moon, and it is called the Moon, and that for this reason, viz., because it is humid, as also because it is capable of receiving the influence and light of the Sun, viz., Sulphur.

9. Salt is that fixt permanent Earth which is in the center of everything, that is incorruptible, and unalterable, and it is the supporter and nurse of the *humidum radicale*, with which it Is strongly mixt. Now this Salt has in it a seed, viz. its Calidum Innatum, which Is Sulphur, and its *humidum radicale*, which is Mercury; and yet these three are not distinct, or to be separated, but are one homogeneal

thing, having upon a different account divers names; for in respect of its heat and fiery substance it is called Sulphur, in respect of it humidity, It is called Mercury, and in respect of Its terrestrial siccity it is called Salt, all which are in Gold perfectly united, depurated and fixed.

10. Gold therefore is most noble and solid of all Metals, of a yellow Color, compacted of principles digested to the utmost height, and therefore fixed.

11. Silver is in the next place of dignity to Gold, and differs from it in digestion chiefly; I said chiefly, because there is some small impurity besides adhering to silver.

12. Now having given some small account of the original matter, first, and second, and manner of the growth of Gold, I shall in the next place set down some Curiosities therein, and preparation thereof. The preparations are chiefly three, viz. aurum potable, which Is the mixtion thereof with other Liquors: Oil of Gold, which is Gold liquid by itself without the mixture of any other Liquor: and the tincture, which is the extraction of the Color thereof.

The Oil of Gold.

13. Dissolve pure fine Gold in aqua regis according to Art (the aqua regis being made of a pound of aqua fortis, and four ounces of Salt Armoniack distilled together by Retort in sand) which clear solution put into a large glass of a wide neck, and upon it pour drop by drop Oil of Tartar made per deliquim, until the aqua regis, which before was yellow, become clear and white; for that is a sign that all calx of Gold is

settled to the bottom; then let It stand all night, and in the morning pour off the clear Liquor, and wash the calx four or five times with common spring water, being warmed, and dry it with a most gentle heat.

14. Note, and that well, that if the heat be too great, the calx takes fire presently like Gunpowder and flies away to thy danger and lose; therefore, it is best to dry it in the sun, or on a stone, stirring it diligently with a wooden spattle. To this calx add half a part of the powder of sulphur; six them together, and in an open crucible let the sulphur burn away in the fire, putting a gentle fire to it at the first, and in the end a most strong fire for the space of an hour, that the calx may in some manner be reverberated, and become most subtle, which keep in a viol close stopped for your use.

15. Then sake a Spirit of Urine after this manner, viz., Take the Urine of a healthy man drinking Wine moderately, put it into a gourd, which you must stop close, and set in horse-dung for the space of forty days, then distill it by a Limbeck in sand into a large Receiver, until all the humidity be distilled off. Rectify this Spirit by cohobation three times, that the Spirit only may rise. Then distill it in sand by a glass with a long neck, having a large receiver annexed, and closed very well to it, and the Spirit will be elevated into the top of the vessel like christal, without any aqueous humidity accompanying of it. Let this distillation be continued, until all the Spirits be risen. These christals must be dissolved in distilled rainwater, and be distilled as before; this must be done six times, and every time you must take fresh rainwater distilled. Then put these christals into a glass bolthead, which close Hermetically, and set in the moderate heat of a Balneum for the space of fifteen days, that they may be reduced into a most clear Liquor. To this Liquor add an equal weight of Spirit of Wine, very well rectified,

and let them be digested in Balneo the space of twelve days, in which time they will be united.

16.   Then take the calx of Gold abovesaid, and pour upon it of these united Spirits as such as will cover them three fingers breadth, and digest them in a gentle heat, until the Liquor be tinged as red as blood; decant off the tincture, and put on more of the aforesaid Spirits, and do as before till all the tincture be extracted; then put all the tincted Spirits together, and digest them ten or twelve days, after which time abstract the Spirit with a gentle heat, and cohobate it once; and then the calx will remain in the bottom like an Oil as red as blood, and of a pleasant odor, and which will be dissolved in any Liquor. Whereof this Oil may be the *succedaneum* of true Gold. If you distil the same solution by Retort in sand, there will come over, after the first part of the menstruum, the tincture with the other part thereof, as red as blood, the earth which is left in the bottom of the vessel being black, dry, spongy and light. The menstruum must be vapored away, and the Oil of Gold will remain by itself, which must be kept as a great treasure: and this is Dr. Anthony's aurum potable.

Four or eight grains of this Oil taken in what manner soever, wonderfully refreshes the Spirits, and works several ways, especially by sweat, and cures all Diseases in young and old.

The True Oil of Gold.

18.   Take an ounce of Leaf-Gold, dissolve it in four ounces of the rectified water of Mercury, expressed page 75[33]. digest them in horse-

---

[33] These page numbers refer to the text of 1662. -pnw

dung the space of two months, then evaporate the Mercurial water, and at the bottom you shall have the true Oil of Gold, which is radically dissolved, another process hereof you say see page 71.

The Child of Gold.

19. Dissolve pure Gold in aqua regia, precipitate it with the oil of sand into a yellow powder, which you must dulcify with warm water, and then dry it; (this will not be fired as *aurum fulminans*) this powder is twice as heavy as the Gold that was put in, the cause of which is the salt of the flints precipitating itself with the Gold. Put this yellow powder into a crucible, and make it glow a little, and it will be turned into the highest and fairest purple that you ever saw, but if it stands longer, it will be brown. Then pour upon it the strongest spirit of salt (for it will dissolve it better than any aqua regis) on which dissolution pour on the best rectified Spirit of Wine, and digest them together, and by a long digestion, some part of the Gold will fall to the bottom like a white snow, and may with Borax, Tartar and salt nitre be melted into a white metal as heavy as Gold, and afterwards with Antimony may recover its yellow Color again; then evaporate the spirit of Salt, and of Wine, and the Gold Tincture remains at the bottom, and is of great virtue.

The Sun of Gold.

20. Take of the aforesaid yellow Calx of Gold, precipitated with Oil of sand, one part, and three or four parts of the Liquor of sand, or of crystal; six them well together, and put them into a crucible in a gentle heat first, that the moisture of the Oil may vapor away (which It will not do easily, because the dryness of the sand retains the moisture thereof, so that it flies away like molten Allum, or Borax) when no more will vapor away, increase your fire, till the crucible be red hot, and the mixture cease bubbling; then put it into a wind furnace, and cover it that no ashes fall into it, and make a strong fire about it for the space of an hour, and the mixture will be turned into a transparent Rubie. Then take it out, and beat it, and extract the tincture with spirit of Wine, which will become like thin blood, and that which remains undissolved, may be melted into a white metal as the former.

The Moon of Gold.

21. Rang plates of Gold over the fume of argent vive, and they will become white, friable, and fluxable as Wax. This is called Magnesia of Gold, as says Paracelsus, in finding out of which (says he) Philosophers, as Thomas Aquinas, and Rupescissa, with their followers, took a great deal of pains, but in vain; and it is a memorable secret, and indeed very singular for the melting of metals, that are not easily fluxed. Now then Gold being thus prepared, and melted together with the Mercury, is become a brittle substance, which must be powdered, and out of it a tincture may be drawn for the transmuting of metals.

The Star of Gold.

22. Take half an ounce of pure Gold, dissolve it in aqua regis, precipitate it with Oil of flints, dulcify the Calx with warm water, and dry it, and so it is prepared for your work. Then take regulus martis powdered, and six it with three parts of salt nitre, both which put into a Crucible, and make them glow gently at first, then give a strong melting fire, and then this mixture will become to be of a purple Color, which then take out, and beat to powder, and add to three parts of this one part of the calx of Gold prepared as before; put them into a wind furnace in a strong crucible, and make them melt as a metal, so will the nitrum antimoniatum in the melting take the calx of Gold to itself, and dissolve it, and the mixture will become to be of an Amethyst Color. Let this stand flowing in the fire till the whole mass be as transparent as a Ruby, which you may try by taking a little out and cooling of it. If the mixture does no flow well, cast in some more salt of nitre. When it is completely done, cast it forth, being flowing, into a brazen Mortar, and it will be like to an oriental Ruby; then powder it before it be cold, then put it into a Viol, and with the spirit of Wine extract the tincture.

This is one of the best preparations of Gold, and of most excellent use in Medicine.

The Rainbow.

23. First make a furnace fit for the purpose, which must be close at the top, and have a pipe, to which a. recipient with a flat bottom must be fitted: When this furnace is thus fitted; put in three or four grains,

not above an ounce, of album fulminens, which as soon as the furnace is hot flies away into the recipient through the pipe like a purple colored fuse, and is turned Into a purple powder; then put in three or four grains more, and do as before, till you have enough flowers of Gold (that which flies not away, but remains at the bottom, may with Borax be melted into good Gold) then take them out, and pour upon them rectified spirit of Wine tartarized, and digest them in ashes till the spirit be colored blood-red, which you must then evaporate, and at the bottom will be a blood-red tincture of no small virtue.

Aurum Fulminans.

Take the purest Gold you can get, pour on it four times as much aqua regia, stop your glass with a paper, and set it in warm ashes, so will the aqua regia in an hour or two take up the Gold, and become a yellow water, if it be strong enough: (be sure that your Gold has no Copper in it, for then your labor will be lost) because the Copper will be precipitated with the Gold, and hinder the firing thereof) then pour on this yellow water drop by drop, pure Oil of Tartar made per deliquium, so will the Gold be precipitated into a dark yellow powder, and the water be clear. Note that you pour not on more Oil of Tartar then is sufficient for the precipitation; otherwise it will dissolve part of precipitated Gold to thy prejudice. Pour off the clear Liquor by inclination, and dulcifie the calx with distilled rainwater warmed. Then set this calx in the Sun, or some warm place, to dry, but take great heed, and especial care, that you set It not in a place too hot; for it will presently take fire and fly away like thunder, not without great danger to the etanders by, if the quantity be great. This is the common way to make album fulminans, and it has considerable difficulties in the preparation. But the best way is to precipitate Gold dissolved in aqua regis by the spirit of Salt Armoniack or of Urine; for by this way the Gold is made purer then by the other, and giveth a far greater crack and sound. Note that the salt of the spirits which is precipitated with the Gold, must be washed away, and the Gold dulcified as before.

A few grains of this being fired give a crack and sound as great as a Musket when it is discharged, and will blow up anything more forcibly far then Gunpowder, and it is a powder that will quickly and easily be fired.

This is of use for Physick as it is in powder, but especially it is used in making the foregoing tincture.

Chapter II.

1. Of Acetus Philsophicum. 2. Of Aqua Mars Scorpio. 3. Of Aqua Mars Subtilltatis. 4. How Yllius Solis Celeetis is made. 5. Row Stella vitae is made. 6. How Filia Lunae is made. 7. How Ignie vitae is made. 8. Of Adjustrix vita. 9. Of Salus vitae. 10. Of Sanguis vitae. 11. Of Amicus vitae. 12. Of Succus vitae. 13. Of aqua Venue, Virgo. 14. Of aqua Mars Aries. 15. Of aqua 301, Cancer. 16. Of aqua Saturn, Libra. 17. Of Medulla vitae. 18. Of aqua Mars Luna. 19. Aqua Mars, Cancer. 20. Aqua Venus, Libra. 21. Aqua Venue, Scorpio. 22. Aqua Sol, Virgo. 23. Aqua Jupiter, Taurus, 24. Aqua Mars Cancer. 25. Aqua Mercury, Virgo. 26. Aqua Jupiter Luna. 27. Puella Sol. 28. Acquisjto Luna. 29. Aqua Luna, Scorpio. 30. Fortuna Major Sol. 31. Rubeus Sol. 32. Puer Sol. 33. Aqua Jupiter. 34. Sol Mars, Aries. 35. Of making spirits. 36. To make a

Vegetable to yield his spirit; and of the wonderful virtues of these waters.

Acetum Philosophicum.

1. Take Honey, Salt melted, of each a pound, of the strongest spirit of Vinegar two pounds; digest them for the space of a fortnight, or more, then distil them in ashes, cohobate the Liquor upon the feces three or four times, then rectify the spirit.

Note that they must be done in a large glass-gourd.

Aqua Martis Scorpio.

2. Take of the best rectified spirit of Wine, with which imbibe the strongest unslaked Lime, until they be made into a paste, then put them into a glass-gourd, and distil off the spirit in ashes: This spirit pour on more fresh Lime, and do as before; do this three or four times, and thou shalt have a very subtle spirit, able to dissolve most things, and to extract the virtue out of them.

Aqua Martis Subtiliatis.

3. Take oil of Olive, Honey, rectified spirit of Wine, of each a pint, distil them all together in ashes, then separate all the flega from the oil,

which will be distinguished by many colors, put all these colors into a Pelican, and add to them the third part of the Essence of Balm, and Sallendine, digest them for the space of a month. Then keep it for use.

This Liquor is so subtle that it penetrates everything.

Filius Soli Celestia is made thus.

4. Take of Cinnamon, Cloves, Nutmegs, Ginger, Zedoary, Galingal, Long-pepper, Citron-pill, Spikenard, Lignum-Aloes, Cububs, Cardasums, Calasus aromaticus, Germander, Ground-pine, Mace, white frankincense, Tormentil, Hermodactyls, aurum potable, the pith of Dwarf-elder, an ounce of each: Juniper Berries, Bay Berries, the seeds and flowers of Mother-wart, the seeds of Smallage, Fennel, Annise, the leaves of Sorrel, Sage, Felwort, Rosemary, Marjoram, Mints, Pennyroyal, Stechados, the flowers of Elder, Roses red, white, of the leaves of Scabious, Rue, the lesser Moonwort, Egrimony, Centory, Fumitary, Pimpernel, Sowhistle, Eyebright, Maidenhair, Endive, red Saunders, Aloes, of each two Ounces, pure Amber, the best Rhubarb, of each two drams, dried Figs, Raisons of the Sun, Dates stoned, sweet Almonds, Grains of the Pine, of each an ounce, of the best aqua vitae to the quantity of them all, of the best hard Sugar a pound, of white Honey half a pound, then add the root of Gentian, flowers of Rosemary, Pepperwort, the root of Briony, Sowbread, Wormwood, of each half an ounce. Now before these are distilled, quench Gold, being made red hot oftentimes in the foresaid water, put therein oriental Pearls beaten small an ounce, and then distill it after twenty-four hours infusion.

This is very Cordial water, good against fainting and infection.

Stella vitae, is made thus.

5. Take of the rind of Citrons dried, Oranges, Nutmeg, Cloves, Cinnamon, of each two ounces; the roots of Flower-deluce, Cyprus, Calamus Aromaticus, Zedoary, Galingal, Ginger, of each half a pound; of the tops of Lavender, Rosemary, of each two handfuls; the leaves of the Bay-tree, Marjoram, Balm, Mints, Sage, Thise, flower of Roses white, Damask, of each half a handful, aurum potable a dram, Rose-water four pints, the best white wine a gallon: Bruise what must be bruised, then infuse them all twenty-four hours, after which distil them.

This is the same virtue as the former.

Filia Lumae Celestis, is made thus.

6. Take of Cloves, Galingal, Cubebs, Mace, Cardasuss, Nutmegs, Ginger, of each a dram, the juice of Celendine half a pint, spirit of Wine a pint, White wine three pints: infuse all these twenty-four hours, and then distil off two pints by a Limbeck.

This water is very good against wind in the stomach and head.

Ignis vitae, is made thus.

7. Take a gallon of Gascoign wine, Ginger, Galingal, Cinnamon, lutsegs, Grains, Anniseeds, Fennel seeds, Carroway seeds of each a dram, aurum potable, an ounce, Sage, red Mints, red Roses, Thime, Pellitory, Rosemary, wild Thime, Casmomile, Lavender, of each a handful: Beat the Spices small, and bruise the Herbs letting them macerate twelve hours, stirring them now and then, distil them by a Limbic or copper still, with its refrigeratory, keep the first pint by itself, and the second by itself.

Note that the first pint will be the hotter, but the second the stronger of the ingredients.

This water is well known to comfort all the principal parts.

Adjutrix vitae.

8. Take of red Poppy-cakes (after the water has been distilled from them in a cold still) not over dried two pound, pour upon them of the water of red Poppy a gallon and half, Canary wine three pints; add to them of Coriander seeds bruised four ounces, of Dill seed bruised two ounces, of Cloves bruised half an ounce, of Nutmegs sliced an ounce, of Rosemary a handful, three Oranges cut in the middle, distil them In a hot still; to the water put the juice of six Oranges, and hang In it half an ounce of Nutmegs sliced, and as much Cinnamon bruised, two drama of Cloves, a handful of Rosemary cut small, sweet Fennel seeds bruised an ounce, of Raisons of the Sun stoned half a pound, being all put into a bag which may be hanged in the water (the vessel being close stopped) the space of a month, and then be taken out and cast

away, the Liquor thereof being first pressed out into the foresaid water, and of aurum potable a dram.

This water is of wonder virtue in Sursets and Plurises, composes the spirits, causes rest, helps digestion if two, or three, or four ounces thereof be drunk, and the patient compose himself to rest.

Salus vitae.

9. Distil green Hyssop in a cold still till you have a gallon and half of the Water, to this put four handfuls of dried Hyssop, a handful of Rue, as much of Rosemary, Horehound, Elecampane-root bruised, and of Horseradish-root bruised, of each four ounces, of Tobacco in the leaf three ounces, Aniseed bruised two ounces, two quarts of Canary wine, let them all stand in digestion two days, then distil them, and in the water that is distilled put half a pound of Raieons of the Sun stoned, of Licorice two ounces, sweet Fennel seeds bruised two ounces and a half, Ginger sliced an ounce and a half, and let them be infused in frigido the space of ten days, then take them out.

This water sweetened with Sugar-candy, and drunk to the quantity of three or four ounces twice in a day, is very good for those that are Ptieical, it strengthens the Lungs, attenuates thick flega, opens obstructions, and is very good to comfort the stomach.

Sanguis vitae.

10. Take of Wormwood bruised eight ounces, the shavings of Hartshorn two ounces, of Peach-flowers dried an ounce, aurum potable a dram, of Aloes bruised half an ounce, pour on these the water of Tansie, Rue, Peach-flowers, and of Wormwood, of each a pint and a half, let them being put into a glass vessel, be digested the space of three days, then distil them; cohobate this water three times.

This water is very excellent against the Worms; it may be given from half an ounce to three ounces, according to the age of the Patient.

Amicus Vitae.

11. Take of ros vitrioli (which is that water that is distilled from Vitriol in the calcining thereof) two quarts, in this put of Rue a handful, of Juniper berries bruised an ounce, of Bay berries bruised half an ounce, Peony berries bruised six dress, Casphire two dress, Rhubarb sliced an ounce, aurum potable two dress, digest these four days in a temperate balneo, then distil them in a glass vessel in ashes, and there will come over a water of no small virtue.

It cures Convulsions in Children especially, it helps also the Vertigo, the Hysterical passion, and Epilepsy, it is very excellent against all offensive vapors and wind that annoys the head and stomach.

It may be taken from two drams to two ounces.

Succus Vitae.

12.   Take of Wormwood, Broom blossoms, of each a like quantity, bruise them, and mix with them some Leaven, and let them stand in fermentation in a cold place the space of a week, then distil them in a cold still till they be very dry: take a gallon of this water, and half a gallon of the spirit of Urine, pour them upon two pounds of dried Broom blossoms, half a pound of Horse Radish roots dried, three ounces of the best Rhubarb sliced, two ounces of sweet Fennel seed bruised, and an ounce and a half of Nutmegs; let them digest a week being put into a glass vessel in a temperate balneo then press the Liquor hard from the feces, put this Liquor in the said vessel again, and to it put three ounces of sweet Fennel seeds bruised, Licorice sliced two ounces, digest them in a gentle heat the space of a week, then pour off from the feces, and of aurum potable, two drams, and keep it close stopped.

This water being drank from the quantity of an ounce to four ounces every morning, and at four o'clock in the afternoon, doth seldom fail in curing the dropsy; it strengthens also the Liver, is very good against gravel in the back, stone, cures the Scurvy, Gout, and such diseases as proceed from the weakness and obstructions of the Liver.

Aqua Venus Virgo.

13. Take of Aniseed three ounces, Cuminseed three dross, Cinnamon half an ounce, Mace, Cloves, Nutmeg, of each a dram, Galingal, three drama, Calamus Aromaticus dried, half an ounce, the dried rind of Oranges two ounces, Bay berries half an ounce, Aurum Potable an ounce.

Let all these being bruised, be macerated in six pints of Mallago wine 48 hours, then be distilled in balneo till all be dry.

This water being drank to the quantity of an ounce or two at a tine do ease the gripings of the belly and stomach, very much.

Aqua Mars Aries.

14.   Take of black cherries bruised with their kernels, a gallon, of the flowers of Lavender three handfuls, half an ounce of white Mustard seed bruised, six these together, then put some ferment to them and let them stand close covered the space of a week, then distil them in balneo till all be dry.

This water being drank to the quantity of an ounce or two or three, doth much relieve the weakness of the head, and helps the Vertigo thereof, as also strengthen the sinews and expel windiness out of the head and stomach.

Aqua Sol, Cancer.

15.   Take the root of the great Buxre, fresh, Swallow wort, fresh, Aurum Potable an ounce, The middle rind of the root of the Ash tree, of each two pounds; cut them small, and infuse them 24 hours, in the best White wine and Rue vinegar, of each five pints, then distil them in balneo till all be dry, put to the water as much of the Spirit of Sulphur per campanam, as will give it a pleasant acidity, and to every pint of

the water put a scruple and a half of Campbire cut small, and tied up in a bag, which may continually hang in the water.

This was a famous water in Germany against the plague, pestilence and Epidemical diseases; it causes sweat wonderfully if two or three ounces thereof be consumed and the patient compose himself to sweat.

Aqua Saturn, Libra.

16. Take of the best Spirit of Wine a gallon, Andromachus trecle, six ounces, Myrrth two ounces, the roots of Colts-foot, three Ounces, Sperma Ceti, Aurum Potable, Terra Sigillata, of each half an ounce, the root of swallow wort, an ounce, Dittany, Pimpernel, Valerian root, of each two drama, Casphire, a dram. Mix all these together in a glass vessel, and let them stand close stopped the space of eight days in the Sun.

Let the Patient drink of this a spoonful or two, and Compose himself to sweat.

Medulla Vitae.

17. Take three pints of Muscadine, and boil it in Sage, and Rue of each a handful till a pint be wasted, then strain it and set it over the fire again, put thereto a dram of long Pepper, Ginger and Nutmeg of each half an ounce being all bruised together: then boil them a little, and put thereto half an ounce of Andromachus trecle, and three drama of

Mithridate, and a quarter of a pint of the best Angelica water, an ounce of Aurum Potable.

This water (which, as says the Author, must be kept as your life, and above all earthly treasure) must be taken to the quantity of a spoonful or two, morning & evening; if you be already Infected, and sweat thereupon, if you be not infected, a spoonful is sufficient, half in the morning and half at night: all the plague time under God (says the Author) trust to this, for there was never man, woman, or child that failed of their expectation in taking 01 it. This is also of the same efficacy not only against the plague, but pox, measles, surfeits, & etc.

Aqua Mars, Luna.

18. Take of Andromachus Treacle, five ounces, the best Myrrh, two ounces and half, the best Saffron half an ounce, Camphire two drams, Aurum Potable an ounce. Mix them together, then pour upon them ten ounces of the best spirit of wine, and let them stand 24 hours in a warm place, then distil them in balneo with a gradual fire, cohobate the spirit three times.

This spirit causes sweat wonderfully, and resists all manner of infection.

It may be taken from a dram to an ounce in some appropriate Liquor.

Aqua Mars, Cancer.

19.   Take of the roots of Bistort, Gentian, Angelica, Tormentil, of each ten drama, Pimpernel ten drama, Bay berries, juniper berries, of each an ounce, Nutmeg, five drams, The shavings of Saffafras two ounces, Zedoary half a drain, Aurum Potable a dram, White Sanders three drama, the leaves of Rue, Wormwood, Scordium, of each half a handful, the flowers of Wall flower, Buglosse, of each a handful and half, intromachus Treachie, Mithridate of each six drams infuse them all in three pints of the best White wine vinegar the apace of eight days in frigido in glass vessels; then distil them In Balneo.

This Spirit is very good to prevent them that are free from infection, and those that are already infected, from the danger thereof, if two or three spoonsful thereof be taken once in a day, with sweating after, for those that are infected, but without sweating for others.

Aqua Veneris, Libra.

20.   Take of the middle rind of the root of Ash bruised, two pounds, Juniper berries bruised, three pounds, Aurum Potable a dram, Venice turpentine that is very pure, two pounds and a half. Put these into twelve pints of spring water in a glass vessel well closed, and. there let them putrefy in horse dung for the space of three months, then distil them in ashes, and there will come forth an oil and a water, separate the one from the other.

Ten or twelve drops of this oil being taken every morning in four or six spoonsful of the said water, dissolves the gravel and stone in the kidneys, most wonderfully.

Aqua Veneris, Scorpio.

21.   Take the juice of Radish, Lemons, of each a pound and a half: Waters of Betony, Tonsey, Saxifrage, and Vervin, of each a pint. Hydromel, and Maimfey, of each two pounds. In these Liquors mixed together, infuse for the space of four or five days in a gentle Balneo, Juniper berries ripe and newly gathered being bruised, three ounces: the seed of Grosel, Bardock, Radish, Saxifrage, Nettles, Onions, Anise, and Fennel, of each an ounce and a half, the four cold seeds, the seed of great Mallows, of each six drama, the Calx of Eggshells, Cinnamon, of each three drama, of Casphire two drama, let all be well strained and distilled in ashes, and afterwards an ounce of Aurum Potable.

Two ounces of this water taken every morning, doth wonderfully cleanse the Kidneys, provoke Urine, and expel the Stone, especially if you calcine the feces and extract the Salt thereof with the said Water.

Aqua Sol, Virgo.

22.   Take Plantain, Ribwort, Bone-wort, wild Angelica, Red-mints, Betony, Egrimony, Sanacle, Blewbottles, Whitebottles, Dandelion, Avens, Honeysuckle leaves; Bramble-buds, Hawthorn buds and leaves; Mugwort, Dasie roots, leaves and flowers; Wormwood, Southernwood, of each one handful: Boil all these in a bottle of White wine, and as much Spring water, till one half be wasted; and when it is thus boiled, strain it from the herbs, and put to it half a pound of honey, and let it boil a little after: then put it into bottles, and keep it for your use.

Mote that these herbs must be gathered in May only, but you may keep them dry, and make your water at any time.

This water is very famous in many Countries, and it has done such cures in curing outward and inward Wounds, Imposthumes, and Ulcers, that you would scarce believe it, if I should recite them to you: also, it is very good to heal a sore mouth.

The Patient must take three or four spoonful's thereof morning and evening, and in a short time he shall find ease, and indeed a cure, unless he be so far declined as nothing almost can recover his. If the round be outward, it must be washed therewith, and linen cloths wet in the same be applied thereto,

Aqua Jupiter, Taurus.

23. Take of Lavender flowers a gallon; pour upon them of the best spirit of wine three gallons: the vessel being close stopped, let them be macerated together in the Sun for the space of six days, then distil them in an Alembic with its refrigeratory, then take of Aurum Potable a dram; the flowers of Sage, Rosemary, Betony; of each a handful, Borage, Buglose, Lilly of the Valley, Cowslips, of each two handfuls; Let all the flowers be fresh and seasonably gathered, and macerated in a gallon of the best spirits of Wine, and mixed with the aforesaid spirit of Lavender, adding then the leaves of Balm, Motherwort, Orange tree newly gathered, the flowers of Stechados, Oranges, Bay berries, of each an ounce. After a convenient digestion let them be distilled again; then add the outward rinds of Citrons six drams, the seed of Peony husked, six drama, Cinnamon, Nutmegs, Mace, Cardamoms, Cubebs, of yellow

Sanders, of each half an ounce, Lignum Aloes one dram, the best Jujube, the kernels taken out, half a pound. Let them be digested for the space of six weeks, then strain & filter the Liquor, to which add of Aurum Potable an ounce, prepared Pearl, two drama, prepared Emerald a Scruple, Ambergris, Musk, Saffron, Red Roses, Sanders, of each an ounce, Yellow Sanders, Rinds of Citrons dried, of each a dram. Let all these spices be tied in a silken bag and hanged in the foresaid spirit,

Aqua Mars, Cancer.

24. Take the leaves of both sorts of Scurvie grass, being made very clean, of each six pounds: let these be bruised, and the juice pressed forth: to which add the juice of Brooklime, Water creases, of each half a pound, of the best White wine, eight pints, twelve whole Lemons cut, of the fresh roots of Briony four pounds, Horse Radish two pounds, of the bark of Winteran, half a pound, of Nutmegs four Ounces; Let them be macerated three days and distilled.

Three or four spoonsful of this water taken twice in a day, cures the Scurvy presently,

Aqua Mercury, Virgo.

25. Take of fresh Castoreum two ounces, flowers of Lavender fresh, half an ounce, Sage, Rosemary, of each two drains, Cinnamon three drama, Mace, Cloves of each a dram, the heat rectified Spirit of Wine,

three pints. Let them be digested in a Glass (two parts of three being empty) stopped close with a bladder and Cork two days in warm ashes; then distil the spirit in balneo, and keep it in a glass close stopped. If you would make it stronger, take a pint of this spirit, and an ounce of the powder of Castoreum; put them into a glass and digest them in a cold place for the space of ten days, and then strain out the spirit.

This spirit is very good against fits of the Mother, passions of the heart which arise from vapors, etc.

Aqua Jupiter, Luna.

26. Take of the leaves of the greater Salladine together with the roots thereof, three handfuls and a half, Rue, two handfuls, Scordium, four handfuls, Dittany of Crete, Carduus, of each a handful and half, root of Zedoary, Angelica, of each three drama, the outward rind of Citrons, Lemons, of each six drama, the flower of Wallgilly-flower, and ounce and half, Red Roses, the lesser Centory, of each two drama, Cinnamon, Cloves, of each three drama, Andromachus his Treacle, three ounces, Mithridate, an ounce and a half, Camphire, two scruples, Trochlaces of Vipers, two ounces, Mace, two drains, Aurum Potable, one ounce, Lignum aloes, half an ounce, Yellow Sanders, a dram and half, the seeds of Carduus, an ounce, Citron, six drains.

Cut those things that are to be cut, and let them be macerated three days in the best Spirit of Wine, and Muscadine, of each three pints and half, vinegar of Wallgilly-flowers, and the juice of Lemons, of each a pint, let them be distilled in a glazed vessel in balneo.

After half the Liquor is distilled off, let that which remains in the vessel be strained through a linen cloth, and vapored away to the thickness of honey, which may be called

This water is a great Cordial, and good against any Infection.

Puella Solis.

27.   Take of Ginger a pound, long Pepper, and black Pepper, of each half an ounce, of Cardamunas three drama, of Grains an ounce, powder them and put them into a glass with half an ounce of the best Camphire, distilled vinegar two pounds, digest them a month, then separate the vinegar by expression, which must putrefy a month, and then be circulated for the space of a week, then filter it, and you have as powerful a Sudorific as was or can be made.

The dose is from a drain to half an ounce, and to be drank in a drought of posset-drink.

Acquisitio Luna.

28.   Take of the juice of the green shells of Walnuts four pounds, the juice of Rue, three pints, Carduus, Mary-Gold, Balm, of each two pints, the root of Butterburre fresh a pound and half, Burre, Angelica, Master-wart, fresh, of each half a pound, the leaves of Scordium, four handfuls, old Andromachus treacle, Mithridate, of each eight ounces, Aurum Potable a dram, the best Canary, twelve pints, the sharpest

Vinegar, six pints, the juice of Lemons, two pints. Digest them two days in horse dung, the vessel being close stopped; then distil them in sand.

Aqua Luna, Scorpio.

29. Take of Sugar candid, one-pound, Canary Wine, six ounces, Rosewater, four ounces, make of these a Syrup, and boil it well, to which add Aurum Potable a dram, of Aqua Celestis, two pints, Ambergryse, Musk, of each eighteen grains, Saffron, fifteen grains, yellow Sanders infused in Aqua Celestis, two drama.

Fortuna Major Sol.

30. Take of the root of Briony, four pounds, the leaves of Rue, Mugwort, of each two pounds, Savin dried, three handfuls, Motherwort, Nippe, Pennyroyal, of each two handfuls, Garden Basil, Crecensian Dittany, of each a handful and half, the rind of yellow Oranges, fresh, four ounces, Myrrh, two ounces, Aurum Potable, an ounce, Castoreum, an ounce, the best Canary wine, twelve pints. Let them be digested four days in a fit vessel, then distil them in balneo.

Rubeus Sol.

31.   Take of the beat Tobacco in leaves, cut small, four ounces, Squils two ounces, Nutmegs sliced half an ounce; put these into three pints of spring water, a pint of White wine vinegar, distil them in a hot Still or Alembic.

If you would have it stronger, you may put this water on fresh ingredients, and distil it again.

A little quantity of this water is a most safe and effectual vomit, and say be taken from the eldest to the youngest, if so be you proportion the quantity to the strength of the Patient.

You say dulcify it with sugar or syrup if you please.

Puer Sol.

32.   Take green Walnuts gathered about Midsummer, Radish roots, of each bruised two parts, of distilled Wine vinegar four parts, digest them five days, then distil them in balneo.

This being taken to the quantity of two spoonful's ox three, causes easy vomiting.

Aqua Jupiter.

33.   Take of Scammony an ounce, Hermodactyls two ounces, the seeds of Broom, of the lesser Spurge, of Dwarf Elder, of each half an ounce, the juice of Dwarf Elder, of wild Asses cucumber, of black

Hellebore, the fresh flower of Elder, of each an ounce and half, Polypodium six ounces, of Sane three ounces, Red sugar eight ounces, **coSinon** distilled Water six pints.

Let all these be bruised, and infused in the water twenty-four hours, then be distilled in balneo.

This water may be given from two drama to three ounces, and it purges all manner of humours, opens all obstructions and is pleasant to be taken, and they whose stomachs loath all other physick, may take this without any offence.

After it is distilled there may be hanged a little bag of Spices in it, as also it may be sweetened with sugar, or any opening syrup.

Sol, Mars, Aries.

34. Take of oil of Cloves well rectified half an ounce, in it dissolve half a dram of Camphire, add to them of the Spirit of Turpentine four times rectified, in which half a drain of Opium has been infused, half an ounce.

A drop or two of this Liquor put into a hollow tooth with some lint, eases the toothache presently.

The Spirit of any vegetable may suddenly, at any time of the year be made thus.

35.   Take of what Herb, Flower, Seeds, or Roots you please, fill the head of the Still therewith, then cover the mouth thereof with a course Canvas, and set it on the Still, having first put into it sack, or low Wines. Then give it fire.

If any time thou wouldst have the spirit be of the Color of its vegetable, then put of the flowers thereof dried a good quantity in the nose of the Still.

To make any vegetable yield its Spirit quickly.

36.   Take of what vegetables you please, whether it be the seed, flower, root, fruit, or leaves thereof, cut or bruise them small, then put them into warm water, put yeast or balm to them, and cover them warm, and let them work three days as doth Beer, then distil them, and they will yield their spirit easily.

Chapter III.

1. Of the Essence of an Herb. 2. Of the appearing of the Idea of an Herb in a Glass. 3. Of a wonderful famous Medicine experienced by the Rosie Crucians. 4. Of its virtue. 5. How to turn Quick—silver into water without sizing anything with it, and to sake thereof a good purgative and Diaphoretick Medicine. 6. Of the fragrant Oil of Mercury. 7. It virtues. 8. Its use. 9. To make a Spirit of Honey. 10.Of the Quintessence of Honey. 11. Of the Oil of Honey. 12. Of the Essence of Honey. 13. Of its virtues. 14. Use. 15. Fortuna Veneris, and of the virtues, and use. 16. Aqua Magnanimitatis. 17. The Famous Restorative of Plato and Pythagoras. 18. Of Paracelsus HOMUNCULUS. 10. The Process. 20. The Second rule. 21. The Third Rule. 22. Of the virtues and use of it. 23. How to make artificial flesh, and of its virtues.

To reduce the whole Herb into a liquor, which may well be called the Essence thereof.

1. Take the whole Herb with flowers and roots, make it very clean, then bruise it in a stone Mortar, put it into a large glass vessel, so that two parts of three may be empty: then cover it exceeding close, and let it stand in putrefaction in a moderate heat the space of half a year, and it will be all turned into a water.

Make an Essence of any Herb, which being put into a glass, and held over a gentle fire, the lively form and Idea of the Herb will appear (?) the Glass.

2.	Take the foregoing water, and distill it in a gourd glass (the (?)nts being well closed) in ashes, and there will come forth a Water and an Oil, and in the upper part of the vessel will hang a volatile (?)t. The oil separate from the water, and keep by itself; with the (?)er purify the volatile salt by dissolving, filtering, and coagulating. The Salt being thus purified imbibe with the said Oil, until it will imbibe no more; digest them well together for a month in a vessel hermetically sealed. And by this means you shall have a most subtle essence, which being held over a gentle heat will fly up into the glass,

(?) represent the perfect Idea of that Vegetable whereof it is the Essence.

A wonderful famous Medicine experienced by the Rosie Crucians

3.	Take Calx of Saturn, or else minimum, pour upon it so much spirit Vinegar as say cover it four fingers breadth; digest them in a warm (?)ce the space of twenty four hours, often stirring them, that the matter settle not too thick in the bottom; then decant the menstruum, pour on more; digest it as before, and this do often until all the saltness be extracted: Filter and clarify all the menstruum being put together, then evaporate it half away, and set the other part in a cold place till it crystalline. These Crystals dissolve again in fresh spirit do often until all the e all the menstruum being put id set the other part in a cold a dissolve again in fresh spirit of Vinegar, filter and coagulate the Liquor again into Crystals, and this do so often, until they be sufficiently impregnated with the salt Armoniack of the Vinegar as with their proper ferment. Digest them in a temperate balneo, that they say be resolved into a Liquor like Oil. Then distil this Liquor in sand in a

Retort, with a large Receiver annexed to it, and well closed, that no spirits evaporate, together with the observation of the degrees of the fire; then there will distil forth a spirit of such a fragrant smell, that the fragrancy of all flowers, and compounded perfumes, are not to be compared to it. After Distillation when all things are cold, take out, and cast away the black feces which is of no use. Then separate the yellow oil which swims on the top of the spirit, and the blood red oil which sinks to the bottom of it: Separate the phlegm from the spirit in balneo. Thou shalt by this means have a most fragrant spirit that even revives the senses; and so balsamical, that it cures all old and new sores inward and outward, and so cordial, that the dying are with admiration revived with it.

4. They that have this Medicine need scarce use any other either for inward or outward griefs.

How to turn Quicksilver into a water without mixing anything with it, and to make thereof a good Purgative and Diaphoretic medicine.

5. Take an ounce of Quicksilver not purified, put it into a bolt head of glass, which you must nip up, set it over a strong fire in sand for the space of two months, and the Quicksilver will be turned into a red sparkling Precipitate. Take this powder, and lay it thin on a Marble in a Cellar for the space of two months, and it will be turned into a water which may be safely taken inwardly; it will work a little upward and downward, but chiefly by sweat.

Note that you may set divers glasses with the same matter in the same Furnace, that so you say make the greater quantity at a time.

I suppose it is the Sulphur which is in the Quicksilver, and makes it so black, that being stirred up by the heat of the fire fixes the Mercury.

A fragrant Oil of Mercury.

6.   Take of Mercury seven times sublimed, and as often revived with unslaked Lime, as much as you please, dissolve it in spirit of Nitre in a moderate heat, then abstract the spirit of Salt, and edulcorate it very well by boiling it in spirit of Vinegar; then abstract the spirit of Vinegar, and wash it again with distilled rain water; then dry it, and digest it two months in a like quantity of the best rectified spirit of Wine you can get. Distil them by Retort, making your fire moderate at the beginning, afterwards increasing it; then evaporate the spirit of Wine in balneo, and there will remain in the bottom a most. fragrant oil of Mercury.

7.   This oil so purifies the blood by sweat and urine, that it cures all distempers that arise from the impurity thereof, as the venereal Disease, etc.

8.   The truth is, they that have this Medicine well made, need but few other Medicines; the dose is four or five drops.

To make a spirit of Honey.

9. Take good strong stale Mead, otherwise called Metheglm, as much as thou please, distil it in a Copper Still or Limbeck, with its refrigeratory, and it will yield a spirit like AQUA VITAE.

The Quintessence of Honey is made thus.

10. Take of the purest Honey two pounds, of Fountain water one pound; boil these together till the water be boiled away, taking off all the scum that rises; then take the honey and put it into a glass, four parts of five being empty, close it well, and set in digestion a whole year, and thou shalt have the Essence of Honey swimming on the top in form of an oil being of as fragrant smell as anything in the world; the flegme will be in the middle, and the feculent matter in the bottom, of a dark Color, and stinking smell.

Some make the Quintessence of Honey after this manner.

11. Take as much Honey as thou pleases of the best, put it into a Gourd of glass, first distil off the phlegm in balneo, then extract the tincture out from what remains, with the said water, then calcine the remaining feces, and extract from thence the salt with the foresaid water, being distilled off from the tincture, calcine the salt and melt it in a Crucible, then let it dissolve in a Cellar; then again evaporate it away, and thou shalt have a most white salt, which let imbibe as much of the tincture as it will; digest them for three months, and thou shalt have an Essence of Honey.

An Essence of Honey may be made thus.

12. Take of Honey well despumated as much as you please, pour upon it as much of the best rectified spirit of Wine as will cover it five or six fingers breadth, digest them in a glass vessel well closed (the fourth part only being full) in a temperate balneo the space of a fortnight, or till the spirit be very well tinged, then decant off the spirit, and put on more till all the tincture be extracted, then put all these tinctures together, and evaporate the spirit till what remains begin to be thicket at the bottom, and of a golden Color.

13. This is a very excellent Essence of Honey, and is of so pleasant an odor, that scarce anything is like to it.

14. It is so cordial, that it even revives the dying, if two or three drops thereof be taken in some cordial water.

Fortuna Veneris.

15. Take of Pismires or Ants (the biggest, that have a sourish smell, are the best) two handfuls, spirit of Wine a gallon, digest them in a glass vessel close shut the space of a month, in which time they will be dissolved into a Liquor, then distil them in balneo till all be dry. Then put the same quantity of Ants as before, digest and distil them in the said Liquor as before: do this three times, then add aroma to the spirit with some Cinnamon.

Note that upon the spirit will float an oil, which must be separated.

This spirit is of excellent use to stir up the Animal spirit; in so much that John Casimire Palsgrave of the Rhene, and Setfrie of Collen, General against the Turks, did always drink of it when they went to fight, to increase magnanimity, and courage, which it did, even to admiration.

This spirit doth also wonderfully irritate them that are slothful to Venery.

It also provokes Urine even to admiration.

It doth also wonderfully irritate the spirits that are dulled, and deadened with any cold distemper.

This Oil doth the same effects, and indeed more powerfully.

This Oil doth, besides what is spoken of the spirit, help deafness exceedingly, two or three drops being dropped into the ear after it is well syringed, once in a day, for a week together.

It helps also the Eyes that have any film growing on them, being now and then dropped into them.

Aqua Magnanimitatis is made thus.

16. Take of Ants or Pismires a handful, of their Eggs two hundred, of Millepedes, i.e. Wood—lice one hundred, of Bees one hundred and fifty, digest all these in two pints of spirit of Wine, being very well impregnated with the brightest Soot. Digest them together the space of a month, then pour off the clear spirit, and keep it safe.

This water or spirit is of the same virtue as the former.

The famous Restorative of Plato and Pythagoras used by Paracelsus.

17. First, we must understand that there are three acceptions of the word homunculus in Paracelsus, which are these.

> 1. Homunculus is an image made in the place or name of any one, that It may contain an Astral and invisible man; wherefore it was made by Numbers.

> 2. Homunculus is taken for an artificial man, made of sperma humanum masculinum, digested into the shape of a man, and then nourished and increased with the essence of man's blood; and this is not repugnant to the possibility of Nature and Art, but is one of the greatest wonders of God, which he ever did suffer mortal man to know. I shall not here set down the full process, because I think it unfit to be done, at least to be divulged; besides, neither this nor the former is for my present purpose.

> 3. Homunculus is taken for a most excellent arcanum, or Medicament, extracted by the Spagyrical Art, from the chiefest staff of the natural life in man, and according to this acceptation I shall here speak of it: But before I show you the process, I shall give you an account why this Medicament is called homunculus, and it is this.

18. No wise man will deny that the staff of life is the nutriment thereof, and that the chiefest nutriment is Bread and Wine, being

ordained by God and Nature above all other things for the sustentation thereof. Besides Paracelsus preferred this nutriment for the generation of the blood and spirits, and the forming thence the Sperm of this homunculus. Now by a suitable allusion the nutriment is taken for the life of man, and especially because it is transmuted into life: and again, the life is taken for the man; for unless a man, be alive he is not a man, but the carcass only of a man, and the basest part thereof, which cannot perfectly be taken for the whole man, as the noblest part may. In as such therefore as the nutriment, or ailment of life, may be called the life of man; this nutriment extracted out of Bread & Wine & being by digestion exalted into the highest purity of a nutritive substance, and consequently becoming the life of man, being so potentially, say Metaphorically be called homunculus.

19. The process, which in part shall be set down allegorically is thus: Take the best Wheat, and the best Wine, of each a like quantity, put them into a glass, which you must hermetically close: then let them putrefy in horsedung three days, or until the Wheat begin to germinate, or to sprout forth, which then must be taken forth and bruised in a Mortar, and be pressed through a linen cloth, and there will. come forth a white juice like silk; you must cast away the feces: Let this juice be put into a glass, which must not be above half full; stop it close, and set it in horsedung as before, for the space of fifty days. If the heat be temperate, and not exceeding the natural heat of a man, the matter will be turned into a spagyrical blood, and flesh, like an Embryo. This is the principal, end next matter, out of which is generated a twofold sperm, viz, of the father and mother, generating the homunculus, without which there can be made no generations, whether human, or animal.

20. From the blood and flesh of this Embryo let the water be separated in balneo, and the air in ashes, and both be kept by themselves. Then to the feces of the latter distillation let the water of the former distillation be added, both which must (the glass being close stopped) putrefy in balneo the apace of ten days, after this distil the water the second time, (which is then the vehiculum of the first) together with the fire, in ashes, then distil off this water in a gentle balneo, and in the bottom remains the fire, which must be distilled in ashes. Keep both these apart. And thus, you have the four Elements separated from the Chaos of the Embryo.

21. The feculent earth is to be reverberated in a close vessel for the space of four days: In the interim distil off the fourth part of the first distillation in balneo, and cast it away; the other three parts distil in ashes, and pour it upon the reverberated earth, and distil it in a strong fire; cohobate it four times, and so you shall have a very clear water, which you must keep by itself: Then pour the air on the sane earth, and distil it in a strong fire; and there will come over a clear, splendid, odoriferous water, which must be kept apart; After this pour the fire upon the first water, and putrefy them together in balneo the space of three days, then put them into a Retort, and distil them in sand, and there will come over a water tasting of the fire: let this water be distilled in balneo, and what distils off keep by itself, as also what remains in the bottom, which is the fire, keep by itself, This last distilled water pour again upon its earth, and let them be macerated together in balneo for the space of three days, and then let all the water be distilled in sand, and let what will rise be separated in balneo, and the residence remaining in the bottom be reserved with the former residence. Let the water be again poured upon the earth, be abstracted, and separated as before, until nothing remains in the bottom, which is not separated in balneo. This being done, let the water which was last

separated be mixed with the residue of its fire, and be macerated in balneo three or four days, and all be distilled in balneo, that can ascend with that heat, and let what remains be distilled in ashes from the fire, and what shall be elevated is aerial; and what remains in the bottom Is fiery. These two last Liquors are ascribed to the two first principles, the former to Mercury, and the latter to Sulphur, and are accounted by R. Crucians, not as elements, but their vital parts, being as It were the natural spirits and soul, which are in them by nature. Now both are to be rectified, and reflected into their center with a circular motion, that this Mercury may be prepared with Its water, being kept clear, and odoriferous, in the upper place, but the Sulphur by itself. Now it remains that we look into the third principle; it the reverberated earth, being ground upon a marble, imbibe its own water, which did above remain after the last separation of the Liquors made in balneo, so that this be the fourth part of the weight of its earth, and be congealed by the heat of ashes into its earth, and let this be done so often, the proportion being observed, until the earth has drank up all its water. And lastly, let this earth be sublimed into a white powder as white as snow, the feces being cast away. This earth being sublimed and freed from its obscurity, is the true Chaos of the Elements; for it contains those things occult, seeing it is the salt of nature, in which they lye hid, being, as it were, reflected in their center. This is the third principle of Paracelsus, and the salt, which is the matrix, in which the two former sperms, viz. of the man and woman, the parents of the homunculus, viz, of Mercury and Sulphur, are to be put, and to be closed up together in a glazed womb, sealed with Hermes Seal, for the true generation of the homunculus, produced from the spagyrical Embryo: and this is the homunculus or great arcanum, otherwise called the nutritive Medicament of Paracelsus.

22. This homunculus or nutritive Medicament, is of such virtue, that presently after it is taken into the body, it Is turned into blood and spirits, If then Diseases prove mortal because they destroy the spirits, what mortal Disease can withstand such a Medicine, that doth so soon repair, and so strongly fortify the spirits, as this homunculus, being as the oil to the flame, into which it is immediately turned, thereby renewing the same? By this Medicament therefore, as Diseases are overcome, and expelled, so also youth is renewed, and gray hairs prevented.

An artificial way to make Flesh.

23. Take of the crumbs of the best wheaten Bread as soon as it comes forth out of the Oven, being very hot, as much as you please, put it into a glass vessel, which you must presently hermetically close. Then set it in digestion in a temperate balneo, the space of two months, and it will be turned into a fibrous flesh.

If any Artist please to exalt it to a higher perfection, according to the Rules of Art, he may find out, how great a nourisher and restorative Wheat is, and what an excellent Medicine it may make.

Note that there must be no other moisture put into the glass besides what is in the bread itself.

Chapter IV.

1. The way to raise a dead Bird to life. 2. Of generating many Serpents of one, etc. 3. To purifie and refine Sugar. 4. To make a Vegetable grow and become more glorious then any of its species. 5, To make a Ballet grow in two or three hours. 6. To make the idea of any plant appear in a glass, as if the very plant it self were there. 7. To make Fir-trees appear in Turpentine. 8. To make Harts-horn appear in a Glass. 9. To make Golden Mountains to appear in a glass. 10. To make the world appear in a Glass. 11. To make four Elements appear in a Glass. 12. To make a perpetual Motion in a Glass. 13. To make a Luminous water that shall give light by night. 14. Of a room that shall seem on fire, if you enter with a Candle. 15. To make a powder that by spitting upon it shall be inflamed. 16. To make a Loadstone draw a Nail out of a post.

The way to raise a dead bird to life, and for the generating many Serpents of one, both which are performed by putrefaction.

1. A Bird is restored to life thus, viz. Take a Bird, put it alive into a gourd glass, and seal it up hermetically, burn it to ashes in the third degree of fire, then putrefy it in horse dung into a mucilaginous flegme, and so be a continued digestion that flegme must be brought to a further maturity (being taken out, and put into an oval vessel of a Just bigness to hold it) by an exact digestion, and will so become a renewed bird: which says Paracelsus Is one of the greatest wonders in Nature, and shows the great virtue of putrefaction.

2.   Cut a Serpent into small pieces, which put into a gourd glass which you must Hermetically seal, up, then putrefy them in horse dung, and the whole Serpent will become living again in the glass, in the form either of worms or spawn of fishes; Now if these worms be in a fitting manner brought out of putrefaction, and nourished; many hundred Serpents will be bred out of one Serpent, whereof everyone will be as big as the first. And as it is said of the Serpents so also many other living creatures may be raised, and restored again.

To purify and refine Sugar.

3.   Make a strong Lixivium of Calx vive, wherein dissolve as much course Sugar as the Lixivium will bear, then put in the white of Eggs (or 2 to every part of the Liquor) being beaten into an oil; stir them well together, and let them boil a little, and there will arise a scum which must be taken off as long as any will arise, then pour all the Liquor through a great Woolen cloth bag, and so the feces will remain behind in the gab, then boil the Liquor again so long till some drops of it being put upon a cold plate, will, when they be cold, be congealed as hard, as salt. Then pour out the Liquor into pots, or molds made for that purpose, having a hole in the narrower end thereof, which must be stopped for one night after, and after that night be opened, and there will a moist substance drop forth which is called Molasses, or Treacle; then with potters clay cover the ends of the pot, and as that clay sinks down by reason of the sinking of the Sugar, fill them up with more clay, repeating the doing thereof till the Sugar shrink no more. Then take it out till it be hard, and dried, then bind it up in papers.

To make a vegetable grow and become more glorious then any of its species.

4.   To reduce any vegetable into its three first principles, and then join them together again being well purified, & put the same into a rich earth, and you shall have it produce a vegetable far more glorious then any of its species.

Note how to make such an essence; look into the first book, and there you shall see the process thereof.

To make a Plant grow in two or three hours.

5.   Take the ashes of Moss, moisten them with the Juice of an old dunghill being first pressed forth, and strained, then dry them a little, and moisten them as before, do this four or five times, put this mixture being neither very dry, nor very moist, into some earthen, or metalline vessel; and in it set the seed of Lettice, Purslane or Parsley (because they will grow sooner than other Plants) being first impregnated with the essence of a vegetable of its own species, (the process whereof you shall find Book 1.) till they begin to sprout forth, then I say, put them in the said earth with that end upwards which sprouts forth: Then put the vessel into a gentle heat, and when it begins to dry, Moisten it with some of the said juice of dung.

You may by this means have a salad grow whilst supper is making ready.

To make the Idea of any Plant appear in a glass, as if the very plant itself were there.

6. The process of this you may see, page 32. and therefore I need not here again repeat it; only remember that if you put the flame of a candle to the bottom of the glass where the essence is, by which it may be made hot; you will see that thin substance which is like impalpable ashes or salt send forth from the bottom of the glass the manifest form of a vegetable, vegetating and growing by little and little, and putting on so fully the form of stalks, leaves and flowers in such perfect and natural wise in apparent show, that any one would believe verily the same to be naturally corporal, when as in truth it is the spiritual idea, endued with a spiritual essence: which serves for no other purpose, but to be matched with its fitting earth, that so it may take unto itself a more solid body. This shadowed figure as soon as the vessel is taken from the fire, returns to its ashes again and vanishes away, becoming a Chaos, and confused matter.

To make Fir-trees appear in Turpentine.

7. Take as much Turpentine as you please, put it into a Retort, distil it by degrees; when all is distilled off, keep the Retort still in a reasonable heat, that what humidity is still remaining may be evaporated, and it becomes dry; Then take this off from the fire and hold your hand to the bottom of the Retort, and the Turpentine that is dried (which is called Colophonia) will crack asunder in several places,

and in those cracks or chips you shall see the perfect effigies or Fir—trees which will there continue many months.

To make Harts-horn seemingly to grow in a glass.

8.   Take Harts-horn broken into small pieces, and put them into a glass Retort to be distilled, and you shall see the glass to be seemingly full of horns, which will continue there so long till the volatile salt come over.

To make golden mountains as it were appear in a glass.

9.   Take of Adders eggs half a pound, put them into a glass Retort, distil them by degrees; when all is dry, you shall see the feces at the bottom turgid and puffed up, and seem to be as it were golden mountains, being very glorious to behold.

To make the representation of the whole World in a Glass.

10.   Take of the purest salt Nitre as much as you please, of Tin half so much, mix them together, and calcine them Hermetically, then put them into a Retort, to which annex a glass receiver, and lute them well together; let there be leaves of Gold put into the bottom thereof, them put fire to the Retort, until vapors arise that will cleave to the Gold:

augment the fire till no more fumes ascend, then take away the Receiver, and close it Hermetically, and make a lamp fire under it, and you will see presented in it the Sun, Moon, Stars, Fountains, Flowers, Trees, fruits, and indeed even all things, which is a glorious sight to behold.

To make four Elements appear in a glass.

11.   Take of the subtle powder of Jet an ounce and a half, of the oil of Tartar made per deliquium (in which there is not one drop of water besides what the Tartar itself contracted) two ounces, which you must Color with a light green with Verdergreece, of the purest Spirit of Wine tinged with a light blue with Indigo, two ounces of the best rectified Spirit of Turpentine colored with a light red, with Hadder two ounces. Put all these into a glass, and shake them together, and you shall see the Jet which is heavy and black fall to the bottom, and represent the earth; next, the Oil of Tartar made green, representing the element of water, falls: upon this swims the blue Spirit of Wine which will not mix with the oil of Tartar; and represents the element of air: uppermost will swim the subtle red oil of Turpentine which represents the element of fire.

It is strange to see how after shaking all these together they will be distinctly separated the one from the other. If it be well done, as it is easy enough to do, it is a most glorious sight.

To make a perpetual motion in a glass.

12. Take seven ounces of Quicksilver, as much Tin, grind them well together with fourteen ounces of Sublimate dissolved .in a cellar upon a Marble the space of four days, and it will become like oil Olive, which distil in sand, and there will sublime a dry substance; then put the water which distils off back upon the earth, in the bottom of the Still and dissolve what you can; filter it, and distil it again, and this do four or five times, and then that earth will be so subtle, that being put into a vial, the subtle atoms thereof will move up and down for ever.

Note that the vial, or glass must be close stopped, and kept in a dry place.

To make a Luminous Water that shall give light by night.

13. Take the tails of Glowworms, put them into a glass still, and distil them in balneo, pour the said water upon more fresh tails of Glowworms, do this four or five times, and thou shalt have a most Luminous Water, by which you may see to read in the dark night.

Some say this Water may be made of the skins of Herrings; and for ought I know it may be probable enough: for I have heard that a group of Herrings coming by a ship in the night has given a great light to all the ship.

It is worth the while to know the true reason why Glowworms, and Herrings and some other such like things should be luminous in the night.

To make a vapor in a chamber, that he that enters into it with a candle shall think the room to be on fire.

14. Dissolve Camphire in rectified aqua vitae, and evaporate them in a very close chamber where no air can get in, and he that first enters the chamber with a lighted candle, will be much astonished; for the chamber will seem to be full of fire, very subtle, but it will be of little continuance.

You must note that it is the combustible vapor, with which the chamber is filled, that takes flame from the candle.

Divers such like experiments as this nay be done, by putting such a combustible vapor into a box, or cupboard or such like, which will as soon as any one shall open them having a candle in his hand, take fire, and burn.

To make a powder that by spitting upon shall be inflamed.

15. Take a Loadstone, powder it, and put it into a strong calcining pot, cover it all over with a powder made of Calx vive, and Cobophonia, of each a like quantity, put also some of this powder under it; when the pot is full, cover it, and lute the closures with potters earth, put them into a furnace, and there let them boil, then take them out and put them into another pot, and set them in the furnace again, and this do till they become a very white and dry Calx. Take of this Calx one part, of salt Nitre being very well purified four parts, and as much Campfire, Sulphur vivian, the oil of Turpentine, and Tartar; grind all these to a subtle powder and searse[34] them, and

put as much Spirit of Wine well rectified as will cover them two fingers breadth, then close them up and set the vessel in horse-dung three months, and in that time they will become an uniform paste: evaporate all the humidity, until the whole mass becomes a very dry stone: then take it out and powder it, and keep it very dry.

If you take a little of this powder and spit upon it, or pour some water upon it, it will take fire presently, so that you may light a match, or any such thing by it.

To fortify a Loadstone that it shall be able to draw a nail out of a piece of wood.

16. Take a Loadstone, and heat it very hot in coals, but so that it be not fired, then presently quench it in the Oil of Crocus Martis made of the best steel, that it may imbibe as much as it can.

Thou shalt by this means make the Loadstone so very strong and powerful, that you may pull out nails out of a piece of wood with it, and do such wonderful things with it that the common Loadstone can never do.

Now the reason of this (as Paracelsus says) is because the Spirit of Iron is the Life of the Loadstone, and this may be extracted from, or increased in the Loadstone.

---

[34] Spelled exactly as in the 1662 printed edition. -pnw

Chapter V.

1. To make Quicksilver Malleable in seven hours. 2. To reduce glass into its first principles, viz. Sand and Salt. 3. To write or engrave upon an egg, or pebble, with wax. 4. To make Pearle. 5. Make Arabian perfume. 6. To make strange Oils and Liquors. 7. To make Steel grow like a Tree. 8. To melt any Metal in the hand without burning of it. 9. Secret observations. 10. To extract a white Milkie substance from the rays of the Moon.

To make Quicksilver Malleable in seven hours.

1.   Take of the best lead, and melt it, and pour it Into a hole, and when it is almost congealed make a hole in it, and presently fill up the hole with quicksilver, and it will presently be congealed into a friable substance; then beat it into powder, and put it again into a bowl of fresh melted lead as before; do this, three or four times, then boil it being all in a piece in Linseed oil the space of six hours; then take it out and it will become malleable.

Note that after this, it may by being melted over the fire be reduced into quicksilver again.

A thin plate of the said Mercury laid upon an inveterate Ulcer takes away the malignity of it in a great measure, and renders it more cureable then before.

A plate of the said Mercury laid upon tumors would be a great deal better repercussive then plates of lead, which Chirugeons use in such cases.

The powder of the friable substance of Mercury before it be boiled in the oil is very good to be strewed upon old ulcers, for it doth much correct the virulence of them.

To reduce glass into its first principles, viz. Sand and Salt.

2. Take bits or powder of Glass as much as you please, as much of the salt which Glass men use in the making of Glass: melt these together in a strong fire: Then dissolve all the melted mass in warm water, then pour off the water and you shall see no Glass, but only sand in the bottom, which sand was that which was in the glass before.

This censures the vulgar opinion, vim, that the fusion of Glass is the last fusion and beyond all reduction.

To write or engrave upon an egg, or pebble, with wax or grease.

3. Make what letters or figures you please with wax, or grease, upon an egg or pebble, put them into the strongest Spirit of Vinegar, and there let them lye two or three days, and you shall see every place about the letters or figures eaten or consumed away with the said Spirit, but the place where the wax or grease was, not at all touched:

the reason whereof is because that the Spirit would not operate upon the said oleaginous matter.

To make artificial Pearls, as glorious as any oriental.

4. Dissolve mother of Pearl in Spirit of vinegar, then precipitate it with Oil of Sulphur per campanam (and not with Oil of Tartar, for that takes away the splendor of it) which adds a luster to it: when it is thus precipitated, dry it, and mix it with whites of eggs, and of this mass you may make Pearls of what bigness or fashion you please: before they be dried you may make holes through them, and when they be dried they will not at all or very hardly be discerned from true, and natural Pearls.

To make a Mineral perfume.

5. Dissolve Antimony, or Sulphur in the Liquor or Oil of flints or pebbles, or Chrystals, of sand, coagulate the solution into a red mass, pour thereon the spirit of urine and digest them till the spirit be tinged, then pour it off, and pour more on, till all the tincture be extracted, put all the tinctures together, and evaporate the Spirit of urine in balneo, and there will remain a blood-red Liquor, at the bottom, upon which pour Spirit of wine, and you shall extract a purer tincture, which smells like garlic; digest it three or four weeks, and it will smell like balm; digest it longer, and it will smell like Musk or Ambergris.

Besides the smell that it has, it is an excellent Sudorific, and cures all diseases that require sweat, as the plague, putrid fevers, Lues venerea, and such like.

The Oil or Liquor of Sand, flints, pebbles, or Chrystals, for the aforesaid preparation, is thus made.

6. Take of the best salt of Tartar being very well, by two or three dissolutions and coagulations purified, and powdered in a hot mortar, one part, of flints, pebbles or crystals being powdered, or small sand well washed, the fourth part, mingle them well together; put as much of this composition as will fill an eggshell into a Crucible set in the earthen Furnace, (expressed page 83.) and made red hot, and presently there will come over a thick, and white spirit, this do till you have enough, then take out of the Crucible whilst it is growing hot, and that which is in it like transparent glass, which keep from the air.

The Spirit may be rectified by sand in a glass Retort.

The Spirit is of excellent use in the gout, stone, ptisick, and indeed in all obstructions, provokes sweat, and urine, and cleanses the stomach, and by consequence effectual in most diseases.

It being applied externally clears the akin, and makes it look very fair.

Take that which remains at the bottom in the crucible, and beat it to powder, and lay it in a moist place, and so it dissolves into a thick fat Oil: and this is that which is called the Oil of Sand, of flints, and pebbles or crystals.

This oil is of wonderful use in medicine, as also in the preparation of all sorts of Minerals.

This oil being taken inwardly in some appropriated Liquor, dissolves tartarous coagulations in the body, and so opens all obstructions. It precipitates metals and makes the calx thereof weightier then oil of Tartar doth.

It is of a golden nature: it extracts colors from all metals, is fixed in all fires, makes fine Crystals, and Borax, and matures imperfect metals into Gold.

If you put it into water, there will precipitate a most fine white earth, of which you may make as clear vessels as are China-dishes.

Note that all sand, flints, and pebbles, even the whitest, have in them a golden sulphur, or tincture, and if a prepared lead be for a tine digested in this oil, it will seem as it were gilded, because of the Gold that will hang upon it, which may be washed away in water. Gold also is found in sand and flints, etc. and if you put Gold into this oil, it will become more ponderous thereby.

To make Steel grow in a glass like a tree.

7. Dissolve Steel in a rectified Spirit of Salt, so shall you have a green and sweet solution, which smells like brimstone; filter it, and abstract all the moisture in sand with a gentle heat, and there will distil over a Liquor as sweet as rainwater; for Steel by reason of its dryness detains the corrosiveness of the Spirit of Salt, which remains in the bottom like a blood red mass, which is as hot on the tongue as fire;

dissolve this red mass, in oil of flints, or of sand, and you shall see it grow up in two or three hours like a tree with a stem and branches; prove this tree at the test, and it yields good Gold, which this tree has drawn from the aforesaid oil of sand, or flints, which has a golden sulphur in it.

To melt any metal in one's hand without burning of the hand.

8. Take a little calcining pot in your hand, make in it a lane or course of the powder of any metal, then upon it lay a lane of Sulphur, Saltpeter and Sawdust, of each a like quantity, mixed together, put a coal of fire to it, and forthwith the metal will be melted into a mass.

An observation upon the beams of the Sun and heat of fire, how they add weight to Minerals and Metalline bodies.

1. Take any Mineral Liquor and set it in an open vessel in the sun for a space, and it will be augmented in quantity, and weight. But some will say that this proceeds from the air: to the which I answer and demands, whether the air had not this impregnation from the sun, and what the air has in itself that proceeds not from the Bun and stars.

2. Put this liquor in a cold cellar, or in a moist air, and you shall find that it increases not in weight, as it doth in the sun, or in the fire (which has in this respect some analogy with the sun) I do not say but haply it might attract some little moisture which is soon exhaled by any small heat.

3. Dissolve any sulphureous and imperfect metal as Iron, Copper, or link, in aqua fortis, or any other acid spirit, then abstract the Spirit from it, make it glowing hot, yet not too hot, that the Spirit may only vapor away, then weigh this metalline Calx, and set it in a crucible over the fire, but melt it not, only let it darkly glow; let it stand so three or four weeks, then take it off, and weigh it again, and you shall find it heaver then before.

4. Set any sulphureous metal, as Iron, or Copper, with sixteen or eighteen parts of Lead on a test made with ashes of wood or bones in a probatory furnace: first weigh the test, copper and lead before you put them into the furnace; let the iron or copper fly away with the lead, yet not with too strong a heat, then take the test out, and weigh it, and you shall find it (though the metals are gone) when it is cold to be heaver then it was when it was put into the furnace with the metals.. The question is now whence this heaviness of all the aforesaid Minerals and metals proceed, if that the heat of the sun and fire through the help of the Minerals and metals be not fixed into a palpable Mineral, and Metalline body.

5. Set a test with lead, or copper in the sun, and with a concave glass unite the beans of the sun, and let them fall on the center of the metal, hold the concave glass in your hand, and let your test never be cold, and this will be as well done in the sun, as in the fire. But this concave must be two foot in Diameter, and not too hollow or deep, but about the eighteenth or twentieth part of the circle, that it may the better cast its beans forth, and it must be very well polished.

6.   Calcine Antimony with a burning glass, and you shall see it smoke, and fume, and be made dryer then before, yet weigh it and be heavier than before.

I shall take in, for the confirmation of all this, a relation of Sir Kenelm Digby concerning the precipitating of the sun beams. I remember (says he) a rare experiment that a Nobleman of much sincerity, and a singular friend of mine, told me he had seen, which was, that by means of glasses made in a very particular manner, and artificially placed one by another, he had seen the sun beams gathered together, and precipitated down into a brownish or purplish red powder. There (says he) could be no fallacy in this operation. For nothing whatsoever, was in the glass, when they were placed, and disposed for this intent; and it must be in the hot time of the year; else the effect would not follow. And of this magistery he could gather some days near two ounces in a day, and it was of a strong volatile virtue, and would impress its spiritual quality into Gold itself (the heaviest and most fixed body we converse withal) in a very short time.

I leave it now to the reader to Judge whether the beams of the sun, and heat of the fire add weight to Minerals, and Metals.

To extract a white Milky substance from the rays of the Moon.

Take a concave glass and hold it against the Moon when she is at the full in a clear evening, and let the rays thereof being united fall upon a sponge, and the sponge will be full of a cold Milky substance, which you may press out with your hand, and gather more. DE-LA-BROSSE is of opinion that this substance is of the substance of the Moon: but I cannot assent to him in that, only this I say, if this experiment were well prosecuted, it might be the key to no small secrets.

Chapter VI.

1. To condense air in summer. 2. How to fix two volatile salts. 3.Of a Rosie Crucian Medicine, and its use and virtues. 4. Another. 5.Of a Cordial Tincture, and its virtues. 6. Another of excellent virtues, and its use. 7. To reduce distilled Turpentine into its body again, and of its use, and virtues. 8. To make the distilled oil out of any herb or flower, or seed in an instant without a furnace. 9. To know what Metal there is in any Ore. 10. A pretty observation upon the Melting of Copper and Tin together. 11. Admirable observations upon the melting salt Armoniack, and calx vive together. 12. A cheap powder like unto aurum fulminans. 13. To make an Antimonial cup, and to cast divers figures of Antimony.

To condense the air in the heat of summer and in the heat of the day, into water.

1.   Fill an earthen vessel unglazed, made pointed downward, and fill it with snow-water (which must be kept all the year) in which is dissolved as much Nitre as the water would dissolve; Let the vessel be close stopped. Hold this vessel against the sun, and the air will be so condensed by the coldness of the vessel that it will drop down by the sides thereof.

How two sorts of volatile Salts will be fixed by Joining them together.

2. Take a strong Lixivium made of unslaked Lime, and evaporate it, and whereas you would expect to find a salt at the bottom, there is none; for all the salt in the Lixivium is vapored away, and the more the Liquor is evaporated, the weaker the Lixivium becomes, which is contrary to other Lixiviums: Also if you take spirit of Vinegar, and evaporate it, you shall find no salt at the bottom. Now if you take the clear Lixivium of Lime, and spirit of Vinegar, of each a like quantity, and mix them together, and evaporate the humidity thereof, you shall find a good quantity of salt at the bottom, which tastes partly hot, and partly acid.

This salt being set in a cold cellar on a marble stone, and dissolved into an oil, is as good as any lac virginis, to clear and smooth the face, and dry up any hot pustules in the skin, as also against the Itch, and old Ulcers to dry them up.

To make an Unguent, that a few grains thereof being applied outwardly, will cause vomiting or looseness, as you please.

3. Take lapis infernalis, mix therewith of distilled oil of Tobacco as much as will make an Ointment: Keep it in a dry place.

If you would provoke vomiting, anoint the pit of the stomach with five or six grains thereof, and the party will presently vomit, and as much, as with taking of a vomit.

If you would provoke to looseness anoint about the navel therewith, and the patient will presently fall into a looseness.

Note that you must give the patient some warm supping's all the time this medicine is working.

Note also, and that especially, that you let not the ointment lye so long as to cauterize the part to which it is applied.

To make a medicine that half a grain thereof being taken every morning will keep the body soluble.

4. Take of the distilled oil of Tobacco, of which let the essential salt of Tobacco imbibe as much as it can. Then with this composition make some Lozenges by adding such things as are fitting for such a form of medicine: Note that you put but such a quantity of this oily salt as half a grain only may be in one Lozenge.

One of these Lozenges being taken every morning, or every other morning, keeps the body soluble, and is good for them as are apt to be very costive in their bodies.

Note that you may put some aromatic ingredient into the Lozenges that may qualify the offensive odor of the oil, if there shall be any.

To make a Cordial, stomachic, and purgative tincture.

5. Make a tincture of Hiera picra with Spirit of wine well rectified, and aromatized with Cinnamon of Cloves.

Two or three spoonsful of this tincture being taken in a morning twice in a week, wonderfully helps those that have weak and foul stomachs;

it opens obstructions, end purges viscosities of the stomach and bowels, cures all inveterate headaches, killed worms, and indeed leaves no impurities in the body, end is very cordial; for it exceedingly helps them that are troubled with faintings. There is nothing offensive in this medicine but the bitterness thereof, which the other extraordinary virtues will more than balance.

Another.

6. Dissolve Scammony in Spirit of Wine, evaporate the one **moity**, them precipitate it by putting Rosewater to it: and it will become most white; for the black and fetid matter will lie on the top of the precipitated matter which you must wash away with Rose-water. Then take that white gum being very well washed, and dry it; if you please you may powder it, and so use it; for indeed it has neither smell nor taste, and purges without any offence, and may be given to children or to any that distaste physick, in their milk or broth, without any discerning of it; and indeed it 40th purge without any manner of gripping's. I was wont to make it up into pills with oil of Cinnamon or Cloves which gave it a gallant smell, and of which I gave a scruple which wrought moderately and without any manner of gripping's; then dissolve it again in Spirit of wine being aromatized with what spices you please, and this keep.

This tincture is so pleasant, so gentle, so noble a purgative that there is scarce the like in the world; for it purges without any offence, is taken without any nauseating, and purges all manner of humours, especially choler, and melancholy, and is very Cordial.

It may be given to those that abhor any medicine, as to children, or those that are of a nauseous stomach.

The dose is from half a spoonful to two or three.

Note it must be taken of itself; for if it be put into any other Liquor, the Scammony will precipitate and fall to the bottom.

After this manner you may prepare Jollap by extracting the gum therefore, and then dissolving it in Spirit of wine.

By this means Jollap would not be so offensive to the stomach, as usually it is; for it is the gum that is purgative, and the earthliest that is so nauseous.

Jollap being thus prepared is a most excellent medicine against all hydropick diseases; for it purges water away without any nauseousness or griping at all.

To reduce distilled Turpentine into its body again.

7.   Take the oil of Turpentine, and the Colophonie thereof (which is that substance which remains in the bottom after distillation) which you must beat to powder. Mix these together and digest them, and you shall have a Turpentine of the same consistency as before, but of a very subtle mature.

Pills made of this Turpentine, are of excellent use in obstructions of the breast, kidneys and the like.

To make the distilled Oil out of any herb, seed or flower in an instant without any furnace.

8.   You must have a long pipe made of tin, which must have a bowl in the middle with a hole in it as big as you can put your finger into it; by which you must put your matter that you would have the Oil of. Set this matter on fire with a candle or coal of fire, then put one end of the pipe into a basin of fair water, and blow at the other end, and the smoke will come into the water; and there will an oil swim upon the water, which you may separate with a funnel.

To prove what kind of metal there is in any Ore, although you have but a very few grains thereof, so that as you cannot make proof thereof the ordinary way with lead.

9.   Take two or four grains (if you have no greater quantity) of any Ore that you have, put to it half an ounce of Venice glass, and melt them together in a crucible, (the crucible being, covered) and according to the tincture that the glass receives from the Ore, so may you Judge what kind of metal there is in the Ore; for if it be a copper Ore, then the glass will be tinged with a sea-green Color. If copper and iron, a glass-green, if iron, a dark yellow; if tin, a pale yellow; if silver, a whitish yellow; if Gold, a fine sky Color; if Gold and silver together, a Smaragdine Color; if Gold, silver, copper, and iron together, an. amethyst Color.

A pretty observation upon the melting of Copper and Tin together.

10.   First make two bullets of red copper of the same magnitude, make also two bullets of the purest tin in the same mold, as the others were made: weigh all four bullets, and observe the weight well: then melt the copper bullets first, upon them being melted put the two tin bullets, and melt them together, but have a care that the tin fume not away. Then cast this molten mixture in the same molds as before, and it will scarce make three bullets, but yet they weigh as heavy as they did before they were melted together.

I suppose the copper condenses the body of the tin, which was very porous, which condensation rather adds then diminishes the weight thereof.

A remarkable observation upon the melting of Salt Armoniack, and Calx vive together.

11.   Take Salt armoniack, and Calx vive, of each a like quantity, mix, and melt them together. Note that Calx of itself will not melt in less than eight hours with the strongest fire that can be made; but being mixed with this salt melts in half an hour, and less, like a metal, with an indifferent fire.

This mixture being thus melted becomes a hard stone, out of which you may strike fire as out of a flint, which if you dissolve again in water, you shall have the Salt armoniack in the sane quantity as before, but fixed.

Note that hard things have their congelation from Salt armoniack, as horns, bones and such like; for little fixed salt can be extracted from them, only volatile and armoniack.

An ounce of any of these volatile Salts, (as of horns, bones, amber and such like) reduced into an acid Liquor by distillation, condenses, and indurates a pound of Oily matter.

An easy and cheap powder like unto Aurum Fulminans.

12. Take of salt Tartar one part, Saltpeter three parts, Sulphur a third part, grind these well together, and dry them, A few grains of this powder being fired will give as great a clap as a musket when it is discharged.

To make an Antimonial cup, and to cast divers figures of Antimony.

13. Take the best crude antimony very well powdered, Nitre, of each a pound, of crude Tartar finely powdered two pounds, mix them well together, and put them into a crucible, cover the crucible, and melt them, and the regulus will fall to the bottom, and be like a melted metal, then pour it forth into a brass mortar, being first smeared over with Oil.

Or,

Take two parts of powdered Antimony, and four parts of powder of crude Tartar; melt these as aforesaid.

This regulus you may (when you have made enough of it) melt again and cast it into what molds you please; you may either make cups, or what pictures you please, and of what figures you please. You may cast it into forms of shillings or half-crowns, either of which if you put it into two or three ounces of wine in an earthen glazed vessel, or glass, and infuse in a moderate heat all night, you may have a Liquor in the morning which will cause vomit: of which the dose is from two drams to two ounces and a half.

Note that in the Wine you may put a little Cinnamon to correct and give a more grateful relish to it.

It is the custom to fill the Antimonial cup with Wine, and to put as much Wine round about betwixt that and the little earthen cup where it stands, and so infuse it all night, and then drink up all that Wine: but I fear, that so much Wine will be too much as being three or four ounces when as we seldom exceed the quantity of two ounces of the infusion of Antimony.

These cups or pictures will last forever, and be as effectual after a thousand times infusion as at first: and if they be broken at any time, (as easily they may, being as brittle as glass) they may be cast again into what forms you please.

Note that he that casts them must be skillful in making his spawde[35], as also in scouring of them, and making them bright afterwards: for if they be carefully handled, they will look even as bright as silver.

The Rosie Crucians give this Rule for the Gout: To be taken in this order.

---

[35] Spelled exactly as in the 1662 edition. -pnw

The Pultasie.

1. Take of Manchet about three ounces, the crumb only, thin cut, let it be boiled in Milk till it grow to a Pulp. Add in the end a drachm and a half of the powder of red Roses, of Saffron ten grains, of oil of Roses an ounce; let it be spread out upon a linen cloth, and applied lukewarm, and continued for three hours space.

The Bath or Fermentation.

2. Take of Sage leaves half a handful, of the root of Humlock sliced six drachms, of Briony roots half an ounce, of the leaves of red Roses two pugile; let them be boiled in a bottle of water, wherein Steel has been quenched, till the Liquor come to a quart; after the straining put in half an handful of Bay-salt: Let it be used with scarlet Cloth or scarlet Wool dipped in the Liquor hot, and so renewed seven times, all in the space of a quarter of an hour, or little more.

The Plastier.

3. Take Emplaistrum Diacalciteos as much as is sufficient for the part you mean to cover; let it be dissolved with Oil of Roses in such a consistence as will stick, and spread upon a piece of Holland, and applied.

Chapter VII.

1. Of a water to Cause hair fallen to grow again. 2. A water to cause hair taken off never to grow again. 3. How to sake another. 4. How to take away spots in the face. 5. A water against scabs. 6. To preserve the sight. 7. Another water. 8. How to restore the sight of. an old san. 9. How to cure the Gout. 10. To cure the Web and spots in the eye. 11. How to cure Tetters, Fistulaes, Cankers, etc. 12. How to cure the redness of the face, and beautify the skin. 13. Another. 14. Another of the same virtue.

A water to cause Hair fallen, to grow again.

1.   Take Mountain-Hyssop, Mountain-Calasint, leaves of Southern-wood, of each two handfuls, Canary-Wine, Urine, Honey, Milk, of each two pounds, Mustard seed half a pound, bruise what is to be bruised, macerate them three days, then distil them in balneo.

A water to cause hair taken off never to grow again.

2.   Take seeds of Henbane bruised two pounds, lay it a while in some moist place, then add great Stonecrop half a pound, distil it according to Art.

Another.

3. Take blood of frogs, Terrae Sigilatae, Sumach, Roses, House-leek, what is sufficient; macerate them together twenty-four hours, then distil them in balneo.

A water to take away spots in the face.

4. Take Asses milk four pounds, White wine one pound, the inside of two new Loaves, twelve Eggs with the shells, Sugar-candy three drachma; mix them well together and distil them.

A water against Scabs.

5. Take Sorrel water two pounds, juice of Plantain, Rose-water, of each four ounces, juice of Lessons two ounces, Lytharge six ounces, Ceruse Sublimate, of each half an ounce, Sulphur vive three drachms; bruise them that are to be bruised; then infuse them 24 hours, and after distil them according to Art.

A water to preserve the sight.

6. Take Fennel, Vervain, Eye-bright, Endive, Betonie, Red Roses, Venus Hair, of each three handfuls. Bruise the herbs and macerate them 24 hours in white wine, (as much as is sufficient) then distil them in a limbeck in balneo.

Another.

7. Take Fennel, Celandine, Sage, Rosemary, Vervain, Rue, of each equal parts. Prepare as it before.

A water to restore the sight decayed.

8. Take Fennel, Celandine, Vervain, Rue, Leaves of Snula, Fullers Teesel, Milfoil, of each one handful; Camphire half a dram, Bruise them and distil them in an Alembic.

A water against the Gout.

9. Take Licorice half a pound, Aniseeds 1. pound, Cinnamon, 3 ounces, Galingale, Ginger, Roots of Iroes, Enula Caspana, Seeds of Fennel, Caraway, Amomum, Amid, Piony, Basil, Savory, Marjoram, of each one ounce, Juniper Berries 2 Ounces, Ground Ivy half a handful, Long Pepper, Calasue, Spikenard, Mace, of each 3 drams;, Valerian 1 drabs, Roots of Angelica half an ounce, Cyprus 4 ounces, Lignum

Aloes half an ounce, Sugar 4 ounces, Maliga line, or strong Ale 32 pounds, Prepare and distil them according to Art.

This water taken inwardly strengthens cold and weak stomachs, and breaks the Stone.

Outwardly applied it eases the Gout, enlarges sinews that are shrunk, & is good against all aches and passions proceeding from melancholy and cold.

A water for the Web and spots in the Eyes.

10.   Take Rue, Plantain, Red roses, Red Poppies, Vervain, Celandine leaves, of each 1-ounce, Red rose water 1 pound and a half, Tutia prepared 1 drachim, Aloes Kepatick an ounce and a half, Cloves 1 ounce, Powder, prepare and distil them according to Art. Drop the water into the Eyes morning and evening.

A water for Tetters, Fistulas, Cankers, etc.

11.   Take strong white Wine Vinegar 8 pounds, Wood Ashes 1 pound, infuse them three days natural, and stir them twice a day, then put thereto unslaked lime 1 pound, let it stand other three days, and stir it as before; when it is well settled, filter off the clear Lee, and put thereto Sal Gemme, Salt Alkali, Salis Vitae, Salt Armoniac, Salt of Tartar, of each one dram. Calx of Eggshells, and Calx vive, of each 1 drachm; grind all these together, and temper then with the said Lee;

put them into a glass alembic and distil them in balneo; give it the first 24 hours no more heat then will make it, and keep it warm: after that distil it off according to Art.

A water against redness of the Face, and to beautify the akin.

12. Take Wild Purslaine, Mallowa, Nightshade, Plantain with the seeds, of each three handfuls. The Whites of 12 Eggs, Lemons number 12; Roch Allum, 4 ounces; prepare and distil them according to Art.

Another.

13. Take Calx of Eggshells, White Coral pulverized, of each 2 ounces, Salt calcinated, and Borax, of each 6 ounces, Gum Tragagant 5 ounces, Roots of white Lillies, number 6, White sop. 8 pounds, Styrax, Calamita, Blzoin, of each 4 ounces. Mix and distil them by Alembic.

Another of the same virtue.

14. Take Wine Vinegar half a pound, Lytharge of Gold one ounce and a half, Ceruse one ounce, Sal Gem six drachms, Roch Alum, half an ounce, Borax, Sulphur vive, Salt liter, of each three drachms, Camphire half a drachm, prepare and distil them according to Art.

Chapter VIII.

1. How to cure inordinate Flux of Tears. 2. Or thus. 3. How to cure red Eyes. 4. How to dense and dry a sharp ulcer. 5. How to make white teeth. 6. To take away the marks of the small pox. 7. To Cicatrize Ulcers. 8. Another thus. 9. To cure Ulcers, 10. Of hollow Ulcers and their cure. 11. Of a Cicatrizing water. 12. Of curing wounds. 13. Another water. 14. To make teeth white. 15. Of the Colic, how to cure it. 16. To cure a cold stomach. 17. Of Sage water. 18. Of Lavender water compounded, and its virtues and use. 19. A pectoral water. 20. Another. 21. Aqua Splenetica & its virtues. 22. Aqua Febrifuga, and its virtues. 23. Aqua Damascena, Odorifera, and its virtues. 214. Aqua Hysterica, and its virtues. 25. Aqua Nephritica. 26. Aqua Apertiva, and their virtues how to use them.

A water against the inordinate Flux of Tears.

1. Take ripe Strawberries as many as you please, set them to digest in Horse dung, fifteen days, then distil them in balneo.

Or thus.

2.   Take Flowers of the white Thorn, leaves or tops of the Willow, Eyebright, of each what suffices, distil them as before.

A water against redness of the Eyes.

3.   Take juice of Celandine, Rue, Vervain, Fennel, of each three ounces, tops and leaves of Roses, of each what suffices, sugar candy three ounces, of the best Tutia, Sanguis Daconis, of each four ounces. Bruise them that are to be bruised, and distill them according to Art.

A water to cleanse and dry a sharp Ulcer.

4.   Take Crude Ilium two ounces, white of Eggs, number fifteen, Juice of Purslane, Plantain, Nightshade, Nicotian, Houseleek, Water of Meadsweet, Trinity grass, Roses, of each four ounces. Labor them well together and draw off the water by an Alembic of glass in balneo.

A water to make Teeth white.

5.   Take Alum six ounces, Common salt three ounces, Myrrh, Mastic, Cloves, of each three drachms. Mix bruise and distill them according to Art.

A water to take away the marks of the Small Pox.

6.   Take Mastic, Myrrh, Aloes Hepatick, lard, Sanguis Draconis, Olibanum, Opopanax, Bdellum, Carpobalsamum, Saffron, Gum Arabick, Liquid Storax, of each two drams and a handful, beat what is to be beaten, then add thereto of clear Turpentine equal weight, distil them according to Art.

A water to Cicatrize Ulcers.

7.   Take red Wine two pounds, Plantain-water half a pound, Rose-water four ounces, Juice of Plantain, Vervain, Shepheard's Purse, Knot-grass, Centaury the less, Comfrey the greater and lesser, of each two ounces. Crude Alum one pound, Cypress Nuts three ounces, Pomegranate flowers hair an ounce, Pomegranate pills three ounces, Gals half an ounce, Bark of the Oak, Sumac, of each five drachms, Turpentine, three ounces, Crude Honey half a pound, Mastic, Olibanum, of each ten drachats, Sarcocol two ounces, Burnt Vitriol, Burnt lead, of each one drachm; Bole Armoniack three ounces, Cassia lignea, half an ounce, Round Birtwort three ounces. Powder what is to be powdered, then mix and distil then.

Another.

8. Take Mastic, Myrrh, Olibanum, Sarcocol, Mummie, of each three drams. Frankincense one ounce, Nutmegs, Cinnamon. Cloves, Cubebs, of each two drams. Cyprus Nuts half an ounce. Flowers, Barks of Pomegranates, of each one dram. Bole Armoniack one ounce; Sanguis Draconis half an ounce, Red Roses three drachms, Roch Allum one pound, Vitriol 7 drachms, Clarified Honey one ounce, aqua vitae a pound and half, White Wine one pound, Juice of Plantain, Nightshade, Comfrey of the greater and lesser, of each four ounces, Water wherein Iron has been quenched four pounds; Powder what is to be powdered; and infuse them all night in aqua vitae, in the morning draw forth the water by Alembic.

A water for Ulcers.

9. Take White Wine four pounds, Plantain. water two pounds, Ilium half a pound, White Copperas five ounces, Crude Honey one pound, Licorice Rasped one-pound, Bole Armoniack five ounces, Camphire an ounce and a half, Mercury sublimed two drachms, Bruise what is to be bruised; and distil them by Alembic.

A water for hollow Ulcers.

10. Take Fountain water, Red Wine, of each two pounds and a half, Red Roses, four ounces, Flowers, Rinds of Pomegranates, of each two ounces and a half, Sumac two ounces; Sage a handful, Comfrey the greater and lesser of each half a handful, Sarcocoll three ounces, Mastic

two ounces, Olibanum one ounce, Honey one pound, water of Turpentine a pound and a half, bruise what is to be bruised, and distil them through a alembic of glass with a gentle fire.

A Cicatrizing Water.

11.   Take water wherein Iron has been quenched four pounds. Aqua Balsami Veri four pounds. Turpentine a pound and a half, Crude Honey one pound, Alum ten ounces, white Copperas five ounces, Bole Armoniack seven ounces, Mercury sublimated half a drachm, leaves of Plantain, Comfrey the greater, middle and lesser, Teasil, Knotgrass, St. John's Wort, of each a handful and a half, Frankincense two ounces, Olibanum, White Sanders, of each half an ounce, Red Roses, a handful and a half, Cassia Lignes, Cinnamon, of each three drachms for the first distillation; then take Turpentine one pound, Mastick three drachms, pure Rozen six ounces, Cinnamon, Cloves; of each two drachms, Pomegranate rinds half an ounce, Cyprus Nuts one ounce and a half, White Copperas two ounces, Alum three ounces, Olibanum four ounces, Sanguis Draconis an ounce and a half, Aqua Balsam Veri one pound, for the second distillation: Afterwards, Take flowers of St. John's Wort, Sage, Rosemary, Carduits Benedictus, Centaury, of each one ounce, Mastic, Red Sanders, of each three drachmas, Wood of Aloes, two serupies, Cubebs one drachm, Aqua Vitae half a pound, Burnt Allun, white Tartar, of each an ounce and a half, Myrrh half an ounce, Earthworms in powder one drachm, the middle Bark of the Oak six ounces, Cassia Lignea three drachma, White Copperas one ounce, Rinds of Pomegranates one drachm, Guajacum four ounces, Carpobalsamum, of each 1 dram, Myrtles, Mummie, of each two

drachms, Borax half an ounce, Cloves two drachms, Tormentil, Gentian, Round Birt-wort, of each two drachms and a half, This is for the last distillation, afterwards add Burds Allun half an ounce, White Copperas two drachma, Mastic one ounce in fine powder, and then keep for use.

A water for hollow Wounds.

12. Take fountain water, Red Wine, of each two ounces and a half, Red Roses, four ounces, Pomegranate flowers, Pomegranate rinds, of each two ounces and a half, Sumac two ounces, Sage one handful, both the Comfries, of each a handful, Alum half a pound, Sarcocol three ounces, Mastic two ounces, Olibanum one ounce, Honey one pound, water of Turpentine a pound and a half. Prepare the ingredients according to Art, and then distil them all together in a glass alembic with a gentle fire.

A water for wounds and Ulcers.

13. Take Calx vive extinct in fountain water eight pounds, Plantain water four ounces, Rosewater two pounds. Heat all these together; afterwards let them stand and clear, pour forth all the clear to the alembic, and put to it Honey two pounds, Alum an ounce, Borax, Mastic, of each three ounces, Olibanum four ounces, the middle Bark of the Oak dried, three ounces, powder what is to be powdered, and distil then according to Art.

A water to make Teeth white.

14. Take the first distilled water of Honey which is white, one pound, Ilium half a pound, Salt liter, white Salt, of each one ounce, Water of Lentisk leaves one pound, Mastic two ounces, White Vinegar, White Vine, of each two ounces. Mix and distil them according to Art, and reserve the water.

A water against the Cholic.

15. Take Muscadel, or Malmsey four pounds, Nutmegs, Galls, of each one drachm, Cinnamon, Cloves, Grains, of each two drachms. Powder the ingredients grossly, and infuse them in the wine 24 hours, then with a soft fire draw off the water according to Art.

A water for a cold Stomach.

16. Take Citron and Orange peels dried, of each two ounces, Rosemary, Mints, of each one handful. Cinnamon, Cloves, Cubebs, Cardamoms, Nutmegs, Ginger, of each a drachm and a half, Sage, Pennyroyal, Thyme, of each one handful, Caraway seeds, Aniseeds, of each four drachms. Bruise what is to be bruised, and infuse then all the space of 24 hours in Canary wine four pints, then distil them in balneo according to Art.

Water of Sage Compounded.

17. Take Sage, Marjoram, Thyme, Lavender, Epithymum, Betony, of each one-ounce, Cinnamon half an ounce, Ireos Roots of Cyprus, Calamus Aromaticus, of each one ounce, Storax, Benjamin, of each a drachm and a half, infuse them four days in four pounds of Spirit of Wine; then distil them in balneo.

Lavender water Compounded.

18. Take flowers of Lavender, Lilly of the Valley, of each 24 handfuls, Peony, Tillia, Flowers of Rosemary Sage, of each half a handful, Cinnamon, Ginger, Cloves, Cubebs, Galingale, Calamus Aromaticus, Mace, Mistletoe of the Oak, of each a drachm and a half, Peony roots one ounce and a half, of the best Wine what suffices, infuse them in the Wine two days, then distil them in balneo.

This water is good against the Falling sickness, Convulsion fits, and the infirmities of the brain.

A Pectorial water.

19. Take the Liver of a Calf, the Lungs of a Fox, of each number 1. Liverwort, Lungwort, Sage, Rue, Hyssop, of each one handful, Roots of Enula; Gladiol, of each half an ounce, Seeds of Anise; Caraway, Fennel, of each half an ounce, Flowers of Borage and Bugloss, of each two drachma, infuse then the space of 24 hours, in rich old Wine what suffices, water of Scabius, Carduns Benedictus of each four ounces, Hyssop two ounces; then distil it in Balneo Mariae.

Another.

20. Take leaves of Scabius, Veronica, of each two handfuls, Venus Hair, Sage, Hyssop, Horehound, Liverwort, Licorice, of each one handful, Flowers of Borage, Buglosse, Violets, of each half a handful, Roots of Enula Campana, Licorice, Flowers of Ireos, of each half an ounce, Aniseeds, Fennel seed, of each one drachm, choice Cinnamon, oriental Saffron, of each half a dram, let them be bruised and cut, be digested in water of Scabius, Veronica, of each one pound, water of Hyssop half a pound, white Wine three pounds, let them digest two days, then distilled in Balneo Mariae; add Sugar candy what suffices.

This water opens the obstructions of the Liver and Lungs, and strengthens them.

Aqua Splenetiva.

21. Take roots of Fern two ounces, roots of Parsley, Polypody, of each an ounce and a bait, roots of Round Birtwort, Lovage, Calamus Aromaticus, Acorus of this water, of each one ounce, chosen Rubarb; barks of Tamarisk, Copperas, Ash, of each half an ounce, Lovage, Seeds of Caraway, Cumin, Anise, of each two drachms, Scolopendria, tops of Wormwood, Fumiterrie, Dodder, leaves of Agrimony, Ceterach, of each a handful and a half: Rich Wine eight pounds, let them be digested two days, and then distil them in Balneo Mariae.

This rather strengthens the spleen, opens and provokes Urine.

Aqua Febrifuga.

22. Take roots of Vipers grass, Cinquefoil, Tormentil, Dictamum, of each six drachms, Seeds of Citron excorticated, Carduus Benedictus, Carduus Mariae, Sorrel, of each half an ounce, of all the Sanders, of each one drachm, of the Cordial flowers, of each one handful, Goats Rue one handful, Hartshorn rasped half an ounce, pour upon them bruised water of Tormentil, Cichorie, Carduus Benedictus, Carduus Mariae, Wild Poppy, of each what suffices; let them be macerated three days in a glass, close shut, afterward add Citrons bruised number six, Juice of Endive, Cardus Benedictus, Plantain, of each one pound, Borage, Scordium, of each half a pound; let them be distilled in Balneo Mariae.

This water is convenient in Fevers, especially malignant Fevers: because it drives away the malignity, and resists putrefaction.

Aqua Damascena Odorifera.

23. Take Ireos Flowers, Cloves, Cubebs, Cinnamon, Grains of Paradise; Calamus Aromaticus, of each one ounce, Marjoram, Thyme, Bay leaves, Rosemary Flowers, Red Roses, of each a handful. Lavender flowers three drachms, of the best Wine three measures; let then be macerated and distilled: to the distilled liquor add Musk half a scruple, Civet six grains.

This water heats, dries, cuts, discusses, and chiefly strengthens the Heart and head.

Aqua Hysterica.

24. Take roots of Dictamnum, seeds of Dancus, of each one ounce; Cinnamon, Cassia Lignea, Balm, of each two scruples, Oriental Saffron one scruple, New Castoreum one scruple and a half; of all these mixt make a powder, to which let be powdered water of Rue two pounds and a half: let them stand in infusion four days and then distil then in Balneo Maria.

Aqua Nephritica.

25. Take roots of Enula Campana, Cammock, Pimpernel, Radish of each one ounce, Parsley, Lovage, of each seven drachms; leaves of Lovage, Parsley, of each one handful, Saxifrage *cum toto* two ounces,

Flowers of Broom, Balm, Rosemary, of each half a handful; Elder one handful, Berries of Juniper, Myrtle, Alcakengie, Aniseeds, of each two ounces, cut then and infuse them the space of eight days in twelve pounds of the best white Wine, then let them be distilled.

This water opens and provokes Urine: the dose is one spoonful.

Aqua Apertiva.

26.   Take roots of Eringo, Vipers grass, Fern, the greater Centaury, of each half an ounce; roots of Fennel, Barks of Copparis, Tamarisk, Ash, of three drachms, Barks of Citrons two drachms and a half; Seeds of Carduus Bendictus, Cichorie, of each half an ounce, seeds of Endive, Cresses, Citrone, Scariol, of each two drachms, Polytricon, Adianthum, Ceterach, Dodder, Scolopendria, Betony, Endive, of each a handful and a half. Tops of Thyme, Epithymum, Hops, Flowers of St. John's Wort, Broone, Borage, Balm, of each one handful, small Raisons, one ounce: Cinnamon one drachm and a half. Stec. Dialac. half a drachm, Carduus Benedictus, Water of Hops, Scolopendria, Paula Betony, of each one-pound, Rhenish Wine two pounds and a half; let them stand two days in a warm place in a vessel close stopped: afterwards distil then in balneo.

This water opens the obstructions of the whole body, but especially of the liver, spleen, and Mesentery.

Chapter IX.

1. How to make the Golden tree of Philosophers. 2. To make the Tree of the Sun. 3. To make Gold grow in the Earth. 4. Of the Golden Marcasite. 5. Of preparing of it. 6. Of the virtues of prepared Gold. 7. Of prepared Silver. 8. Of Beata's Medicine. 9. Beata's green Oil of silver. 10. To make oil of silver. 11. To make a liquor of silver, that it shall make the glass wherein it is so exceeding cold, that no man is able for the coldness thereof to hold it in his hand any length of time. 12. How to make silver as white as snow. 13. Of Silver Trees. 14. Of preparing Philosophers Gold and silver. 15. The process of the Terrestrial Holy Celi. 16. The Process of the Pantarva. 17. The Process of the Posie Crucian Medicines, and of their dissolving Gold. 18. The Process of the Panarea, and Hermes Medicines.

To make Gold grow in a glass like a Tree which is called the golden Tree of the Philosophers.

1. Take of Oil of Sand as much as you please, pour upon it the sane quantity of Oil of Tartar *per deliquim*, shake them well together that they be incorporated and become as one Liquor of a thin consistence, then is your Menstruum or Liquor prepared. Then dissolve Gold in aqua regia, and evaporate the Menstruum and dry the Calx in the fire, but make it not too hot, for it will thereby lose its growing quality; then take it out and break it into little bite, not into powder, put those bits into the aforesaid Liquor (that they may lay a fingers breath the one from the other) in a very clear glass. Keep the

Liquor from the air, and you shall see that those bits of the calx will presently begin to grow; first they will swell, then they will put forth one or two stems, then divers branches and twigs so exactly, as that you cannot choose but exceedingly wonder. This growing is real, and not imaginary only. Note that the glass must stand still, and not be moved.

The Tree of the Sun,

2. Calcine fine Gold in aqua regis, that it become a calx, which put into a gourd glass, and pour upon it good and fresh aqua regia, and the water of gradation, so that they cover the calx four fingers breadth; This Menstruum abstract in the third degree of fire until no more will ascend. Pour this distilled water on it again and abstract it as before, and this do so often till you see the Gold rise in the glass, and grow in the form of a Tree having many bows and leaves.

To make Gold grow and be increased in the earth.

3. Take leaves of Gold, and bury them in the earth which looks towards the East, and let it be often soiled with man's urine, and doves' dung, and you shall see that in a short time they will be increased.

The reason of this growth I conceive may be the golds attracting that universal vapor and sperm that comes from the center through the earth (as has been spoken in the anatomy of Gold) and by the heat of putrefaction of the dung purifying and assimilating it to itself.

A remarkable observation upon a golden Marcasite.

4. There is found a certain stone in Bononia, which some call a golden Marcasite, some a solarie Magnes, that receives light from the sun in the day time, and gives it forth in the dark. About this there has been much reasoning amongst Philosophers, as whether light be really a body, or any kind of substance, or any accident only, and whether this stone had any Gold in it or not, and what it did consist of. He that first discovered it, thought that he had found a thing that would transmute metals into Gold, (by which it appears that there seemed to be something of Gold in it or something more glorious then Gold) but his hopes were frustrated by a fruitless labor, notwithstanding which I conceive there might be some immature or crude Gold in it; for crude Gold is a subject (being there is some life in it) that is most fit to receive the influences of the sun according to the unanimous consent of all Philosophers, and therefore is by them not only called Solary, but 501 (i.e.) the sun itself.

5. It is prepared for the receiving of light thus, it is calcined two ways, first it is brought into a most subtle powder with a very strong fire in a crucible; secondly, being thus brought into a powder, is made up into cakes as big as a dollar, or a piece of eight, either with a common water alone, or with the white of an egg; put those cakes being dried by themselves into a Wind Furnace S.S.S.,[36] with coals, and

calcine them in a most strong fire for the space of four hours. When the furnace is cold, take them out, and if they be not sufficiently calcined the first time, (which is known by their giving but little light,) then reiterate the calcination after the same manner as before, which is sometimes to be done thrice, that is the best which is made with the choicest stones that are clean, pure, and diaphanous, and gives the best light. With this being powdered you may make the forms of divers animals of what shapes you please, which you must keep in boxes, and they will receive light from the sun in the day tine, give light in the night, or in a dark place, which light will vanish by degrees.

The virtues of the aforesaid preparations of Gold, and their virtues and use.

6.   With the aforesaid preparations the Ancients did not only preserve the health and strength of their bodies, but also prolonged their lives to a very old age, and not that only, but cured thoroughly the Epilepsy, Apoplexy, Elephantiasis, Leprosy, Melancholy, Madness, the Quartan, the Gout, Dropsy, Pleurisy, all manner of Fevers, the Jaundice, Lues Venerea, the Wolf, Cancer, Ioli Me Tangere, Asthma, Consumption, the Stone, stopping of Urine, inward Impostumes, and such like diseases, which most men account incurable. For there is such a potent fire lying in prepared Gold, which doth not only reassume deadly humours, but also renews the very marrow of the bones, and raises up the whole body of man being half dead.

---

[36] S.S.S. means layer upon layer. -pnw

They that use any of these preparations for any of the foregoing diseases, must betake themselves to their bed for the apace of two or three hours, and expect sweating to ensue; for indeed it will send forth sweat plentifully, and with ease, and leave no impurity or superfluity in the whole body. Note that they must take it for ten days together in appropriated Liquors.

Let young men that expect long life, take any of the aforesaid preparations once a month, and in the morning; but they must abstain from meat and drink, till the evening of the same day; for in that time that matter will, be digested into the radical humor; whereby the strength of the body is wonderfully increased, beauty doth flourish most wonderfully, and continues till extreme old age.

Let old men take it twice in a month, for by this means will old age be fresh till the appointed time of death.

Let young women and maids take it once in a month after their menstrua, for by this means they will look fresh and beautiful.

Let women that are in travel take it, and it will help and strengthen them to bring forth without much pain, notwithstanding many difficulties.

Let it be given to women that have past the years of their menstrua once or twice in a month, and it will preserve them very fresh, and many times cause their menstrua to return, and make them capable, again of bearing children.

It cures the plague, and expels the matter of a carbuncle by sweat most potently.

When I say that this, or it will do thus or thus, I mean any one of the forenamed preparations; viz. aurum potable, Oils, or Tincture of Gold.

The preparations of silver in general.

7. All the several preparations of Gold may, except that of aurum fulminans, be applied to silver, of which being thus prepared the virtues are inferior to those of Gold, yet comes nearer to them than those of any other matter whatsoever, or howsoever prepared.

Note that silver has some peculiar preparations which neither Gold nor any other metal are capable of.

Beata's Gift.

8. Take fine Silver, and dissolve it in twice so much rectified Spirit of Nitre, then abstract half of the said spirit in sand; let it stand a day or two in a cold place, and much of the Silver will shoot into Crystals, and in oft doing most of it.

These Crystals are very bitter, yet may be made into pills, and taken inwardly from three grains to twelve; they purge very securely and gently, and Color the lips, tongue, and mouth black. If in this dissolution of Silver, before it be brought to Crystals, half so much Mercury be dissolved, and both shoot together into Crystals, you shall have a stone not much unlike to Alum. This purgeth sooner, and better, and is not so bitter; it colored the nails, hair, skin, if it be dissolved in rainwater, with a lovely brown, red, or black, according as you put more or less thereof.

Take of the aforesaid Crystals of silver, mix with them a like quantity of pure Saltpeter well powdered, then put this mixture into the distilling vessel, at the bottom of which must be pondered coals to the thickness of two fingers breadth, then make a strong fire, that the vessel and coals be red hot; put in a drachm of the aforesaid mixture, and it will presently sublime in a silver fume into the recipient, which being settled, put in more, and so do till you have enough. Take out the flowers, and digest them ii the best alcholimated spirit of Wine, that thereby the tincture nay be extracted, which will be green.

Beata's Green Oil of Silver.

9. Take of the abovesaid Crystals of silver one part, of spirit of Salt armoniack two or three parts, digest them together in a glass with a long neck, well stopped, twelve or fourteen days, so will the spirit of Salt armoniack be colored with a very specious blue Color; pour it off, and filter it, then put it into a small Retort, and draw off most of the spirit of Armoniack, and there will remain in the bottom a grass-green Liquor. Then draw off all the spirit, and there will remain in the bottom a Salt, which may be purified with spirit of Wine, or be put into a Retort, and then there will distil off a subtle Spirit, and a sharp Oil.

This green Liquor is of great use for the gilding of all things presently.

If you take common rainwater distilled, and dissolve and digest the aforesaid Crystals of silver for a few days, you shall after the appearance of divers colors find an essence at the bottom, not so bitter as the former, but sweet, and in this Liquor may all metals in a gentle heat by long digestion be maturated, and made fit for medicine; but note that they must first be reduced into salts, for then they are no more dead bodies, but by this preparation have obtained a new life, and are the metals of Philosophers.

To make Oil of Silver per deliquum.

10. Take of the aforesaid Salts, or Crystals of silver, and reverberate them in a very gentle fire, then put them into a Cellar on a Marble stone, and they will in two months' time be turned into a Liquor.

To make a Liquor of Silver, that shall make the glass wherein it is so exceeding cold, that no man is able for the coldness thereof to hold it in his hand any long time.

11. Take the aforesaid salt of Silver, pour upon it the spirit of salt Arnoniack, and dissolve it thoroughly, and it will do as abovesaid.

With a glass, being full of this Liquor, you may condense the air into water in the heat of the summer, as also freeze water.

To make Silver as white as snow.

12. Take of the calx of Silver made by the dissolution of it in AQUA FORTIS, dulcify it, and boil it in a Lixivium made of Soap-ashes, and it will be as white as any snow.

To make the Silver-tree of the Philosophers.

13.   Take four ounces of aqua fortis, in which dissolve an ounce of fine silver then take two ounces of aqua fortis, in which is dissolved half an ounce of argent vive; mix these two Liquors together in a clear glass with a pint of pure water, stop the glass very close, and you shall see, day after day, a tree to grow by little and little, which is wonderful pleasant to behold.

To preserve Philosophers Gold and Silver.

14.   I have set down several vulgar preparations of Gold and Silver, and of almost all things else, I shall now crave leave to give an account of some Philosophical preparations of the Philosophers Gold and Silver. For indeed the Art of preparing of then is the true Alchemy, in comparison of which all the Chemical discoveries are but Abortive, and found out by accident, viz., by endeavoring after this. I would not have the world believe, that I pretend to the understanding of them, yet I would have them know, that I am not incredulous as touching the possibility of that great philosophical work, which many have so much labored after, and many have found. To me there is nothing in the world seems more possible, and whosoever shall without prejudice read over my *Harmony of the World*, shall almost, whether he will or no (unless he resolves not to believe anything though never so credible) be convinced of the possibility of it. What unworthiness God saw in Gold more in other things, that he should deny the seed of

multiplication (which is the perfection of the creatures) to it, and gives it to all things besides, seems to me to be a question as hard to be resolved, yea, and harder than the finding out the Elixir itself, in the discovering of which the greatest difficulty is, not to be convinced of the easiness thereof. If the preparations were difficult, many more would find it out then do (says Sendivogius) for they cast themselves upon most difficult operations, and are very subtle in difficult discoveries, which the Philosophers never dreamed of. Nay, says the aforesaid Author, if Hermes himself were now living, together with the subtle-witted Geber, and most profound Raymund Lully, they would be accounted by our Chymists not for Philosophers, but rather for learners. They were ignorant of those so many distillations, so many circulations, and so many other innumerable operations of Artists, nowadays used, which indeed men of this age did find out and invented out of their book; Yet there is one thing wanting to us which they did, vim, to know how to make the Philosophers stone, or physical tincture, the processes of which, according to some Philosophers, are these.

The Process of the terrestrial Bali Call.

15. Take the mineral Electrum, being immature and made very subtle, put it into its own sphere, that the impurities and superfluities may be washed away, then purge it as much as possibly you can with STIBIUM, after the Alchymistical way, lest by its impurity thou suffer prejudice; then resolve it in the stomach of an Estridge, which is brought forth in the earth, and through the sharpness of the Eagle is contorted in its virtue.

Now when the Electrum is consumed, and has after its resolution received the Color of the Marigold, do not forget to reduce it into a spiritual transparent essence, which is like to true Amber; then add half so much as the Electrum did weigh before its preparation of the extended Eagle, and oftentimes abstract from It the stomach of the Estridge, and by the means the Electrum will be made more spiritual. Now when the stomach of the Estridge is wearied with labor, it will be necessary to refresh it, and always to abstract it. Lastly, when it has again lost its sharpness, add the tartarated quintessence, yet so, that it be spoiled of its redness the height of four fingers, and that pass over with it. This do so often till it be of itself white, and when it is enough, and you see that sign, sublime it; so, will the Electrum be converted into the whiteness of an exalted Eagle, and with a little more labor be transmuted into deep redness, and then it is fit for medicine.

The process of the Pantarva; and Projection according to the Rosie Crucians.

16. Take of our Earth through eleven degrees eleven grains; of our Gold, and not of the vulgar, one grain; of our Lune, not of the vulgar, two grains; but be thou admonished that thou take not the Gold and Silver of the vulgar, for they are dead, but take ours which are living, then put them into our fire, and there will thence be made a dry Liquor:

First the Earth will be resolved into water, which is called the Mercury of Philosophers, and in that water, it will resolve the bodies of the Sun and Moon, and consume them, that there remain but the tenth part with one part, and this will be the Humidum Radicale Metallicum.

Then take the water of the salt Nitre of our Earth, in which there is a living stream if you dig the pit knee deep, take therefore the water of it, but take it clear, and set over it that Humidum Radicale, and put it over the fire of putrefaction and generation, but not such as was that in the first operation. Govern all things with a great deal of discretion, until there appear colors like to the tail of a Peacock; govern it by digestion of it, and be not weary, till these colors cease, and there appear throughout the whole a green Color, and so of the rest; and when thou shalt see in the bottom ashes of a fiery Color, and the water almost red, open the vessel, dip in a feather, and smear over some iron with it; if it tinge, have in readiness that water which is the Menstruum of the World, (out of the sphere of the Noon so often rectified, until it can calcine Gold) put in so much of that water as was the cold air which went in, boil it again with the former fire until it tinge again.

The Rosie Crucian Universal Medicine, and a way how to dissolve Metals.

17.   Take the matter, and grind it with a physical contrition, as diligently as may be, then set it upon the fire, and let the proportion of fire be known, vim, that it only stir up the matter, and in a short time, that fire, without any other laying on of hands, will accomplish the whole work, because it will putrefy, corrupt, generate, and perfect, and make to appear the three principal colors, black, white, and red: And by the means of our fire, the medicine will be multiplied, if it be joined with the crude matter, not only in quantity, but also in virtue. Withal they might therefore search out this fire (which is mineral, equal, continual, vapors not away, except it be too much stirred up, partakes

of Sulphur, is taken from elsewhere then from the matter; pulls down all things, dissolves, congeals, and calcines, and is artificial to find out, and that by a compendious and near way, without any cost, at least very small, is not transmuted with the matter, because it is not of the matter) and thou shalt attain thy wish, because it doth the whole work, and is the key of the Philosophers, which they never revealed.

The process of the Panarea and Hermes Medicines, and the Art of projection of the Elixir.

18.   True without all falsity, certain and most true; that which is inferior is as that which is superior, and that which is superior is as that which is inferior; read my *Harmony of the World*, for the accomplishment of the miracles of one thing. And as all things were from one, by the mediation of one, so all things have proceeded from this one thing by adaptation. The Father therefore is the Sun, and the Mother thereof the Moon, the Wind carried it in its belly, the Nurse thereof is the Earth.

The Father of all the perfection of the whole World is this: the virtue thereof is entire, if it be turned into earth: Thou shalt separate the earth from the fire, the subtle from the thick, sweetly, with a great deal of judgement. It ascends from the earth up to heaven, and again descends down to the earth, and receives the powers of superiors and inferiors. So, you have the glory of the whole world; therefore, let all obscurity fly from thee; This is the strong fortitude of the whole fortitude, because it shall overcome everything that is subtle, and penetrate every solid thing, as the world is created: Hence shall wonderful adaptations be, whereof this is the manner, wherefore I am called

Hermes Trismegistus, having three parts of the philosophy of the whole World. It is complete, what I have spoken of the operation of the Sun.

These Medicines are good against all Diseases.

Now if you know the first Matter, you have discovered the Sanctuary of Nature, there is nothing between you and these treasures, the Mountain of Diamonds, the Youth and his Medicines, and all the powers of astromancy and geomancy are at your command; but you must open the door; if your desire lead you on to the practice. Consider well with yourself what manner of man you are, and what it is you would do; for it is no small matter you have resolved, to be a co-operator with the spirit of the living God and to minister to him in his work of generation: Have a care therefore that you do not hinder his work; for if your heat exceeds the natural proportion, you have stirred the wrath of the moist natures, and they will stand up against the central fire, and the central fire against teem, and there will be a terrible division in the chaos: but the sweet Spirit of Peace, the true eternal Quintessence, will depart from the Elements, leaving both them and you to confusion; neither will he apply himself to the matter, as long as it is in your violent destroying hands: take heed therefore, least you turn partner with the serpent, for it is the Devils design from the beginning of the world, to set Nature at variance with herself, that he may totally corrupt and destroy her; *Ne tu augeas fatum*, do not further his designs, many men will laugh at this; but on my word, I speak nothing but what I have known by very good experience, therefore believe me, for my own part, it was ever my desire to bury these secrets in silence, or to print them out in shadows, but I have spoken thus clearly and openly out of the affection I bear to some, who have deserved much more at my bands: True it is, I intended sometimes to expose a greater work to the world, which I promised in my *Temple of Wisdom*; but I have been since acquainted with the world, and I found it base and unworthy. I fear not Man and his noise is nothing to me; I seek not his applause, and so I end the fifth Book.

Book VI.

The Rosie Cross Uncovered,

and

The Places, Temples, holy Houses, Castles, and invisible Mountains of the Brethren discovered and communicated to the World, for the full satisfaction of Philosophers, Alchemists, Astromancers, Geomancers, Physicians and Astronomers.

Whereunto is added,

A Bar to stop Thomas Street from his impudent Attempts, and mad clambering up to Astronomy; to which is demonstrated, that his *Tabula Corolina* is all false, and that he belies his Authors, notwithstanding he was mine years studying his own admired Experience.

To my much-honored Friends, Thomas Temple of Bourton upon the Water in the Country of Glocester Esquire, Page to Prince Rupert, and Gentleman of the Kings Privy Chamber.

And

Christopher Rodd of Hereford Esq; and in Clifford's-Inn, one of the Attourneys of the Kings Bench.

All Celestial and terrestrial Happiness be wished.

Gentlemen,

As boyish airs please trivial Ears, so they kiss the fancy and betray it; but behold without flattery or expectation of gain, I give you an unheard of piece of Rosie Crucian philosophy and physick, I do not cry Hail first, and after crucify; I present it to you, because you are two guards of safety; and if you except it not, I shall not therefore be angry, but question me self for this presumption, to come so plain before wisdom and virtue; you gave me the first encouragement, and my philosophy returns to you for Patronage; I know your abilities to discern, and knowledge to defend; you have art and candor, let the one judge, let the other excuse.

June 9. Your most humble Servant

1662.

John Heydon.

An Apologue for an Epilogue.

I shall here tell you what Rosie Crucians are, and that Moses was their Father, and he was Θεῦ παῖς; some say they were of the order of Elias, some say the Disciples of Ezekiel; others define them to be

Ὑπάρχης τῦ πανηγέμιν☉, ὥσπερ μεγάλε βασιλέως ὀφθαλμὸς κ̀ ὦτα, ἀφορῶσας πάντα κ̀ ἀκέεσας

i.e., The Officers of the Generalissimo of the world, that are as the eyes and ears of the great King, seeing and hearing all things; they are Seraphically illuminated, as Moses was, according to this order of the elements, Earth refined to Water, Water to Air, Air to Fire; so of a man to be one of the heroes, of a heroes a demon, or good genius, of a genius a partaker of Divine things, and a companion of the holy company of unbodied Souls and immortal Angels, and according to their vehicles, a versatile life, turning themselves, Proteus-like, into any shape.

But there is yet Arguments to procure Mr. Walyord, and T. Williams, Rosie Crucians by election, and that is the miracles that were done by them, in my sight; for it should seem Rosie Crucians were not only initiated into the Mosaical Theory, but have arrived also to the power of working miracles, as Moses, Elias, Ezekial, and the succeeding Prophets did, as being transported where they please, as Habakkuk was from Jewby to Babylon, or as Philip, after he had baptized the Eunich, to Azotus, and one of these went from me to a friend of mine in Devonshire, and came and brought me an answer to London the same day, which is four days journey; they taught me excellent predictions of Astrology, and Earthquakes; they slack the Plague in

Cities; they silence the violent Winds and tempests; they calm the rage of the Sea and Rivers; they walk in the Air, they frustrate the malicious aspects of Witches; they cure all Diseases; I desired one of these to tell me whether my Complexion were capable of the society of my good genius? When I see you again, said he, I will tell you, which is, (when he pleases to come to me, for I know not where to go to him) When I saw him, then he said, Ye should pray to God; for a good and holy man can offer no greater nor more acceptable Sacrifice to God, then the oblation of himself, his soul.

He said also, that the good genii are as the benign eyes of God, running to and fro in the world, with love and pity beholding the innocent endeavors of harmless and single-hearted men; ever ready to do them good, and to help them; and at his going away he bid me beware of my seeming friends, who would do me all the hurt they could, and cause the Governors of the Nations to be angry with me, and set bounds to my liberty: which truly happened to me, as they did indeed: Many things more he told me before we parted, but I shall not name them here.

For this Rosie Crucian Physick or Medicines, I happily and unexpectedly light upon in Arabia, which will prove a restoration of health to all that are afflicted with that sickness, which we ordinarily call natural, and all other Diseases, as the Gout, Dropsy, Leprosy and Falling-sickness; and these men may be said to have no small insight in the body, and that Walford, Williams, and others of the Fraternity now living, may bear up in the same likely Equipage, with those noble Divine Spirits their Predecessors; though the unskillfulness in men commonly acknowledge more of supernatural assistance in hot unsettled fancies, and perplexed melancholy, then in the calm and distinct use of reason; yet for mine own part, but not without submission to better judgements, I look upon these Rosie Crucians above all men truly inspired, and more than any that professed or pretended themselves so, this sixteen hundred years, and I am ravished with admiration of their miracles and transcendent mechanical inventions, for the solving the phenomena in the world; I may without offence therefore compare them with Bezaliel and Aboliab, those skillful and cunning workers of the Tabernacle, who, as Moses testifies, were filled with the Spirit of God, and therefore were of an excellent understanding to find out all manner of curious work.

Nor is any more argument, that these Rosie Crucians are not inspired, because they do not say they are; then that others are inspired, because they say they are; which to me is no argument at all; but the suppression of what so happened, would argue much more sobriety and modesty; when as the procession of it with sober men, would be suspected of some piece of melancholy and distraction, especially in these things, where the grand pleasure is the evidence and exercise of reason, not a bare belief, or an ineffable sense of life, in respect whereof there is no true Christian but he is inspired; but if any more zealous pretender to prudence and righteousness, wanting either leisure or ability to examine these Rosie Crucian Medicines to the bottom, shall notwithstanding either condemn them or admire them, he has unbecomingly and indiscreetly ventured out of his own sphere, and I cannot acquit him of injustice or folly: Nor am I a Rosie Crucian, nor do I speak of spite, or hope of gain, or for any such matter; there is no cause, God knows, I envy no man, be he what he will be, I am no Physician, never was, nor never mean to be: what I am it makes no matter as to my profession.

Lastly, these holy and good men would have me know, that the greatest sweet and perfection of a virtuous soul, is the kindly accomplishment of her own nature, in true wisdom and divine love; and these miraculous things that are done by them, are, that that worth and knowledge that is in them may be taken notice of, and that God thereby may be glorified, whose witnesses they are; but no other happiness accrues to them from this, but that hereby they may be in a better capacity of making others happy.

Spittlefields this

10th. of May, 1662.

John Heydon.

The Rosie Cross Uncovered: The Sixth Book.

God, because he was good, did not grieve to have others enjoy his Goodness, (that is, to be and to be well) meaning to make a World, full of all kinds of everlasting and changeable things; First made all, and blended them in one whole confused mass and lump together, born up by his own weight, bending round upon itself.

Then seeing it lay still, and that naught could beget and work upon itself; he sorted out, and sundered a way round about, a fine lively Piece (which they call heaven) for the male mover and working; leaving still the rest as gross and deadly, which moves in opposition to light, and is called darkness, the reward of the wicked; and below this lies the female, to receive the working and fashioning, which we term the four beginnings (or Elements) earth, water, air and fire; And thereof springs the Love which we see get between them, and the great desire to be joined again and coupled together.

Then, that these might be no Number of Confusion in doing causes, but all to flow from one head, as he is One, he drew all force of working and virtue of begetting into one narrow and round compass, which we call Sol; from thence he Bent out, spread and bestowed all about the world, both above and below, which again meeting together, made one general light, heat, nature, life and soul of the World, the cause of all things.

And because it becometh the might, wisdom and pleasure of God to make and rule the infinite variety of changes here below, and not evermore one self-same thing: He commanded that (one light in many) to run his eternal and restless Race to and fro, this way and that way, that by their variable presence, absence and meeting they might fitly work the continual change of flitting Creatures. So, Virgil sings: Thus, translated by Eugenius Theodidactus.

> And first the Heavens, Earth, and liquid Plain,
> The Moons bright Globe, and Stars Titanian,
> A Spirit fed within, spread through the whole,
> And with the huge heap mixt infused a Soul:
> Hence Man, and Beasts, and Birds derive their strain,
> And Monsters floating in the marbled Main.
> These seeds have fiery vigor, and a birth
> Of Heavenly race, but clogged with heavy Earth.

Now there are a kind of men, as they themselves report, named Rosie Crucians, a divine Fraternity that inhabit the Suburbs of Heaven, and these are the Officers of the Generalissimo of the World, that are as the eyes and ears of the great King, seeing and hearing all things: they say these Rosie Crucians are seraphically illuminated, as Moses was, according to this order of the Elements, Earth refined to Water, Water to Air, Air to Fire. So, of a man to be one of the heroes, of a heroes a daemon, or good genius, of a genius a partaker of Divine things, and a Companion of the holy Company of unbodied Souls and immortal Angels, and according to their Vehicles, a versatile life, turning themselves, Proteus-like, into any shape.

But the richest happiness they esteem is the gift of healing and medicine; it was a long time, great labor and travel before they could arrive to this Bliss above set; they were at first poor Gentlemen that studied God and Nature, as they themselves confess; (saying) seeing the only wise and merciful God in these latter days has poured out so richly his mercy and goodness to mankind, whereby we do attain more and more to the perfect knowledge of his Son Jesus Christ and Nature; that justly we may boast of the happy time, wherein there is not only discovered unto us the half part of the World, which was heretofore unknown and hidden; but he has also made manifest unto us many wonderful and never heretofore seen works and Creatures of Nature, and moreover has raised men, endued with great wisdom, which might partly renew and reduce all Arts (in this our Age, spotted and imperfect) to perfection.

So finally, man might thereby understand his own nobleness and worth, and why he is called microcosmos, and how far knowledge extends in nature.

Although the rude World herewith will be but little pleased, but rather smile and scoff thereat; also the pride and covetousness of the Learned is so great, it will not suffer them to agree together; but were they united, they might out of all those things, which in this our age God doth so richly bestow upon us, collect the Book of Nature, or a perfect method of all other Arts, whereof this is the chief; and therefore called R. C. *Axiomata*, But such is their opposition that they still keep, and are loath to leave the old course esteeming Porphory, Aristotle and Galen, yea and that which has but a mere show of learning, more than the clear and manifest light and truth; who if they were now living, with much joy would leave their erroneous doctrines. But here is too great weakness for such a great work.

And although in theology, physick, and the mathematics, the truth doth oppose it itself; nevertheless, the old enemy by his subtilty and craft doth show himself in hindering every good purpose by his instruments and contentions (wavering people.) To such an intent of a general reformation, the most godly and seraphically illuminated Father, our Brother, C.R. a German, the chief and Original of our Fraternity, has much and longtime labored, who by reason of his poverty (although a Gentleman born, and descended of Noble Parents) in the fifth year of his Age was placed in a Cloister, where he had learned indifferently the Greek and Latin tongues, (who upon his earnest desire and request) being yet in his growing years, was associated to a Brother P.A.L. who had determined to go to Apania.

Although his brother died in Cyprus, and so never came to Apamia, yet our brother C.R. did not return but shipped himself over, and went to Damasco, minding from thence to go to Apamia but by reason of the feebleness of his body he remained still there, and by his skill in physick, he obtained much favor with the Isemalits. In the meantime he became by chance acquainted with the wise men of Damcar in Arabia, and beheld what great wonders they wrought, and how Nature was discovered unto them, hereby was that high and noble spirit of brother C.R. so stirred up that Apania was not so much now in his mind as Damcar; also he could not bridle his desires any longer, but made a bargain with the Arabians that they should carry him for a certain sum of money to Damcar; this was in the 16th. year of his Age, when the wise received him (as be himself witnessed) not as a Stranger, but as one whom they had long expected; they called him by his name, and showed him other secrets out of his Cloister, whereat he could not but nightly wonder.

He learned there better the Arabian tongue: so that the year following he Translated the Book M. into good Latin, and I have put it into English, wearing the Title of *The Wisemans Crown*; whereunto is added, *A New Method of Rosie Crucian Physick*. This is the place where he did learn his Physick and Philosophic how to raise the dead; for example, as a snake cut in pieces and rotted in dung, will every piece prove a whole snake again, & etc. and then they began to practice further matters, and to kill birds that are bred by force of seed and conjunction of male and female, and to burn them before they are cold in a glass, and so rotted, and then enclosed in a shell, to hatch it under a Hen; and restore the same; and other strange proofs they made of Dogs, Hogs, or Horses, and by the like kindly corruption to raise them up again, and renew them: And at last they could restore, by the same course, every Brother that died to life again, and so continue many Ages; the rules you find in the fourth book,

Let me speak a word (although I am no Rosie Crucian) of this matter and manner of restoring of a man; Let us call it before reason, and consider what that seed is that makes man, and the place where he is made: what is all the work, is it anything else but a part of man (except his mind) rooted in a continual, even, gentle, moist, and natural heat? Is it not like that the whole body, rotted in like manner, and in a womb agreeable, shall swim out, at last quicken, and arise the same thing? As Medea found true upon Jason's father, and made him young again, as Tully says, recoquendo. And Hermes was after this manner raised from death to Life; so was Virgil the Poet: but the Spanish Earl failed, through the ignorance of his Friend the artist that mistook the heat, moisture, and temper of the work, as you heard in the third book.

But i cannot tell, i will neither avow nor disavow the matter; nature is deep, and wonderful in her deeds, if they be searched to the bottom, and may suffer this, but not religion. But to our R.C. who learned his mathematics here, whereof the world has just cause to rejoice, if there were more love, and less envy. After three years he returned again with good consent, shipped himself over Sinus Arabicus into Egypt; where he remained not long, but only took better notice there, of the Plants and Creatures, of Mineral Medicines, the famous aurum potable, that cures all diseases in body and mind, and of the Oil of Gold. Then he sailed over the whole Mediterranean Sea, for to come unto Fezo where the Arabian had directed him. And it is a great shame unto us that wise men, so far remote the one from the other, should not only be of one opinion, hating all contentious writings; but also, be so willing and ready, under the Seal of Secrecy to impart their secrets to others.

Every year the Arabians and Africans do send one to another, inquiring one of another out of their Arts, if happily they have found out some better things; or if experience had weakened their reasons, yearly there came something to light, whereby the mathematica, chisir and magir (for in those are they of fez most skillful) were amended; as there is nowadays in Germany no want of learned men, cabbalists, physicians, astrologers, geomancers, and philosophers, were there but love and more kindness among them, or that the most part of them would not keep their secrets: as we Germans likewise might gather together many things, if there were the like unity: and desire of searching out of secrets amongst us.

After two years, Brother C.C. departed the City Fez, and sailed with many costly things into Spain, hoping well; he so well and so profitably spent his time in Travel, that the learned in Europe would highly rejoice with him, and began to Rule, and order all their Studies, according to those sound and sure foundations: Be therefore conferred with the learned in Madrid, showing them the Errors of Sodom and Gomorrah, and how the faults of the Church by Episcopacy, and the whole Philosophia Moralis was to be amended.

But because their acceptance happened to him contrary to his expectation, being then ready bountifully to impart all his Arts and Secrets to the Learned, if they would have but undertaken to write *The True and Infallible Axiomata*, which he knew would direct them, like a globe or circle, to the only middle point and centrum, and (as it is usual among the Arabians) it should only serve to the wise and Learned for a Rule, that also there might be a society in Canaan which should have Gold, Silver, and precious Stones, sufficient for to bestow them on Kings for their necessary uses, and lawful purposes: with which such as be Governors might be brought up to learn all that which God has suffered man to know.

Brother C.R. after many Travels, and his fruitless true instructions, returned again into Germany, and there built a neat and fitting habitation, upon a little hill or mount, and on the Hill, there rested always a cloud; and he did there render himself visible or invisible, at his own will and discretion. In this house he spent a great time in the mathematics, and made many fine Instruments, *Ex omnibus hujus artis partibus.*

After five years came into his mind the wished return of the children of Israel out of Egypt, how God would bring them out of bondage with the Instrument Moses. Then he went to his Cloister, to which he bare affection, and desired three of his brethren to go with him to Moses, the chosen servant of God Brother G.V. Brother l.A. and Brother I.O. who besides, that they had more knowledge in the Arts, then at that time many others had, he did bind those three unto himself, to be faithful, diligent, and secret; as also to commit carefully to writing what Moses did; and also all that which he should direct and instruct them in., to the end that those which were to come, and through especial Revelation should be received into this fraternity, might not be deceived of the least syllable and word.

After this manner began the Fraternity of the Rosie Cross, first by four persons, who died and rose again until Christ, and then they came to worship as the Star guided them to Bethlehem of Judea, where lay our Savior in his mother's arms; and then they opened their Treasure and presented unto him gifts, gold, frankincense, and myrrh, and by the Commandment of God went home to their habitation.

These four waxing young again successively many hundreds of years, made a magical language and writing, with a large dictionary, which we yet daily use to Gods praise and glory, and do find great wisdom therein; they made also the first part of the book M. which I will shortly publish by the Title of *The Wisemans Crown.*

Now whilst Brother C.R. was in a proper womb quickening, they concluded to draw and receive yet others more into their Fraternity: To this end was chosen Brother R.G. his deceased fathers brothers SON; Brother B. a skillful Painter, O. their Secretary, and P.D. another Brother elected by consent; and E.F. all Germans, except l.A. so in all they were nine in number, all bachelors and of vowed virginity; by those was collected a volume of all that which man can desire, wish or hope for.

After such a most laudable sort they did spend their lives; and although they were free from all diseases and pain, yet notwithstanding they could not live and pass their time appointed of God: So, they all died, at the death of our Lord and Savior Jesus Christ, and their Spirits attended him into glory. Now the second row of these men by many were called the Wise men of the East; and eighty-one years the Secrets of this Fraternity were concealed.

Now the true and fundamental Relation of finding the memory of the Fraternity of the Rosie Cross is this. A learned man in Germany, went to find out the wise men of the East into many Countries, but could never hear of any of them: So being provided of Gold and Silver, Medicines, Tinctures and Telesemes, he chose a Master of Numbers A. to be his Companion: and finding an old strange habitation, then they set themselves to alter this building, in which renewing, he lighted upon the memorial Table, which was cast in Brass, and contained all the names of the Brethren, with some few other things; this he transferred to another more fitting Vault with great joy; for he had never heard of this Fraternity, being all dead eighty one years before his time. In this Table stuck a great nail, somewhat strong, so that when it was with force drawn out, it took with it a stone and a piece of thin wall, or plastering of the hidden door, and so, unlooked for, uncovered the door; wherefore we did with joy and longing throw down the rest of the wall, and cleared the door, upon which was written in great Letters, Post 81. Annos Patebo, with the year of our Lord under it.

Wherefore we gave God thanks, and let the rest that same night; in the morning following we opened the door, and there appeared to our sight a Vault of seven aides and corners, every side five foot broad, and the height of nine foot. Although the Sun never shined in this Vault, nevertheless it was enlightened with another Sun which had learned this of the Sun, and was situated in the upper part in the center of the ceiling; in the midst, instead of a Tomb-stone, was a round Altar, covered over with a Plate of Brass, and thereon was this engraved.

A. C. R. C. HOC UNIVERSI COMPENDIUM UNIUS MIHI SEPULCHRUM FERI.

Round about the first circle or brim stood.

JESUS MIHI OMNIA.

In the middle were four Figures, enclosed in four Circles, whose circumscription was.

1. NEQUAQUAM VACUUM.
2. LEGIS JUGUM.

3. LIBERTAS EVANGELII.

4. DEI GLORIA INTACTA.

This all clear and bright, as also the seventh side, and the 2. KEPTAGONI: so, we kneeled all down together, and gave thanks to the sole Wise, sole Mighty, and sole Eternal God, who has taught us more than all man's wit could have found out, and praised be his holy Name: This VAULT we parted into three parts, the upper part or ceiling, the wall or side, the ground or floor.

Of the upper part you shall understand no more of it at this time, but that it was divided according to the seven sides in the Triangle, which was in the bright Center: but what therein is contained, you shall, God willing, (that are desirous of our Society) behold the same with your own eyes; but every side or wall is parted into ten squares, everyone with their several Figures and Sentences, as they are truly showed, and set forth Concentratum here in this Book.

The bottom again is parted in the triangle, but because therein is described the power and rule of the inferior Governors, we leave to manifest the same, for fear of the abuse by the evil and ungodly world. But those that are provided and stored with the heavenly Antidote, they do without fear or hurt tread on the head of the old and evil Serpent, which this our Age is well fitted for. Every side or wall had a door for a Chest, wherein there lay divers things, especially all the Works of CR. how he and his Brethren raised each other to Life again: in those Books were written of their going to Bethlehem to worship our Savior Jesus Christ, and of the Itmerariun, and Vitam of C.R. in another Chest were Looking—glasses of divers virtues; as also in other places were little Bells, and Rings, which if any man put upon his finger, he seemed now in green, then in white and blue, red and bloom, and all manner of colors; thus will his Garments change into a pure Color every moment: there were burning Lamps, and wonderful artificial Songs, which they had kept ever since God spoke to Moses in the Mount: They kept the old Testament carefully, and expected Christ to be born; and chose forty five more to bear witness to the incredulous World and superstitious Sects, that Christ is the Son of God, and was crucified at Jerusalem; and left these Brethren all the wonderful Works of God, and the Acts of Moses and the Prophets, to the end, that if it should happen, after many hundreds of years, the Order or fraternity should come to mothing; and if Tyrants should burn the old Testament, which they bear witness to be the Word of God, that then they might by this only Vault be restored again.

And there is another Vault or Habitation of the Brethren in the West of England, and there is recorded all the New Testament, and every Chapter explained.

Now as yet we had not seen the dead body of our careful and wise Father in the German-Hill; we therefore removed the Altar aside, there we lifted up a strong Plate of Brass, and found a fair and worthy body whole and unconsumed, as the same is here, lively counterfeited with all the Ornaments and Attires; in his hand he held a Parchment book divided into two parts, the first was the old Testament, and every Chapter interpreted, and the other is the Book I, which next unto the Bible is our greatest treasure, which ought to be delivered to the censure of the world. At the end of this Book attended this following Elogium.

C.R. of C. Ex Nobili atque splendida Germanae R.C. Familia oriundus, vir sui seculi Divinis revelationibus, Subtilissimis Imaginationibus, Indefessis Laboribus ad Coelestia atque humana Mysteria, arcanave admissus, postquam suam (quam Arabico & Africano, Itieribus collegerat) plusquam regiam atque imperatoriam Gazam suo secul.o nondum Convenietem posteritate eruendam cusiodivisset, & jam suarum Artium, ut & nominis fidos ac conjunctissimos Heredes instisuisset, mundum Minutum omnibus Motibus Magno illi respondentem Fabricasset, hocque tandem Praeteritarum, Praesentium & futurarum rerum Compendio extracto, Centenario Major, non morbo (quem ipse nunquam Corpore expertus erat, numquam alios infestare sinebat) ullo pellente, sed Spiritu Del evocante, illuminatam animam (inter Fratuum amplexus & ultima Oscula) Fidelissimo Creatori Deo reddidisset, Pater dilectissimus, Fra. suanissimus, Preceptor Fidelissimus, amicus integerrimus, à suis ad 1400. Annos hic absconditus est.

Underneath they had subscribed themselves.

1. Fra. I. A. Fra. C. H. Fra. I. H. Electione fraternitatis Caput.

2. fra. G. V. M. P. C. S.

3. Fra. P. C. Junior haeres S. Spiritus.

4. Fra. B. M. P. A. Pictor & Architectus.

5. Fra. G. O. F. H. M. P. I. C. A. M. Cabballsta F. W. N.

Q. A. Z. B. X. O. N. P. E. D. L. F. K. M. Z. A. S. C. P.

Secundi Circuli.

1. Fra. T. H. Successor, Ira. P. A. Mathematicus.

2. Ira. I. O. Successor, Ira. A. D.

3. Ira. P. P. Successor Patris C. R. C. cum Christo Triumphant.

At the end was written.

Ex Deo nascimur, in Jean Morimur, per Spiritum Sanctum reviviscimus.

At this day the Rosie Crucians that have been since Christ, say, their fraternity inhabits the West of England; and they have likewise power to renew themselves, and wax young again, as those did before the birth of Jesus Christ, as you may read in many Books.

And Dr. F. says, somewhere there is a Castle in the West of England, in the earth, and not on the earth, and there the Rosie Crucians dwell, guarded without walls, and possessing nothing, they enjoy all things; in this Castle is great Riches, the Halls fair and rich to behold, and the Chambers are made and composed of white Marble; at the end of the Hall there is a Chimney, whereof the two Pillars that sustain the Mantle-tree, are of fine Jasper, and the Mantle is of rich Calcedony, and the Lintel is made of fine Emeralds trailed with a wing of fine Gold, and the grapes of fine Silver, and all the Pillars in the Hall are of red Calcedony, and the pavement is of fine Amber.

The Chambers are hanged with rich clothes, and the benches and bedsteads are all of white Ivory, richly garnished with precious stones; the Beds were richly covered; there are Ivory Presses, whereon are all manner of Birds cunningly wrought, and in these Presses are Gowns and Robes of most fine Gold, and most rich Mantles, Furred with Sables, and all manner of rich Garments.

And there is a Vault, but it is bigger than that in Germany, which is as clear, as though the Sun in the midst of the day had entered in at ten windows, yet it is seven score steps underground: And there are ten

Servants of the Rosie Crucians, fair young men: And C. B. reports this; when I first came to the Society (says he) I say a great Oven with two mouths, which did cast out great clearness, by which four young men made Paste for Bread, and two delivered the Loaves to other two, and they sit them down upon a rich cloth of silk; then the other two men took the Loaves, and delivered them unto one man by two Loaves at once, and he did set them into the Oven to bake, and at the other mouth of the Oven, there was a man that drew out the white Loaves and Pasts, and before him was another young man, that received them, and put them into baskets, which were richly painted.

C. B. went into another Chamber eighty one Cubits from this, and the Rosie Crucians welcomed him; for he found a Table ready set, and the cloth laid, and there stood Pots of Silver, and Vessels of Gold, bordered with precious Stones and Pearl, and Basins and Ewers of Gold to wash their hands; then we went to dinner; of all manner of Flesh, Fowl, and Fish, of all manner of Meat in the world, there they had plenty, and pots of Gold garnished with precious Stones full of Wine: This Chamber was made of Chrystal, and painted richly with Gold and Azure, and upon the walls were written and engraved all things past, present, and to come, and all manner of golden Medicines for the diseased, as you read in the Preface: upon the Pavement was spread abroad Roses, Flowers, and Herbs sweet-smelling above all savors in the world; and in this Chamber were divers Birds flying about, and singing marvelous sweetly.

In this place have I a desire to live, if it were for no other reason, but what the Sophist sometimes applied to the Mountains, *hos primum sol sulutat, ultimosque deserit. quis locum non amet, dies longiores habentem.* But of this place I will not speak any more least the Readers should mistake me, so as to entertain a suspicion that I am of this Order.

Tobias Williams, Noah Walford, Ira. H. W. V. C. B. I. and these in all are thirty-six, that bear witness of Christ.

And Ira. N. chose C. B. for his Successor, saying, I have long expected your coming; in this place you shall live, and we will teach you all things, and you shall learn our *Axiomata*.

First, you must, as we do, profess Medicine, and cure the sick, and that Gratis.

2. You shall not be constrained to wear one certain kind of Habit, but may therein follow the custom of the Country.

3. Every year upon the day C. you shall meet us in this House, S. Spiritus, or write the cause of your absence, and when I am dead lay me in a glass, and renew me according to Nature to live again, as you are taught by us.

4. And you must look about for a worthy person, who after your decease must succeed you.

5. The word R. C. must be your Mark, Seal, and Character.

6. Our Fraternity shall be concealed seven years, and no more. And thirty of the Brethren departed; only four and the Brethren T. W. and N. W. remained with the Father Ira. R. C. I. A. and their servants a whole year, and T. W., died, and Father I. A. put him in a glass, and buried him for renewing his life.

After few years there will be a general Reformation both of Divine and Human things, according to our desire, and the expectation of others: For its fitting, that before the Rising of the Sun, there should appear and break forth Aurora, or Divine Light in the sky, and so in the meantime some few, which shall give their names, may join together, thereby to increase the number and respect of our Fraternity, and make a happy and wished for beginning of our Philosophical Canons prescribed to us by our brother R. C. and be partakers with us of our treasures, (which never can fail or be wasted) in all humility and love to be eased of this worlds labor, and not walk so blindly in the knowledge of the wonderful works of God.

But that also every Christian may know of what Religion and belief we are, we confess to have the knowledge of Jesus Christ, among his Disciples, end he is the Son of God, and was crucified for Mankind at Jerusalem, him did our Eyes see and worship, being guided by a star. And episcopacy is the best form of Church Government, being most clear and purely professed, and cleansed from factious Presbyterians, Cromwellian Anabaptists, Jesuitical Quakers, and false prophets.

Also, we use two Sacraments as they are instituted with all forms and ceremonies of the first renewed Church in England, we acknowledge Carolus Magnus Secundus, for our Christian head: and in Politia, we acknowledge the Protestant Empire and Quartam Monarchiam for our Government; albeit we know what Alterations be at hand, 1663, 1664, 1665, 1666, 1667, 1668, 1669, and would fain impart the same with all our hearts to other Godly Learned men.

Notwithstanding our writings which is in our hands no man (except God alone) can make it Common, nor any unworthy Person is able to bereave us of it; but we shall help with secret aid, this so good a cause, as God shall permit, or hinder us: for our God is not blind as the Heathens fortuna, but is the Churches Ornament, and the honor of the Temple: Our Philosophy of numbers also is not a new invention, but as Adam after his Fall has received it, and as Moses and Solomon our Men used it; also she ought not much to be doubted of, or contradicted by other opinions, or meanings; but seeing the Truth is peaceable, brief and always like herself in all things, and especially accordingly with Jesus In Omni Parte and all members: And. as he is the Image of the Father, so is she his Image; it shall not be said this is true according to Philosophy, but true according to theology; and wherein Plato, Aristotle, Pythagoras, and others did hit the mark, and wherein Enoch, Abraham, Moses, our Men, and Solomon did excel; but especially wherewith that wonderful book the Bible agrees, all that same concurs together, and makes a Sphere or Globe, whose total parts are equidistant from the Center, as hereof more at large, and more plain shall be spoken in Christianly Conference.

But now concerning (and chiefly in this our age) the ungodly, and accursed Gold making, which has gotten so much the upper hand, whereby under Color of it, many Renegades and Roguish People do use great Villainies, and cozen and abuse the credit which is given them, yea nowadays men of discretion do hold the transmutation of Metals to be the highest Point and Fastigium in Philosophy, this is all their intent and desire; and that God would be most esteemed by them, and honored, which could make great store of Gold, and in abundance, the which with unpremeditated Prayers, they hope to obtain of the All-knowing God, and searcher of all hearts; we therefore do by these present public testimony, that the true Philosophers are far of another mind, esteeming little the making of Gold, which is but a Paragon; for besides that they have a thousand better things. And we say with our loving Fore-fathers, Phy. Aurum, Nisi Quantum Aurum; for unto them the whole Nature is detected; he doth not rejoice, that he can make Gold, and that as says Christ, the angels and devils are obedient unto him, but is glad that he sees the Heavens open, and the Angels of God ascending and descending, and his name written in the Book of Life.

Also, we do testify that under the name of Crymia many Books and Pictures are set forth in Contumeliam Gloriae Del, as we will name in their due season, and will give to the Pure hearted a Catalog or Register of them; and we pray all learned men to take heed of The Aurum Chemicum Britannicum, published by Elias Ashmole Esquire, and such kind of Books as these; for the Enemy never rests, but sows his weeds till a stronger one does root it out.

To conclude, the Rosie Crucians say, pearl helps swooning's, and withstands the Plague of Poison, and that smarage and jacinth helps the Plague, and heals the wounds of venomous stings.

The water of Nile makes the woman of Egypt quick of conceit and fruitful, and sometimes they bear seven children at a Birth, and this is Saltpeter-water: There is a wonderful virtue in the Oil of Tobacco: in the tincture of Saffron, in the flower of Brimstone, in Quicksilver, in Common Salt, and Copperas, molten and made a water, kills the poison of the Toadstool; and juice of Poppy, Amber, which is no stone, but a hard clammy Juice, called Bitumen, eases the Labor of women, and the falling sickness in children.

Now for Metals, if it be true, which all men grant, that precious stones in that hard and ungentle fashion, show such virtue and power of healing, what shall the mixtures of all these Metals under a fortunate Constellation made in the Conversion of their own planets do, which they call electrum, sigil, or telesme, saying, it will cure the Cramp, Benumbing palsy, Falling-sickness, Gout, Leprosy, Dropsy, if it be worn on the heart finger; others they make to cause beauty in Ladies, & etc.

The third perfume of R. C. is compounded of the Saphirick earth, and the Æther, if it be brought to its full exaltation, it will shine like the

Daystar in her fresh Eastern glories; it has a fascinating attractive faculty; for if you expose it to the open Air, it will draw to it Birds and Beasts, and drive away evil Spirits. Astrum Solis, or the R. C. Mineral Sun is compounded of the Æther, and a bloody, fiery-spirited earth; it appears in a Gummy Consistency, but with a fiery, hot, glowing complexion, it is substantially a certain purple, animated, Divine Salt, and cureth all manner of Venereal distempers, Consumptions, and diseases of the Mind.

We give another Medicine, which is an Azure, or Sky-colored water, the Tincture of it is light and bright, it reflects a most beautiful Rainbow; and two drops of this water keeps a man healthy; in this water lies a blood red earth of great virtue.

The other Medicine is the Heavenly Luna and Moon of the Mine, a very strange stupefying substance: it is not simple but mixed: The Æther, and a subtle white Earth are its Components: and this makes it grosser, then the Æther itself; it appears in the form of an exceeding white oil, but in very truth a certain vegetant, flowing, smooth, soft salt, and this renews youth, and causes wisdom and virtue.

The Pantarva of Rosie Crucians is a water, and no stone; it after-night discovers a fire as bright as day; and if you look on it in the day time, it dazzles the eye with certain gleams or Coruscations; for in it is a Spirit of admirable power to long Life, Wisdom, and Virtue: Now I will show who taught these Secrets, and showed me these things. Walking upon the plain of Bulverton Hill to study Numbers and the nature of things, one evening, I could see between me and the light, a most exquisite Divine beauty; her frame neither long nor short, but a mean descent stature; attired she was in thin loose Silks, but so green that I never saw the like, for the Color was not earthly, in some places it was fancied, with Gold & silver Ribbands, which looked like the Sun and

Lilies in the field of grass; her head was overcast with a thin floating Tiffany; which she held up, with one of her hands, and looked as it were from under it; her eyes were quick, fresh, and Celestial, but had something of a Start, as if she had been puzzled with a sudden occurrence.

From her vail did her looks break out, like Sun beans from a Mist, they ran disheveled to her Breast, and then returned to her cheeks in curls and rings of Gold; her hair behind her was rolled to a curious Globe, with a small short spire flowered with purple and sky Color knots; her Rings were pure entire Emeralds, for she valued no Metal, and her pendants of burning Carbuncles. In brief her whole habit was youthful and flowery, it smelt like the East and was thoroughly aired with rich Arabian Diapasms; this and no other was her appearance at that time.

But whilst I admired her perfections, and prepared to make my address, she prevents me with a voluntary approach; here indeed I expected some discourse from her, but she looked very seriously and silently in my face, takes me by the hand and softly whispers, My love I freely give you, and with it these tokens, my Key and Signet, the one opens, the other shuts, be sure to use both with discretion; as for the mysteries of the Rosie Cross, you have my Liberty to peruse them all; there is not anything here, but I will gladly reveal It to you, I will teach you the virtue of Numbers of Names, of Angels and Genii of men; I have one precept to command to you, and this it is, you must be silent; you shall not in your writings extend my allowance; remember that I am your love, and you will not make me a Prostitute. But because I wish you serviceable to those of your own disposition, I here give you an Emblematical Type of my Sanctuary, viz. The *Axiomata* of the R. C. The secrets of Numbers, with a full privilege to publish it. This is all, and now I am going to the invisible Region, amongst the Ætherial

Goddesses, let not the Proverb take place with you, out of sight, out of mind; remember me and be happy.

Now I asked her if she would favor me with her name; to this she replied very familiarly, as if she had known me long before, My dear friend H. I have many Names, but my best beloved is EUTERPE.

Observe in your R. C. *Axiomata* that the genuine time of impression of Characters, Names, Angels, Numbers, and Genii of men, is, when the principles are spermade and callalo; but being once coagulated to a perfect body; the time of stellification is past. Now the R. C. in old time used strange Astrological Lamps, Images, Rings, and Plates, with the numbers and names engraved, which at certain hours would produce incredible extraordinary effects. The common Astrologer he takes a piece of Metals, another whining Associate he helps him with a Chrystal Stone, and these they figure with ridiculous Characters, and then expose them to the Planets, not in an Alkemusi, but as they Dream they know not what, when this is done, all is to no purpose; but though they fail in their practice, yet they believe they understand the *Axiomata* of Numbers well enough. Now my beloved J. U. that you may know what to do, I will teach you by Example; Take a ripe grain of Corn that is hard and dry, expose it to the Sun beams in a glass or any other vessel, and it will be a dry grain forever; but if you do bury it in the Earth, that the Nitrous Saltish moisture of the Element may dissolve it, then the Sun will work upon it, and make it spring and sprout to a new body; it is just thus with the Common Astrologer; he exposes to the Planets a perfect Compacted body, and by this means thinks to perform the Rosie Crucian Gamaea, and marry the Inferior and Superior worlds.

It must be a body reduced into SPERM, that the Heavenly Feminine moisture, which receives and retains the Impress of the Astral Agent, may be at liberty, and immediately exposed to the Masculine tire of Nature. This is the ground of the Beril; but you must remember, that nothing can be stellified without the joint Magnetism of three Heavens; what they are you know already. When she had thus said, she took out of her bosom two miraculous Medals with Numbers and Names on them, they were not Metalline, but such as I had never seen; neither did I conceive there was in Nature such pure and glorious substances; In my Judgement, they were two Magical Telesms; but she called them Saphiricks of the Sun and Moon. These miracles Euterpe commended to my perusal, and stopped in a mute Ceremony; for I was to be left alone; she looked upon me in silent smiles, mixed with a pretty kind of sadness, for we were unwilling to part; but her hour of Translation was come, and taking as I thought our last leave, she passed before my eyes into the Æther of nature; excusing herself as being sleepy, otherwise she had expounded them to me; I looked, admired, and wearied myself in that Contemplation; their complexion was so heavenly, their continuance so mysterious, I did not well know what to make of them, I turned aside to see, if she was still asleep; but she was gone, and this did not a little trouble me. I expected her return, till the day was quite spent, but she did not appear: at last, fixing my eyes on that place, where she sometimes rested, I discovered certain pieces of Gold, full of Numbers and Names, which she had left behind her, and hard by a Paper folded like a Letter. These I took up, and now the night approaching, the evening Star tinned in the West; when taking my last survey of her flowery pillow I parted from it in these verses.

Pretty Green Bank, farewell & mayst thou wear
Sun-beams, and Rose, and Lilies all the year;
She slept on thee, but needed not to shed

Her Gold, 'twas pay enough to be hen bed:
Thy Flowers are Favorites; for this lov'd day
They were my Rivals, and with her did play;
They found their heav'n at hand, and in her eyes
Enjoy'd a Copy of their absent skies.
Their weaker paint did with true Glories Trade,
And mingled with her cheeks, one posie made;
And did not her soft skin confine their Pride,
And with a screen of Silk her flowers divide;
They had suck'd life from thence, and from her heat
Borrow'd a soul to make themselves complete.
O happy Pillows though thou art laid even
With dust, she made thee up almost a heaven;
Her breath rain'd Spices, and each Amber Ring
Of her bright locks, etrew'd Bracelets are thy Spring;
That Earths not poor, did such a Treasure hold,
But thrice enrich'd with Amber, Spice and Gold.

Thus much at this time, and no more am I allowed by my Mistress Euterpe to publish: Be therefore, gentle Reader admonished that with me you do earnestly pray to God, that it please him to open the hearts and ears of all ill—hearing people, and to grant unto them his blessing, that they may be able to know him in his Omnipotence, with admiring contemplation of Nature, to his honor and Praise, and to the Love, Help, Comfort and strengthening of our neighbors; and to the restoring of all the diseased, by the Medicines above taught.

I had given you a more large account of the Mysteries of Nature, and the Rosie Cross: but whilst I studied Medicines to cure others, my dear Sister Anne Heydon died, and I never heard she was sick (for she was

100 miles from me) which puts an end to my writings, and thus I take my leave of the world, I shall write no more, you know my Books by Name, and this I write (that none may abuse me) by printing books in my Name, as Cole does Culpepper's. But return to my first happy Solitudes.

FINIS.

The Rosie Crucian Prayer

to God.

Jesus Mihi Omnia.

Oh thou everywhere and good of all, whatsoever I do, remember, I beseech tree, that I am but dust, but as a vapor sprung from earth, which even thy smallest breath can scatter; you have given me a soul, and laws to govern it; let that eternal rule, which thou didst first appoint to sway man, order me; make me careful to point at thy glory in all my ways; and where I cannot nightly know thee, that not only my understanding, but my ignorance may honor thee, Thou art All that can be perfect; Thy Revelation has made me happy; be not angry, O Divine One, O God the most high Creator, if it please thee, suffer these revealed secrets, Thy Gifts alone, not for my praise, but to thy Glory, to manifest themselves. I beseech thee most gracious God, they may not fall into the hands of ignorant envious persons, that cloud these truths to thy disgrace, saying, they are not lawful to be published, because what God reveals, is to be kept secret. But Rosie Crucian Philosophers lay up this Secret into the bosom of God, which I have presumed to manifest clearly and plainly. I beseech the Trinity, it may be printed as I have written it, that the Truth may no more be darkened with ambiguous language. Good God, besides thee nothing is. Oh, stream thy Self into my Soul, and flow it with thy Grace, thy Illuminations, and thy Revelation. Make me to depend on Thee: Thou delights that Man should account Thee as his King, and not bide what Honey of Knowledge he has revealed. I cast myself as an honored of Thee at thy feet. O establish my confidence in Thee, for thou art the fountain of all bounty, and canst not but be merciful, nor can you deceive the humbled Soul that trusts Thee: And because I cannot be defended by thee, unless I live after thy Laws, keep me, O my souls sovereign, in the obedience of thy Will, and that I wound not my Conscience with vice, and hiding thy Gifts and Graces bestowed upon me; for this I know will destroy me within, and make thy Illuminating Spirit leave me: I am afraid I have already infinitely swerved from the Revelations of that Divine Guide, which you have commanded to direct me to the Truth; and for this I am a sad Prostrate and Penitent at the foot of thy Throne; I appeal only to the abundance of thy Remissions. O my God, I know it is a mystery beyond the vast Souls apprehension, and therefore deep enough for Man to rest in safety in. O thou Being of all Beings, cause me to work myself to thee, and into the receiving arms of thy paternal Mercies thru myself. For outward things I thank thee, and such as I have I give unto others, in the name of the Trinity freely and faithfully, without hiding anything of what was revealed to me, and experienced to be no Diabolical Delusion or Dream, but the Adjectanenta of thy richer Graces; the Mines and deprivation are both in thy hands. In what you have given me I am content. Good God ray thy self into my Soul, give me but a heart to please thee, I beg no more than you have given, and that to continue me, unconcerned and unpitied honest. Save me from the Devil, Lusts, and Men, and for those fond dotages of Mortality, which would weigh down my Soul to Lowness and Debauchment; let It be my glory (planting myself in a Noble height above them) to condemn them. Take me from my self, and fill me with thee. Sum up thy blessings in those two, that I may be nightly good and wise; And these for thy eternal Truths sake grant and make grateful.

FINIS.

A Word from the Publisher

Thank you for purchasing this book from The R.A.M.S. Library of Alchemy. During his lifetime, Hans Nintzel dedicated himself to the identification, acquisition, study, retyping and, when necessary, translation of what he considered to be the most important known works on Alchemy. Hans was assisted by his sparse network of fellow Alchemists, all members of the Restorers of Alchemical Manuscripts Society (R.A.M.S.). I was an active member of R.A.M.S.

Hans provided copies of the R.A.M.S. works as photocopies. My goal is to publish all of them as professionally printed books.

The works from the original R.A.M.S. Library are republished by R.A.M.S. Publishing Company in the collection, "The R.A.M.S. Library of Alchemy," with permission of the Estate of Hans W. Nintzel.

If you have a work on Alchemy that you believe should be a part of the R.A.M.S. Library, please contact me through R.A.M.S. Publishing Company.

Philip N. Wheeler

https://ramsalchemy.jimdo.com/

The R.A.M.S. Library of Alchemy

The study and practice of Alchemy was extremely important to Hans W. Nintzel. He assembled this Library over a period spanning more than three decades, guided by his teacher Frater Albertus. The R.A.M.S. Library of Alchemy includes all of the most valuable Alchemical texts that Hans painstakingly located, acquired, retyped, and translated during his lifetime, with help from other R.A.M.S. members.

The following is a list of the volumes that are currently available. Volumes that contain works from multiple authors may have only the principle author or editor listed.

http://ramsalchemy.jimdo.com

Volume	Title	Author or Editor
1	Twelve Keys of Basilius Valentinus	Basilius Valentinus
2	Triumphal Chariot of Antimony	Basilius Valentinus
3	His Secret Book	Artephius
4	The Golden Work	Hermes Trismegistus
5	Three Works of Ripley	George Ripley
6	Four Works of Paracelsus	Paracelsus
7	Bacstrom's Notebooks, Part 1	Sigismund Bacstrom
8	Bacstrom's Notebooks, Part 2	Sigismund Bacstrom
9	Summa Perfectionis	Geber (Abu Musa Jabir ibn Hayyan)
10	The Five Centuries	Rudolph Glauber
11	The Greater and Lesser Edifyer	Johann Grashoff
12	Chemical Secrets and Experiments	Sir Kenelm Digby
13	The Turba Philosophorum	Arisleus
14	Das Aceton	Christian Becker
15	The Art of Distillation	John French
16	Non-Violent Destruction of the Atom	Nintzel & Wheeler
17	Philosophical Furnaces	Rudolph Glauber

18	The Last Will and Testament	Basilius Valentinus
19	TBD	
20	TBD	
21	Alchemical Symbols, 4th Edition	Philip N. Wheeler
22	The Book of Formulas	John Hazelrigg
23	18 Short Tracts	Hans W. Nintzel
24	Bacstrom's Notebooks, Part 3	Sigismund Bacstrom
25	A Discourse on Fire and Salt	Blaise Vignere
26	The Mineral Work	Johan Hollandus
27	The Vegetable Work	Johan Hollandus
28	Lamspring's Process	Lamspring
29	The Book of Abraham the Jew	Abraham Eleazar
30	Five Short Works of Glauber	Johann Glauber
31	The Metamorphosis of the Planets	Johannes Monte-Snyder
32	Four Works of Roger Bacon	Roger Bacon
33	The Golden Chain of Homer	Homerus, Kirchweger, Nintzel, Wheeler
34	Alchemy Rediscovered and Restored	Archibald Cochren
35	Aurifontina Chymica	John Houpreght
36	The Golden Fleece, 2nd Edition	Salomon Trismosin
37	The Transmutation of Base Metals into Gold and Silver	David Beuther
38	Sanguis Naturae	Christopher Grummet
39	A Revelation of the Secret Spirit	Giovanni Lambi
40	The Holy Guide, 2nd Edition	John Heydon
41	TBD	
42	Secreta Alchymiae	Kalid Persica
43	The Golden Treatise of Hermes	Hermes Trismegistus
44	Potpourri of Alchemy, Part 1	Hans W. Nintzel
44	Potpourri of Alchemy, Part 2	Hans W. Nintzel
46	TBD	
47	Selected Chemical Universal and Particular Processes	Alexius von Ruesenstein
48	TBD	
49	TBD	
50	Transcendent Magic	Eliphas Levi

John Heydon

Colophon

The Bird King from Dr. Dee's Manuscript,

used on the cover of this book.

Publishing Company

www.ingramcontent.com/pod-product-compliance
Lightning Source LLC
Chambersburg PA
CBHW062317220526
45469CB00008B/2539